T0177482

Philosophical Foundations of Climate Change Policy

Philosophical Foundations of Climate Change Policy

JOSEPH HEATH

OXFORD
UNIVERSITY PRESS

OXFORD
UNIVERSITY PRESS

Oxford University Press is a department of the University of Oxford. It furthers
the University's objective of excellence in research, scholarship, and education
by publishing worldwide. Oxford is a registered trade mark of Oxford University
Press in the UK and certain other countries.

Published in the United States of America by Oxford University Press
198 Madison Avenue, New York, NY 10016, United States of America.

© Oxford University Press 2021

Library of Congress Cataloging-in-Publication Data
Names: Heath, Joseph, 1967– author.
Title: Philosophical foundations of climate change policy / Joseph Heath.
Description: New York, NY : Oxford University Press, [2021] |
Includes bibliographical references and index.
Identifiers: LCCN 2020055999 (print) | LCCN 2020056000 (ebook) |
ISBN 9780197567982 (hardback) | ISBN 9780197568002 (epub) |
ISBN 9780197568019 | ISBN 9780197567999
Subjects: LCSH: Climatic changes—Government policy. |
Climatic changes—Philosophy.
Classification: LCC QC903.H427 2021 (print) | LCC QC903 (ebook) |
DDC 363.738/74561—dc23
LC record available at https://lccn.loc.gov/2020055999
LC ebook record available at https://lccn.loc.gov/2020056000

DOI: 10.1093/oso/9780197567982.001.0001

3 5 7 9 8 6 4 2

Printed by Integrated Books International, United States of America

Contents

Preface

This work was originally brought together as a series of lectures, presented in May 2018 at the Department of Philosophy at Bayreuth University. I would like to extend a special thanks to Julian Fink and his colleagues at Bayreuth, both for the warm reception and intellectually stimulating environment, and also for providing me the impetus to assemble my various thoughts on this question into something more closely resembling a coherent position.

People who have helped me in various ways with the development of this work include Idil Boran, Virginie Maris, Marc Davidson, Chad Horne, James Brandt, Hamish Russell, Rachel Bryant, Katherine Browne, Bruce Chapman, Julian Fink, Axel Gosseries, Xavier Landes, Lukas Meyer, Dan Moller, Nils Holtung, Stephen Gardiner, Kian Mintz-Woo, Lukas Tank, and Arthur Ripstein. Some of our exchanges were adversarial, and so the people listed should not be held responsible for the contents of this work, or for the failure to discourage me from making certain claims. Naturally I owe a special debt to those who read the entire manuscript or provided research assistance. Special thanks to Jovy Chan for completing the index.

Portions of this work have been presented at Queen's University, Duke University, Erasmus University Rotterdam, Brown University, Université de Montreal, Stockholm School of Economics in Riga, Harvard University, University of Maryland, SUNY Buffalo, and of course Universität Bayreuth. Special thanks again to members of the Department of Philosophy at Bayreuth for inviting me to present the 2018 Wittgenstein Lectures, on which this book is based. Thanks as well to the several dozen talented students at Bayreuth, who attended every lecture and spent two hours each afternoon pressing me on the various points made. I would also like to thank students at the University of Toronto, both in philosophy and in public policy, for discussion of these topics over the years. Finally, I would like to thank the Social Sciences and Humanities Research Council of Canada, the Trudeau Foundation, as well as the Jackman Humanities Institute at the University of Toronto for financial support, which has funded both research for the book and release time for its preparation.

This book includes modified presentation of material that has appeared previously in the following publications: "The Structure of Intergenerational Cooperation," *Philosophy and Public Affairs*, 41:1 (2013): 31–66; "Climate Ethics: Defending a Positive Social Time Preference," *Journal of Moral Philosophy*, 14:4 (2016): 436–462; and my book *The Machinery of Government* (New York: Oxford University Press, 2020).

Introduction

I would like to begin with both an apology and an explanation for the title of this book: *Philosophical Foundations of Climate Change Policy*. This is something of a mouthful and, at the same time, not particularly inspiring. I have chosen it, however, because of my desire to communicate clearly both the nature and the scope of the work. This is not a book about climate change per se, but rather about how governments should respond to climate change. It is, in other words, concerned with climate change policy. The argument, however, is somewhat backward compared to many other contributions made by philosophers in this field. Rather than starting with a philosophical view and then working out its implications for climate change policy, I start rather with climate change policy—or more specifically, with the range of policy options that can seriously be contemplated under anything like our current circumstances—and I work back *from this* to a discussion of the philosophical view that one should hold.[1] This is based on my conviction that a plausible normative-philosophical theory (such as a "theory of justice") should have the capacity to generate policy recommendations within the space of feasible alternatives and, thus, should be able to help policymakers both to select the right response and to defend it with serious arguments.

By "feasible" here, I am not talking about *political* feasibility.[2] I am merely talking about feasibility under the ordinary constraints that force us humans to organize our societies with the basic institutional structures that we have. These structures include, most importantly, the organization of production and distribution through a market economy, state intervention to correct a variety of market failures, as well to achieve a less-than-perfect measure of redistribution, and, finally, the *absence* of any supranational authority with independent coercive power (i.e. world government).[3] These are features of the world that we live in that may change someday, but that are unlikely to change within the time frame required to address the problem of climate change. They also serve as constraints on the policy space, in the sense that no one with any real political power, and no one who is in a position to advise anyone with real political power, takes seriously proposals that would require

Philosophical Foundations of Climate Change Policy. Joseph Heath, Oxford University Press. © Oxford University Press 2021. DOI: 10.1093/oso/9780197567982.003.0001

changing any of these basic structural features of the current world system. (To pick just one example, while abolishing capitalism might help to solve the problem of climate change, it is not on the menu of current policy options.)

Most of the normative views held by philosophers, unfortunately, have practical implications that fall so far outside the space of feasible policy alternatives that they are unable to make any productive contribution to the debate over what should be done about climate change. This has not always been apparent to those who hold these views, and so part of the task of this work will be to show how implausible the policy prescriptions are that flow from many of these popular normative positions. The problem, I should note, is not that these theories are excessively "utopian" or "ideal" (and thus, I am not intending to make a contribution to recent debates over "ideal" and "nonideal" theory).[4] To see why this is not the issue, consider utilitarianism, which is extremely idealized, in the sense that it requires that each individual set aside self-interest and act in a way that maximizes total happiness. It therefore demands both complete impartiality and total self-abnegation, something that is deeply in tension with several aspects of human psychology. And yet utilitarianism is also an endlessly productive source of policy-relevant advice. Indeed, it is so good at generating practical recommendations, under any circumstances, that it manages to monopolize the policy debate in many domains. The reason is that, while specifying an ideal that is obviously unobtainable, its fundamental principle also provides a complete ranking of every state of the world that falls short of this ideal. As a result, the theory is able to generate recommendations under any set of constraints.

By contrast, there are many normative theories that, despite being somewhat less idealized in their initial formulation, fail to generate policy-relevant advice, because they have no way of handling imperfect states of affairs, or cannot be applied under constraint. To take a particularly clear example, consider Ronald Dworkin's conception of equality.[5] Dworkin was emphatic that equality was not only the supreme principle of justice, but in fact the *only* principle of justice that we need. He was an equality monist.[6] And yet the specific conception of equality that he endorsed—that of envy-freeness—while allowing him to identify equal allocations of resources, was incapable of ranking imperfectly equal allocations. This is a peculiarity of the envy-freeness standard, that while it can identify equal allocations by the absence of envy, it does not quantify envy, and so is unable to describe any unequal allocation as being more or less equal than some other.[7] As a result, this conception cannot help us with any real-world problems, because as soon as there is *some* inequality, it offers nothing more than a global condemnation of all such states of affairs. Unlike utilitarianism, which generates a ranking of every possible state, Dworkin's egalitarianism has only two statuses: perfect and imperfect.[8] It therefore has, quite literally, nothing to say about the relative merits of different imperfect states of affairs.

Few philosophical views are as self-evidently limited. In most cases the problem is that they are formulated in abstraction, and so when applied to real-world problems generate recommendations that are too extreme to be translatable into policy. For example, John Rawls's well-known difference principle recommends assigning *lexical* priority to the interests of the worst-off representative individual.[9] This means that one is obliged to ignore completely the effects that a policy would have on everyone else in society, until such time as one has maximized the benefit to the worst-off. Then, and only then, can one look to see what the effects are on the *second* worst-off person, and so on. This procedure is obviously one that can be applied in many different imperfect circumstances, but if one were to work through its implications for any practical question, like the design of a tax code, the recommendations that it generates would be so extreme that they could hardly be taken seriously. Given that we live in an electoral democracy, for instance, it is difficult to know how anyone could go about implementing a principle of justice that recommends ignoring the interests of the overwhelming majority of the population. And yet the most prominent critique of Rawls's principle, in the philosophical literature, has been that it is too lax, or that he makes improper concessions to self-interest, because of his willingness to concede that too much redistribution of the social product might undermine the incentives that individuals have to create that product.[10]

In this intellectual environment, contact with the realm of public policy can have the salutary effect of bringing philosophical speculation back down to earth. Bridging the two domains, however, presents certain challenges. The philosopher Jonathan Wolff is one of the few to have addressed this topic explicitly. His book *Ethics and Public Policy* arose from his experience being asked to serve as an "ethics" adviser on a number of government committees in the UK.[11] On his first assignment, dealing with questions of animal welfare, he was asked to provide a summary of the "state of knowledge" within his discipline. He found himself feeling uncomfortable reporting on the views being debated. "On the whole, philosophers seemed to defend views that were so far from current practice as to seem, to the non-philosopher, quite outrageous. The idea that society could adopt any of the views put forward seemed almost laughable. To put it mildly, from the point of view of public policy the views were unreasonable and unacceptable."[12]

One can see a similar problem quite clearly in the case of climate change, where the philosophical discussion has been occurring in a discursive space that is almost completely separate from the policy debate. Practical-minded philosophers have noticed this and been troubled by the disconnect. The way that they express their concern, however, has usually been by complaining that the policy debate is dominated by economists. This is often accompanied by the suggestion that economists have been using some sort of dirty trick to acquire this influence. The most commonly voiced suspicion is that economists have been misrepresenting themselves as purveyors of purely scientific or technical advice, and therefore as neutral experts.[13] Thus the central strategy, among philosophers, by which to elbow aside economists and establish their own place at the table, has been to insist that climate change is a moral problem, and that, as such, it cannot be addressed without consulting the special expertise of "climate ethicists."[14]

If one examines the policy literature, however, one can see the problem with this suggestion. First and foremost, it has always been fairly clear to everyone involved that the problem of anthropogenic climate change has an important moral dimension. One struggles to find an economist of any importance who has ever denied this. Economists simply use a normative vocabulary that is different from the one favored by philosophers and environmental ethicists. The real difference, and the reason that economists have achieved such influence over policymakers, is that they always have something useful to say about the relative merits of the policy options that are actually on the table. Philosophers, by contrast, have a habit of rejecting all the options, then criticizing the construction of the table. There is something to be said, intellectually, for having people around who adopt such radical stances, questioning the paradigm, problematizing the taken-for-granted, waking people from their dogmatic slumbers, and so on. But it is a bit rich for those who adopt these gratifying stances to complain about having insufficient influence over policy, when the theories they are advancing would require a complete structural transformation of society, and possibly human nature as well, in order to be implemented.

The problem of inapplicability is, unfortunately, more pronounced for deontological moral theories than it is for consequentialist ones. Again, because consequentialism is all about producing good outcomes, it lends itself quite naturally to a ranking of outcomes even under adverse conditions. This is obvious in the case of utilitarianism, but it is true of pretty much any consequentialist theory that treats all outcomes as morally commensurable. Furthermore, there is not that much difference between moral consequentialism and the normative vocabulary employed by economists. As a result, philosophers who embrace some form of consequentialism as a comprehensive moral theory—John Broome is the most prominent example—have had little trouble engaging with the nuts and bolts of the policy debates over climate change.[15] Consequentialism, however, is a minority view among philosophers. Especially in political philosophy, neo-Kantian views of one sort or another, largely inspired by the work of John Rawls, are far more widely held.

Kant, of course, was somewhat exceptional in his willingness to act on principle, without regard for the consequences. Modern deontologists inevitably have more moderate views—e.g. they are, for the most part, willing to lie if doing so could prevent a murder. Nevertheless, there is still a strong tendency to want to rule things out "categorically," which can interfere with a judicious weighing of the anticipated consequences. This becomes a challenge when it comes to thinking about climate change policy, which is all about making trade-offs, in some cases very difficult ones. Consider T. M. Scanlon's contractualism, which is widely regarded by philosophers as a credible alternative to utilitarianism. Scanlon considers it an attractive feature of his view that it rejects aggregationism (crudely put, the idea that the needs of the many can outweigh those of the few). While this may have certain advantages in abstraction, or in the stylized examples that philosophers enjoy inventing, it makes the view very difficult to apply in real-world circumstances.

Take, for example, the philosophical puzzle known as "the numbers problem." John Taurek made the observation, back in 1977, that in a typical "lifeboat" scenario, in which it is possible to save the lives of some, but not all, the best way to decide who lives and who dies will be to hold a lottery, in which each individual is given an equal chance of surviving.[16] If, however, one applies this same principle to the situation in which one has a choice between saving one person and saving a group of people, it recommends tossing a coin, in order to decide which to save, since doing so gives each person an equal chance of being saved (i.e. 50 percent). This doesn't seem quite right—almost everyone is of the view that one should save the larger number in this case.

For the most part, philosophers who were aware of this puzzle considered it little more than a curiosity, until Scanlon announced that he had a solution to it, which involved saving the larger group, but did not require giving greater weight to the fact that they were more numerous.[17] Scanlon's argument, however, proved so unpersuasive that it led to widespread re-evaluation of the challenge. What was once regarded as a curiosity came to be seen as a very grave problem. In subsequent years, an extraordinary amount of ingenuity came to be deployed, by philosophers trying to provide an intuitively correct solution to the numbers problem, without just giving in to the idea that the needs of the many, in this case, really do outweigh the needs of the one.[18]

With all deference to those involved in these debates, I think it is fair to say that the jury is still out on whether this variant of contractualism is able to respond to the challenge.[19] Consider then the unenviable circumstance of those who would like to apply a Scanlonian contractualist framework to the problem of climate change.[20] Intergovernmental Panel on Climate Change (IPCC) reports that try to assess potential damages from climate change are sobering documents.[21] Scenarios are being considered that, in the long term, involve large-scale collapse of agriculture that could lead to the starvation of millions, sea-level rise that could displace hundreds of millions, loss of as much as half of the earth's biodiversity, and so on. These are not inevitable consequences, but nevertheless, they are of a moral gravity that necessarily commands our utmost attention and concern. The policies that we adopt now will have extremely serious consequences for our descendants. Given this fact, how can we expect any policymaker to evaluate the options using a normative framework that struggles to explain why it is better to save five people from drowning than it is to save one? The economist's utilitarianism looks good by comparison—even if it is a devil, at least it is the devil we know.

Of course, if the damages were all on the side of climate change *effects*, then the rigidity of these deontological views might not be such a problem. What makes the current set of policy questions so difficult is that much of the quality of life that we enjoy—and much of the prospective gains in quality of life that billions of people in poor countries hope someday to enjoy—depends upon processes that emit greenhouse gases (GHGs), first and foremost the consumption of fossil fuel. To pick just one example, the current human population is sustained only because of the extraordinary increase in agricultural productivity enabled by the Haber-Bosch process, which extracts nitrogen directly from the atmosphere, where it can subsequently be used in the production of synthetic fertilizers. About 40 percent of current food production depends upon this process, which is estimated to consume 2 percent of world energy supply.[22] Without it, the amount of agricultural land under cultivation would have to quadruple (increasing the land under cultivation from less than 15 percent of global ice-free landmass to almost 50 percent).[23] This would, of course, be ecologically catastrophic, would require massive displacement of human populations, and would also exacerbate climate change. So far however, there is no efficient technological alternative to the use of fossil fuel (primarily natural gas) in the extraction of nitrogen from the atmosphere.[24] This is why modern agriculture is sometimes described as a process that transforms fossil fuel into food. Absent technological progress, we are—for essentially Malthusian reasons—locked into producing the emissions currently associated with this process.

A similar story could be told about the production of steel or aluminum, the use of concrete in construction, and the fossil-fuel dependence of aviation or shipping. These are the building blocks of economic development, the success of which impacts the lives of billions of people. Again, to pick just one example, it is estimated that 10 percent of all deaths in India are linked, in one way or another, to poor sanitation.[25] Much of this is due to inadequate or nonexistent sewage and toilet facilities—approximately 450 million people in that country still engage in "open defecation," creating a situation that could plausibly be described as a public health catastrophe. And yet the installation of lavatories and sewage treatment for this many people is an enormous project, completion of which will necessarily involve the use of metal, concrete, and plastics on a massive scale, not to mention the energy used in pumping water. India, however, gets more than half its energy from burning coal—the consumption of which has been increasing at a rate of over 5 percent per year in that country. This is obviously contributing to a different sort of catastrophe. In this context, environmental policy may seem like a matter of deciding which catastrophe is the least bad. Ideally, however, it involves deciding how best to balance the short-term gains that can be achieved through increased energy consumption with the long-terms costs of accelerated climate change. It is not an exaggeration to say that millions of lives depend upon this question being answered correctly.

Many of these dilemmas are a consequence of the speed with which economic development has been occurring around the world. As is often pointed out, more people have been lifted out of poverty, more quickly, in the past 25 years than at any other time in human history. This has made it impossible to ignore the contribution made by developing countries to the problem of climate change. In these countries, however, the trade-off between carbon abatement and development is obvious. Back in the 20th century, in the run-up to the signing of the Kyoto Accord, it was easier to characterize climate change as a consequence of rich countries polluting the atmosphere, to the detriment of poor countries. This in turn made it possible for philosophers and environmentalists to shrug off the costs of carbon abatement, as though combating climate change would only require paring back the "luxury" consumption of affluent Westerners. This stance is no longer credible. Over the decade ending in 2016, European Union countries managed to achieve combined emissions reductions from 5,215 to 4,303 megatons of CO_2-equivalent. During that same period, China increased its emissions from 8,353 to 12,750 megatons, which is to say, over the course of a decade China *added* more emissions than the EU countries—including France, Germany, and the United Kingdom—produce as a whole.[26] And yet, because of the enormous benefits that flow from economic development in China, it is impossible to talk about climate change policy without taking seriously the trade-offs involved.

 In Canada, where I live and work, various features of the economic and political situation have made it such that the trade-offs involved in climate change policy are also difficult to ignore. It is, in fact, not an accident that my own work has focused so resolutely on the policy dilemmas that are generated by the climate change problem. Canada has proven oil reserves of approximately 167 billion barrels (10 percent of the global total), over 164 billion barrels of which are located in the Alberta tar sands.[27] Extraction of this oil is extremely energy-intensive (pressurized steam is used to liquify the bitumen and separate it from the sand), as a result of which carbon emissions in the province of Alberta are at least six times higher than the national average. These emissions are produced almost entirely through the extraction and refinement process; they do not include the oil itself, which is almost all exported. Because of these exports, however, Canada essentially does not have to worry about its balance of trade. Domestically, the industry generates enormous wealth, which has knock-on effects throughout the economy. Trends in economic inequality, for instance, are much less worrisome in Canada than in the United States, in part because the resource economy generates a great deal of well-paid blue-collar work (and has absorbed much of the workforce freed up by the decline in manufacturing). The demand for unskilled labor also facilitates immigrant and refugee integration in myriad ways. As a result, much of the backlash against "globalism" that has wracked Europe and the United States has passed Canada by.

Thanks to these economic facts, there is a great deal at stake in the policy debates over climate change in Canada. It is a basic principle of politics that the only way to achieve change is through a confluence of ideas and interests. It is not sufficient merely to have the best ideas; there must also be some relatively large constituency that stands to gain from the implementation of those ideas and that can be mobilized politically. Mere ideological commitment is normally insufficient—not only is it motivationally weak, but it typically only mobilizes small numbers of people. This is something that academics find deeply counterintuitive, and often slightly repulsive—the fact that in order to implement their ideas, it is necessary to build alliances with interest groups, often ones who have only an instrumental commitment to those ideas.

In the United States, it is extremely difficult to find any large group that stands to gain from any carbon abatement undertaken in the next few decades. By contrast, it is extremely easy to identify and mobilize people who would lose—ranging from petrochemical companies to SUV drivers. That, combined with governance failure in the US political system, has resulted in the state being almost completely dominated by interests opposed to climate change mitigation. Thus the confluence of ideas and interests that would be necessary to force any significant changes in policy is practically impossible to achieve. One perverse consequence of this state of affairs is that it has freed US academics from any obligation to think in practical terms about the consequences of their ideas. They are so completely cut off from political influence that it quite literally *does not matter* what they think. As a result, they have very little reason to think carefully about the policy consequences of their views. This can generate a self-reinforcing dynamic, in which the more impractical their ideas become, the less likely anyone is to consult them on matters of policy, which in turn reduces whatever incentive they may have had to moderate their ideas.

Throughout most of Western Europe the opposite state of affairs prevails. Here there are practically no fossil fuel interests. Indeed, European dependence upon Russian natural gas imports is an economic liability, a geopolitical weakness, and a political embarrassment. So for the most part, European states (including the UK) have a great deal to gain and little to lose from decarbonization. But because of this, environmental ideas often encounter little resistance, no matter how extreme. Academics—including philosophers—are often consulted on matters of policy, but there is seldom much challenge to their views. Thus ideas that are completely impractical on a global scale get a pass domestically because there is no constituency that has a strong interest in challenging them.[28] As a result, much of the academic debate is focused on choosing which shade of radical green politics the various participants find most congenial.[29]

Canada, by contrast, is in the peculiar situation of reproducing domestically many of the same conflicts of interest that exist globally. There are two major poles of political power in the country—the "ethnic block" of francophone voters in Quebec and the "oil patch" interests in Alberta.[30] Control of the federal government (and the position of prime minister) passes back and forth between these two regional power centers. Through a coincidence of geography, the province of Quebec derives 96 percent of its electricity from hydroelectric power (with wind power being the second-largest source, providing 2 percent). As a result, people in that province have nothing to lose, and a fair bit to gain, from an aggressive carbon abatement policy. By contrast, an effective climate change mitigation regime would almost completely destroy the economy of Alberta, and significantly depopulate that province.[31] (In another unfortunate accident of geography, cattle ranching is the most important agricultural sector in Alberta. Among major agricultural products, beef is by far the worst offender, in terms of emissions intensity.) As a result, the configuration of economic interests within the country creates something of a "knife edge" equilibrium, where things could easily go either way—toward American-style intransigence or toward European-style regulation. Under these circumstances, it is much easier to make the case that ideas matter. Any climate change policy proposed is going to be very controversial and will be aggressively challenged in both public and academic debate. Because of this, it is important that one have solid arguments in defense of one's positions, and that one be able to ground these through appeal to very broadly shared norms.

This political context also has a tendency to discipline philosophical speculation, in several different ways.[32] First, it makes it impossible to ignore the fact that when we debate these questions, people's livelihoods are at stake. To pick just one example, if the Alberta tar sands operations were to close, literally thousands of people who have worked and saved to buy a home would suddenly find themselves, not only unemployed, but holding property that was very close to worthless. When the plan is to deprive someone of both savings and livelihood, it is important to have good arguments in defense of that position—ideally, arguments that even those most negatively affected should be able to see the force of. Second, it pushes one to recognize the seriousness of the governance challenge posed by these conflicts. It is one thing to say, "This oil is going to have to stay in the ground"; it is something else entirely to force people to leave it there. Canada's proven oil reserves are worth almost US$10 trillion at the time of writing, a sum that is about six times the total GDP of the country. There is nothing to be gained by understating the amount of coercion that will be required in order to bar access to such a resource and, thus, the high level of political legitimacy that the imposition of a climate change mitigation policy would require.

Both of these considerations speak in favor of an approach to the problem grounded in a liberal theory of justice. (This is a central feature of the "philosophical foundations" that I will be recommending.) I will discuss this issue in greater detail in Chapter 1; for now I would simply like to observe two points that support this approach. First, because liberal political philosophy has always been preoccupied with the question of when state coercion is justifiable (in a way that, for instance, many traditions of moral philosophy have not), it is an appropriate framework for thinking about climate change policy. Second, and closely connected to the first point, liberal theories of justice have incorporated into their design a desire to formulate arguments that will speak to as broad a constituency as possible. This has traditionally been a response to domestic pluralism, but when one turns to a global problem like climate change, the need for arguments that have the widest possible appeal—and that presuppose the least in terms of shared culture or values— becomes even more pressing.

One of the major attractions of the economist's traditional utilitarianism is that it aspires to be a liberal theory in this sense. The central weakness, however, is that it is far too willing to sacrifice the interests of some in order to benefit others. Politically, of course, arguments with this structure tend not to be very effective. (Try telling people in Alberta that it is okay to destroy their economy because people in Quebec will benefit!) The question is whether it is *reasonable* to expect this sort of sacrifice from people. When imposing real hardship on some, is it sufficient justification to point to the gains of others? In other words, is the *mere* fact that someone else will experience a gain sufficient to justify the imposition of the loss on some person? Many have felt that it is not. Consider what Rawls wrote (in the context of his discussion of the "strains of commitment"). The problem with utilitarianism, he thought, was that it was too ready to sacrifice the "life prospects" of some, as a matter of routine policy:

The sacrifices in question are not those asked in times of social emergency when all or some must pitch in for the common good. The principles of justice apply to the basic structure of the social system and the determination of life prospects. What the principle of utility asks is precisely a sacrifice of these prospects. Even when we are less fortunate, we are to accept the greater advantages of others as a sufficient reason for lower expectations over the whole course of our life. This is surely an extreme demand.[33]

Rawls's argument raises a number of very delicate questions about what concessions to self-interest are relevant to the formulation of principles of justice, which I do not intend to address here. Whether the view can ultimately be justified or not, it is clearly a consequence of the utilitarian principle that it licenses these sorts of "extreme demands." The question is whether, in order to justify a climate change mitigation policy, we are obliged to turn to a normative theory that makes extreme demands of us. For if there were a normative theory that made less extreme demands and yet provided a satisfactory, or at least broadly acceptable, solution to the problem, then there would be a great deal to be said in favor of leading with such a theory (particularly given the difficulty of achieving international agreement). The recourse to full-blooded utilitarianism, of the sort that economists sometimes naively espouse, should be reserved for desperate circumstances; to the extent that we make use of it at all, it should never be the opening move.

Rawls of course aspired to produce a conception of justice that would be less motivationally demanding, based on the idea of a social contract. One of the attractive features of this theory, he felt, was that it would ensure that no one came out a net loser from their interactions with others. Society was to be conceived of as a "cooperative venture for mutual advantage," which would make "possible a better life for all than any would have if each were to try to live solely by his own efforts."[34] The principles of justice, he argued, were essentially a set of rules for determining an acceptable institutional structure for this system of cooperation.

This conceptual framework seems like a much more attractive one for thinking about climate change policy. After all, climate change is widely regarded as a collective action problem, which implies that the solution to it will involve the creation of a system of cooperation. A normative theory that offers specific guidance on the organization of cooperative schemes would appear to be ideally suited to the challenges of this policy domain. Unfortunately, while Rawls provided a very compelling way of framing the general question of how a theory of justice should be formulated, the specific theory that he proposed to guide the exercise was not entirely compelling. As a result, more orthodox Rawlsians (as I will attempt to show), who stick close to the letter of Rawlsian doctrine, have not succeeded in making any particularly helpful contributions to the debate over climate change policy.

What I intend to develop and apply in this book are a set of general contractualist ideas, which are of recognizably Rawlsian provenance, but which lie conceptually upstream from the more specific principles of justice that Rawls articulated (such as his "difference principle"). These more general ideas provide a perfectly adequate and compelling basis for thinking about problems such as climate change. Indeed, many of these are *already implicit* in the way that policy options are being debated in a wide range of domains. In previous work, I have described these ideas as constituting a sort of "minimally controversial contractualism."[35] My central ambition here is to show how this way of thinking can provide appropriate philosophical and normative foundations for our approach to climate change policy.

The basic framework is relatively simple. Human beings live in societies that are governed by a moral code, the details of which have a great deal of cultural specificity, and so vary from time to time and place to place. Whether or not there is a universal morality underlying all of this variation, existing human societies are characterized by what Rawls calls "the fact of pluralism," in relation to one another, and usually internally as well. People disagree about fundamental moral questions, and there is no recognized procedure for resolving these disputes. And yet despite all of this disagreement, people also find themselves quite frequently in a position where they are able to engage in interactions that would be mutually beneficial, in the sense that each could be made better off from his or her own perspective. Achieving these benefits requires that they *cooperate* with one another. Yet unfortunately, given the underlying pluralism, the "first order" moral codes will typically underdetermine the choice of cooperative arrangement. Not only are there many different ways to organize a cooperative interaction, but there is potential for conflict over how the benefits and burdens of cooperation should be allocated to those involved. Thus the parties require a set of normative principles to bring about agreement on the specific modalities of cooperation.

It has become conventional, following Rawls, to describe these abstract normative principles as constituting a "theory of justice." Furthermore, the basic problem—of picking out a cooperative arrangement that will be acceptable to all—can be broken down analytically into two components. On the one hand, the interaction generates mutual benefit, and the parties have a common interest in ensuring that these benefits are maximized. On the other hand, certain arrangements may be more to the benefit of some parties than others, and so there is a potential conflict of interest over how the benefits and burdens of cooperation should be distributed. The specifically contractualist idea, then, is that the principles that govern choice in these two dimensions should be ones that favor arrangements that no one has any good reason to reject. With respect to the "common interest" dimension, this constraint generates the Pareto-efficiency principle, which states that an outcome is to be preferred if it makes at least one person better off without making anyone worse off. To see how this follows, one need only consider outcomes that violate the Pareto principle. If an outcome is Pareto-inefficient, this means that at least one person could be made better off without making anyone worse off. This gives that person obvious grounds for complaint, whereas if one were to shift to the Pareto-efficient state, this would eliminate the complaint, while not giving anyone else fresh grounds for complaint.

Things are a bit more difficult with respect to the "conflict of interest" dimension, because this aspect of the interaction is win-lose, and so one might be tempted to conclude that the problem is insoluble, because any movement in either direction will necessarily generate a complaint from someone. There is, however, good reason to favor allocations that divide the benefits and burdens equally, which is to say, those that give everyone exactly the same amount. To see this, consider what happens when one deviates from equality. This immediately generates a "place-switching" complaint, where the person who gets less will demand the portion being assigned to the one who gets more. So while the equal allocation is still vulnerable to self-serving objections from those who want arrangements more favorable to their own interests, it nevertheless neutralizes a specific type of complaint. Of course, finding the equal allocation raises a number of difficult technical questions, depending upon how complex the system of cooperation is, and what form the benefits and burdens take. Nevertheless, the idea of finding an allocation that neutralizes the desire to switch places provides powerful orientation to the task.

One need not think that these two principles are so powerful as to compel rational agreement in all cases to see that they will, nevertheless, have a strong tendency to minimize conflict. While they are unlikely to silence all objections, efficiency and equality do eliminate two very obvious sources of complaint. They can also be generalized to produce a ranking of imperfect states of affairs, using some notion of "intensity" or "seriousness" of complaint. Again, there are technical challenges involved in this, but the task is not insuperable. The overall result is typically a *prioritarian* theory of justice, which permits deviations from equality in order to achieve gains in welfare, but insists that these gains in welfare must be larger the further removed one is from equality.[36] Institutional arrangements that satisfy these constraints are often referred to as "fair." Thus my central philosophical contention will be that we should be thinking about climate change policy in terms of the fairness of our existing institutional arrangements and the level of emissions that they permit.

It is worth emphasizing that a minimal conception of justice developed along these lines is generally not put forward as a complete theory of morality. People have a variety of "thick" moral commitments, which govern many other areas of life beyond just cooperative interactions. Among philosophers, David Gauthier is almost alone in having suggested that a contractual theory of this sort was the sum total of morality, and even then he qualified this claim in various ways. The standard contractualist view follows David Hume, who drew a distinction between the "natural" and the "artificial" virtues, where the latter were a set of traits specifically cultivated to encourage cooperation in areas where individuals could derive mutual benefit from normatively regulated interaction.[37] Under such a framework, there might be certain moral duties that impose basic requirements of respect and charity toward all persons, regardless of circumstance, whereas duties of justice would be more narrow, applying only to the outcome of cooperative interactions, but at the same time imposing much stricter demands.

There are a variety of different ways of drawing a distinction along these lines. Rawls, for instance, distinguished between the "political" conception of justice that should govern the basic institutional structure of society—and is thus focused on assigning the benefits and burdens of cooperation—from the "comprehensive moral views" held by individual citizens. This implies that the "political" obligations we have toward other people—duties that flow from our conception of justice—are not exhaustive, we may have moral obligations as well. Thus neither Rawls nor the minimally controversial contractualist is committed to the view that, outside of cooperative interactions, we are normatively unconstrained in our relations with other people. The duties of justice specified by the contractualist framework are simply the most important of the duties owed to others and, in particular, those whose enforcement is most easily justified.

My objective in this book is to show that minimally controversial contractualism provides all of the normative resources we require for thinking about climate change *policy*—understood as the set of obligations that are to be coercively imposed—even if it does not provide the basis for a complete environmental philosophy. Furthermore, philosophers have tended to put greater emphasis on the equality dimension of the climate change problem while underestimating the importance of the efficiency dimension. I will argue that most of the heavy lifting, in our response to climate change, is done by the efficiency principle. (For those who are familiar with my other work, I should note that this argument is an application of the "market failures" approach.[38] The latter is intended as a contribution to business ethics, where the efficiency principle is applied to the question of self-regulation by firms. Part of the attraction of the market failures approach, however, is that it provides a unified account of, not just business ethics, but also markets, firms, corporate law, and regulation. This book is about regulation, not self-regulation, and so it falls outside the purview of business ethics. But those who follow this literature will notice that I am using the same normative framework, this time to call for a regulatory solution to a market failure.)

I began by saying that this is not really a book about climate change, but rather about climate change policy. Even this turns out to be not quite true. It is not really about climate change policy, but rather about the philosophical foundations of climate change policy. More polemically, one might say that it is about the type of philosophical view that one should hold—in particular, the approach that one should adopt toward questions of justice—if one wants to have sensible things to say about climate change policy. An unfriendly way of describing the project would be to say that I am using the climate change issue as a way of scoring points in a philosophical debate that, ultimately, does not have all that much to do with climate change. There is a grain of truth to this, but it is not the entire story.

My basic view, as I have explained, is that the climate change policy debate is currently dominated by economists, because so far they are the only ones with consistently useful advice to offer on the question. I consider this unfortunate in one respect, which is that economists typically subscribe to a type of utilitarianism that I consider deeply mistaken. For the most part, however, this does not make all that much difference, because climate change is a collective action problem, and both utilitarianism and contractualism have very similar things to say about how one should proceed within the space of Pareto improvements. Thus the fact that most economists subscribe to the wrong normative theory does not lead them too far astray on the policy question. There is, however, one issue on which having the right philosophical theory does matter, and that is the issue of the social discount rate. Traditionally, this has been something of an esoteric issue, but the publication of the *Stern Review on the Economics of Climate Change* in 2006 put it suddenly at the center of debates over climate change policy.[39]

When it comes to thinking about our obligations to people who have not yet been born, consequentialist theorists are sorely tempted to "expand the circle" of moral concern to include not just actual people, but "future people" as well. This leads quite directly to a number of very serious conceptual difficulties, which philosophers have spent many years worrying about. In the case of climate change policy, it leads to a rather important conundrum over the appropriate social discount rate. In this case, consequentialism encourages the extremist view that it is impermissible to discount future costs and benefits.[40] Absent some manipulation, this generates a discount rate that is implausibly low for policy purposes. And yet since this conclusion seems to be the inescapable verdict of normative theory, many economists have been tempted to abandon the normative perspective entirely and to adopt some positive indicator, such as the prevailing interest rate, as a basis for choosing a discount rate. This leads to a rate that is implausibly high. Thus the choice of discount rate has come to be seen as a serious dilemma.

I am not in a position to resolve this issue, but I would like to suggest a dramatically different normative approach to the problem, one that does not require "expanding the circle." This is based on the minimally controversial contractualism that I am recommending. In order to get to this debate, however, it is necessary to provide a more general justification for the way that the question is framed. Discounting only arises as an issue when one attempts to calculate the "social cost of carbon," which is necessary, in turn, to determine the appropriate price for carbon. There is a great deal that one must accept in order to see that carbon pricing is the best approach to resolving the problems posed by anthropogenic climate change. Thus I will begin with the most basic presuppositions that structure the policy debate, in order then to work toward the more specific, and complex, normative issues that arise within that framework. There is a great deal that philosophers can learn from taking the policy debate seriously, by recognizing and coming to respect the intellectual and moral seriousness of its structuring presuppositions. And yet I hope to show that, in the end, there is still something that policy experts can learn from thinking about the central issues in a more philosophically well-grounded manner. Thus there is some back-and-forth between the policy perspective and the more general normative theory being proposed. While thinking about the policy framework imposes important constraints on the range of plausible philosophical theories of justice, different philosophical theories within the space of plausible alternatives do have different policy implications, and so getting the philosophy right remains an important pre-condition for getting the policy right.

Finally, I should mention that I will be ignoring entirely arguments and issues that have been put forward by those who are acting in bad faith. There is a great deal of demagoguery surrounding the climate change issue, particularly in North America. In Canada, for instance, carbon pricing is opposed by both the right-wing (Conservative) and the left-wing (New Democratic) political parties. The Conservatives routinely appeal to the costs that it would create for the economy, while ignoring entirely the benefits. The New Democrats oppose it on the nonsensical grounds that it would impose the abatement costs on "ordinary Canadians" instead of corporations, who are taken to be the real malefactors. One could spend a great deal of time and energy sorting through the confusions and lies that structure the *political* debate over climate change policy. This is, it seems to me, something of a waste of time in an academic publication, since those who have been misled by such claims are not likely to be among the readers of this book. While I have made some effort in this introduction to situate my project politically, my objective in doing so has been merely to explain why I am adopting what many would regard as a "backward" approach to the evaluation of philosophical positions. The remainder of the discussion will be focused entirely on the policy debate, which is a discussion carried out among those who maintain a good-faith commitment to solving the problem of climate change, but have a number of unanswered questions about what should be done or what constitutes a reasonable response. The considerations adduced here should be thought of as ones that would inform the deliberations, not of the minister of the environment, who is beholden to political interests, but rather of the deputy minister (or permanent secretary), in the Ministry of the Environment, who is charged with advising the minister on what "best policy" would be in this domain.

1

False Starts

It is not every day that economists and politicians actively solicit the opinion
of moral philosophers in order to help decide controversial policy questions.
The task of formulating an appropriate response to the problem of global
climate change is one of those rare instances.[1] There are a number of different options, within the space of available policy alternatives, and selecting
among them requires answering a number of difficult normative questions.
Furthermore, these questions are generally agreed to be *ineliminably* normative, as there is no way of reducing them to scientific, technical, or pragmatic
questions. Thus policymakers have turned to philosophers—environmental
ethicists in particular—for counsel on how to proceed. Unfortunately, the
responses they have received have been, for the most part, quite unhelpful.
Environmental ethicists have been unable, or more often unwilling, to answer the specific normative questions that arise within the policy space—
such as "How high should carbon taxes be?" or "How should we trade off
present costs against future benefits?"—because, almost without exception,
they reject the entire conceptual framework that structures the policy discussion. As a result, the standard response to the specific questions posed has
been a refusal to answer, combined with blanket condemnation of the entire
way of thinking about the problem.[2] Moreover, what is considered a fairly
mainstream position within the environmental ethics literature is so radical,
from the perspective of the standard policy framework, that there is essentially no room for productive engagement between the positions. Perhaps
the most significant example is that environmental ethicists are usually opposed to economic growth, whereas the policy debate is structured by the
assumption that economic growth will continue, and that it is desirable. One
might think that the latter point is obvious—as Darrel Moellendorf puts it,
"If mitigating climate change involves slowing the process by which billions
of people will climb out of poverty, perhaps people in the future are better off
with less mitigation."[3] And yet Moellendorf is one of the few philosophers to
take seriously this concern. The dominant tendency has been to assume that
economic growth is a malign force that must be ended.[4] This creates such a

Philosophical Foundations of Climate Change Policy. Joseph Heath, Oxford University Press. © Oxford University Press
2021. DOI: 10.1093/oso/9780197567982.003.0002

profound divide that it becomes almost impossible to have any productive conversation about real-world policy issues.

There is some irony in this, since environmental ethics arose as part of a broader movement within philosophy, starting in the 1960s, aimed at developing different fields of applied philosophy (bioethics, business ethics, etc.), in order to show how everyday practice could be enriched through philosophical reflection and analysis.[5] The objective was to prove the practical value of philosophical methods. Despite these early ambitions, however, environmental philosophy quickly became immersed in a discussion of what amounted to a metaphysical question, concerning the nature of value, and in particular, the possibility of developing a system of "non-anthropocentric" value to inform our thinking about environmental questions. The practical upshot of this tended to take the form of calls for a new "environmental consciousness," the transcendence of "materialism," or a new way of "being-in-the-world" that would produce social changes so fundamental that they would do away with the practices that were generating everyday environmental problems like pollution or habitat destruction.[6]

While many have found inspiration in the wisdom of environmental ethicists, such as Aldo Leopold's injunction to "think like a mountain," the downside is that, when it comes to addressing specific questions, such as how the burden of carbon abatement should be distributed in a global climate treaty, the theoretical framework fails to provide much guidance. (Indeed, it is not even obvious that a "mountain" would care about climate change.) Given the urgency of the problem of global warming, and the rather self-evident impracticality of traditional environmental philosophy, it was not long before a new generation of theorists began to weigh in on the question. These philosophers were not trained in applied ethics, but rather in political philosophy, and they brought a distinctly new sensibility to the discussion. Under the influence of John Rawls, their first impulse was to set aside the metaphysical questions that had preoccupied environmental ethicists, and to treat the climate change problem as an issue of justice. Central figures in this "second wave" of environmental philosophers include John Broome, Simon Caney, Stephen Gardiner, Catriona McKinnon, and Henry Shue.[7]

In certain respects, the theoretical tools deployed by these second-wave environmental philosophers were more appropriate to the task at hand, since the policy debate is essentially concerned with how *states* should respond to the problem of global climate change, and in particular, what legal restrictions should be imposed on greenhouse gas emissions. This is not really an "ethics" question, or a question of personal morality, it is instead an issue of *justice*, as the latter has been conceived of by political philosophers. Thus there would appear to be much greater potential for productive engagement between the participants in this second-wave philosophical discussion and those involved in the policy debates. Unfortunately, this has also largely failed to materialize. The reasons for failure, in this case, are more complex. Many of them are due to the old problem of the hammer's tendency to see everything as a nail. In this case, philosophers trained to think about issues of distributive justice have been inclined to regard climate change as just another "cutting the cake"-style distribution problem. And since they have thought very little about efficiency questions, there has been a remarkable tendency to ignore the fact that climate change policy also involves an optimization problem. More perniciously, there has been a concerted attempt to avoid the optimization issue, by producing theoretical frameworks that preempt the need to make any trade-offs in climate change policy. This is reflected in the rather hypertrophied aversion to cost-benefit analysis that pervades the philosophical literature.

The net effect of this has been a tendency, among both "traditional" and "second wave" climate ethicists, to fall back on a kind of scientism with respect to policy questions. It is sometimes suggested, for instance, that a global "carbon budget" can be calculated based on a target warming level, and then the appropriate policy response can be chosen in a technocratic manner.[8] This allows one to evade the difficult normative questions that arise when attempting to determine a just emissions level, and to focus purely on the distributive justice issue of dividing up the carbon budget. Unfortunately, there are some questions that science alone cannot answer. For instance, when contemplating the problem of global climate change, it does not count as a helpful contribution simply to say that we should stop emitting greenhouse gases, or that we should try to phase them out as quickly as possible. Although there is no doubt a great deal of wasteful consumption of fossil fuels, the fact remains that a significant fraction of our current consumption is extremely valuable. As Moellendorf has observed, much of the poverty in the world today is in fact "energy poverty."[9] The crucial questions in climate change policy debates all involve balancing the very real benefits of fossil fuel consumption against the damages that are being caused. Particularly since China has surpassed the United States as the world's largest emitter of greenhouse gases, the trade-offs between economic development and climate change mitigation can no longer be ignored.

My plan in this book is to promote a rapprochement between the philo-sophical and the policy discourse. My strategy for doing so is to work through, systematically, what I take to be the major barriers that have prevented more fulsome participation, on the part of philosophers, in the policy discussion. There are, in my view, five such barriers. First, the policy debate presupposes a stance of liberal neutrality. As a result, it does not privilege any particular set of environmental values over other concerns, such as economic develop-ment, or even just consumer satisfaction. Second, there is the assumption of ongoing economic growth, along with a commitment to what is some-times called a "weak sustainability" framework when analyzing the value of the bequest being made to future generations. Third, there is the analysis of climate change as fundamentally a collective action problem, not an issue of distributive justice. Fourth, there is the acceptance of cost-benefit analysis, or more precisely, the view that a carbon-pricing regime should be guided by our best estimate of the social cost of carbon. And finally, there is the view that when this calculation is undertaken, it is permissible to discount costs and benefits, depending on how far removed they are from the present. All five of these positions are, in my view, both reasonable and defensible. More importantly, despite being anathema to many environmental philosophers, they are not even considered controversial in policy circles—they are, rather, structuring presuppositions of the debate. Of course, being presuppositions, they are often not articulated as perspicuously as they might be, much less given an explicit defense. As a result, it is possible to mistake them for mere prejudices.[10] Thus my primary objective, in the discussion that follows, will be to make explicit these presuppositions and to offer a philosophical defense of them.

The end result, it should be noted, might easily appear to be a defense of economism, insofar as I think that the basic conceptual apparatus and normative approach being used by climate economists, such as William Nordhaus and Nicholas Stern, is correct.[11] At the same time, I hope to offer different *reasons* for thinking that it is correct. In particular, I in-tend to show that it can be defended on the basis of minimally controver-sial contractualism, rather than the usual sort of welfare consequentialism that economists presuppose.[12] (To put it more crudely, I intend to show that one need not be a utilitarian, or even sympathetic to utilitarianism, in order to accept the way that economists have been thinking about the problem. Rawlsian and post-Rawlsian liberals can, and indeed should, help them-selves to the basic position as well.) Finally, it is worth noting that, although

I have a few suggestions, I do not expect to resolve any of the major normative questions that arise in the policy debate. My objective is primarily to defend the way that the debate is framed. This may seem like an overly modest ambition, until one considers how profound and visceral the opposition to this framing is. In the same way that mainstream views in environmental philosophy generate policy recommendations that are too radical to be seriously contemplated, defending what amount to mainstream views in the policy debate involves taking positions that are philosophically radical, at least with respect to the current climate of opinion. Thus I would hope that my defense of these positions will be of some interest. In any case, if I succeed in promoting greater dialogue, or intellectual engagement, between the various parties, I would count that as a worthwhile achievement.

1.1. Traditional Environmental Ethics

Open a textbook on environmental ethics, and the first topic discussed is usually some version of the contrast between "intrinsic" and "instrumental" value—which is often taken to underlie the distinction between "deep" and "shallow" ecology. This preoccupation with the metaphysics of value is reflected as well in the fact that many of these texts feature prominent discussion of the question of animal rights—an issue that, superficially at least, would appear to have little in common with environmental ethics.[13] What ties them together is the desire to provide a framework for thinking about moral questions that is not bound by the constraints of "anthropocentrism." This is based on the observation that, historically, philosophical reflection on morality has focused primarily on the way that people treat one another. This emphasis has tended to obscure, or at least relegate to secondary status, the question of how people treat animals, as well as other aspects of the natural world. What many proponents of animal rights share in common with environmental ethicists is the desire to "expand the circle" of moral concern, by showing that more than just humans possess "moral standing," or are "morally considerable."[14]

When stated at a high level of generality, it is not difficult to see the point that is being made. Providing a more precise specification, however, has turned out to be quite difficult. The traditional way of doing it, in the environmental ethics literature, has been to introduce a distinction between two ways in which an entity can have value. It may possess instrumental value, insofar as it is able to serve the satisfaction of human interests, or it may possess intrinsic value, which is defined in various ways, but can be understood intuitively as value that it possesses "all on its own," or independent of its capacity to serve human interests. On the basis of this distinction, anthropocentrism is then defined as the view that only humans, or human attributes (such as states of human consciousness), possess intrinsic value, while the rest of nature has only instrumental value (e.g. Katie McShane writes, "Anthropocentrism . . . is the view that the nonhuman world has value only because, and insofar as, it directly or indirectly serves human interests").[15] On this view of things, all that needs to be done, in order to defeat anthropocentrism, is to show that nature possesses intrinsic value. This is then thought to provide the basis for "expanding the circle" of moral concern to include elements of nature.[16]

The argument that is often appealed to, in order to demonstrate the existence of intrinsic value in nature, is Richard Sylvan's "last man" argument.[17] Suppose that humans become infertile, and so the human population slowly ages and passes away. At some point there is only one man left alive. Would it be morally permissible for him, before he dies, to lay waste to nature, wantonly burning forests and killing animals? If the answer is no, then this shows that nature must possess more than just instrumental value, since by hypothesis, once the last man dies, these animals and forests can be of no further use to any human. On this basis, the need for a new, non-anthropocentric ethic is established, an ethic that will attempt to articulate more clearly, and that will encourage us to show greater respect for, the value inherent in nature.

The problem with the argument is that, while it may serve as a good test for uncovering non-instrumental value in the world around us, it does not necessarily move us outside the sphere of human valuation and concern. Imagine that the last man, instead of burning down forests, were to go on a rampage through the Louvre, smashing statues and destroying paintings. Would we not also say that he acts wrongly? Most people agree that a museum rampage would be equally wrong.[18] This is not surprising, since we normally regard aesthetic value as a type of non-instrumental, or intrinsic, value. And yet these artworks are human artifacts, with aesthetic qualities that only humans respond to. So the mere fact that value is intrinsic does not make it non-anthropocentric. On the contrary, intrinsic value can be seen as just another *mode* of human valuation, one that is not necessarily associated with, or conferred upon, nature.[19] Humans value things in different ways. One way that we can value them is for their use, or their ability to satisfy our interests and desires. Another way we can value them is for their non-instrumental qualities, such as their inner harmony and beauty. But the latter form of value is still anthropocentric, in the sense that it is conferred upon objects in the world by humans, and it may involve properties that only humans care about.

This is sufficiently obvious that it may lead one to wonder how the point could have been overlooked for so long. The most likely explanation is that early environmental ethicists, under Peter Singer's influence, took something like 19th-century utilitarianism as a template for understanding moral theories in general. The medieval worldview, one may recall, was one in which nature was suffused with goodness. Everything was taken to have its own particular "good," corresponding to the divine intention that informed its creation. One of the most revolutionary features of the scientific revolution was that it produced broad-based skepticism about this conception of the good as something embodied in nature (accompanied by a devaluation of the pattern of explanation that Aristotle referred to as "final causation").[20] This led some to search for a new, scientifically respectable form of natural good. Utilitarianism takes as its point of departure the suggestion that, whatever doubts one might have about goodness in nature, it seems unreasonable to deny that human happiness (or pleasure) is good, while suffering (or pain) is bad. If one goes on to assert that this is the *only* thing that is intrinsically good, then everything else will be good only insofar as it leads to the production of human happiness. If human happiness consists in the satisfaction of our desires, then it follows that anything valuable, other than human happiness itself, will be valuable only instrumentally, insofar as it can serve as a means to the satisfaction of human desires, and thus to the promotion of our happiness (or the reduction of unhappiness). Seen from this perspective, an anthropocentric theory of morality does seem equivalent to the claim that nature has only instrumental value (and Singer's claim, that this framework arbitrarily privileges human happiness, or pleasure, over that of other creatures, seem plausible).

It is not difficult to see, however, that many other moral theories disrupt these equivalencies. Consider a Humean theory of value. Hume was more thoroughly skeptical than the utilitarians were about the concept of natural goodness. On his view, absolutely nothing in the world is good in itself.[21] The world just is what it is. Human beings, however, experience a variety of sentiments in reaction to the world, and this leads us to value states of affairs in various ways. (Thus moral properties "may be compared to sounds, colours, heat and cold, which, according to modern philosophy, are not qualities in objects, but perceptions in the mind.")[22] In modern parlance, the claim is that various forms of value are "response-dependent" properties of states of affairs.[23] Despite being grounded in our reactions, however, these values need not be instrumental. When we watch a caribou being killed and eaten by a pack of wolves, we might find it upsetting, and so describe it as "horrific" or "cruel," and yet there is no implication that its badness is in any way linked to the frustration of our own desires or interests. It is bad for that particular caribou. Furthermore, the question of who or what has moral "standing" does not really arise, since one does not have to be a member of the circle of moral agents in order for one's existence to have value, or one's suffering to register as bad. Indeed, the whole question of "standing" is not really coherent on this view, suggesting that the concept is tied to an essentially utilitarian way of thinking about morality and moral valuation. So while Hume's view is thoroughly anthropocentric, in the sense that it regards all values as dependent upon human valuation, it has no tendency to instrumentalize nature.

Traditional environmental ethics posits three dichotomies, which are then assumed to line up with one another: anthropocentric/non-anthropocentric, instrumental/intrinsic, and artificial/natural. Yet as we have seen, there is no necessary alignment across the pairs, and different normative theories mix and match the qualities in various ways. The assumption that they must line up is typically based on either confusion or unstated postulates that arbitrarily narrow the range of options. The first step to clarifying these issues involves distinguishing two quite different senses of the terms "anthropocentric" and "non-anthropocentric." It is one thing to say that an object in the world has value independent of any *use* it may have to humans, and quite another to say that it has value independent of any human *valuation*. This distinction, however, is one that has been persistently confused by environmental ethicists. This confusion has been noted already in the literature, but the point bears repeating.[24] Many environmental ethicists are frankly inconsistent in the way that they construe anthropocentrism. J. Baird Callicott, for instance, states that "an anthropocentric value theory (or axiology), by common consensus, confers intrinsic value on human beings and regards all other things, including other forms of life, as being only instrumentally valuable, i.e., valuable only to the extent that they are means or instruments which may serve human beings."[25] In later work, however, he defines anthropocentrism as the view that "there exists no value independent from human experience."[26] He does not appear to notice the shift in the definition (which is significant, since the later definition classifies a Humean theory of value as anthropocentric, while the earlier one does not). In other cases, the two notions are run together in a single phrase. Eric Katz defines anthropocentrism as "the idea that human interests, human goods, and/or human values are the focal point of any moral evaluation of environmental policy, and the idea that these human interests, goods, and values are the basis of any justification of an environmental ethic."[27] To say that policy should be based on human values is quite different from saying that it should be driven by human interests, because one can endorse the metaethical thesis, that the goodness of the natural world is a consequence of human valuation, without endorsing any particular view of what this goodness consists in. In particular, one need not endorse the view that the goodness consists in some capacity to serve human interests.

The fact that various supposedly non-anthropocentric conceptions of the good still move within the circle of human valuation is particularly apparent in arguments that rely on Singer's "expanding the circle" style of reasoning (variously known as "moral expansionism" or "moral extensionism").[28] Singer's arguments all follow a familiar pattern.[29] He begins with a property P, which ascribes some value according to conventional moral criteria. He notes that we consider it uncontroversial to apply P to some class of entities (circumstances, events, etc.) X. These constitute the canonical cases. He then observes that there is some other class Y, which is like X in all morally relevant respects (or differs from X only in ways that are irrelevant from the moral point of view), and so on pain of inconsistency, we are obliged to apply P to all entities in Y as well. Thus the value judgment is extended from the canonical to a set of non-canonical cases. (The somewhat more dubious version of the argument takes some class X, to which we apply P, observes that there is no way of drawing a clear distinction between X and Y with respect to the morally relevant properties, and so concludes that, since there is no principled basis for restricting the application P to X, it must be applied to Y as well. In the animal rights literature this is often referred to as the "argument from marginal cases.")[30] This is how Singer gets from "human suffering is bad" and "animals suffer just like humans," to "the suffering of all sentient creatures is bad."[31]

This method is obviously anthropocentric, because it takes a canonical set of human values or human moral judgments as its point of departure, then merely extends the scope of application. Furthermore, it is difficult to see how a theory that was not anthropocentric in this way would work. As Tim Hayward has observed, "The ineliminable element of anthropocentrism is marked by the impossibility of giving meaningful moral consideration to cases which bear no similarity to any aspect of human cases. The emphasis is on the 'meaningful' here: for in the abstract one could of course declare that some feature of the nonhuman world was morally valuable, despite meeting no determinate criterion of value already recognized by any human, but because the new value is completely unrelated to any existing value it will remain radically indeterminate as a guide to action."[32] In other words, what constitutes the circle that Singer wants to expand is conventional human morality. His ambition is to expand it so that all of the judgments and motivations that we apply in the familiar human cases will be extended to the new ones. He is not trying to introduce a set of new, disjoint circles, representing genuinely nonhuman values—that would be "multiplying the circles," not expanding the familiar one.

Environmental ethicists, of course, have wanted to go much further than Singer when it comes to expanding the circle, taking it beyond the realm of sentient creatures. Indeed, the problem for environmental ethicists has been not so much expanding the circle as finding some place to stop. There has been a steady progression outward, with each proposed stopping point being subjected to the same argumentative pressure. Singer, for instance, is concerned with the suffering of sentient creatures, but what about marginal cases, like lobsters or sea cucumbers? How many ganglia must there be in a nerve cluster before we declare it to be a "brain"? And what about ancient redwoods, great baobabs, or tropical rainforests? Do they not deserve some consideration?[33] And if we are concerned about plants, then what about soils, coral reefs, or water systems?[34] In the same way that the traditional utilitarian, who is concerned only with human suffering, stands accused of "speciesism," or a chauvinistic concern for humans, the more broad-minded utilitarian, who shows a concern for all animal suffering, can be (and has been) accused of "sentientism," or of chauvinistic concern for fellow animals.[35] Singer's basic argument can be iterated. In the same way that it is difficult to draw a principled distinction between X and Y, there may also be some Z, which is difficult to distinguish clearly from Y. Should it not be part of the circle as well? As Warwick Fox has shown, this basic argumentation pattern has generated an increasingly expansionist ethic, moving from sentience ethics, to a biological or autopoetic ethics, to an ecosystem ethics, and ultimately to a "cosmic" ethics.[36]

With each iteration of the argument, the expansionist forces accuse their opponents of being narrow-minded, chauvinistic, or anthropocentric in their thinking, essentially fetishizing a set of human preoccupations. Thus a contest begins to see who can be the most open-minded. The concern for the suffering of individual creatures, for instance, can easily be made to seem like a parochial human concern, when compared to broader worries about what is good for a species. And yet the concern about species is also rather narrow, since the species boundary is much more important to animals than it is to other life forms, including plants. And in any case, species are just a part of a much larger set of natural systems. How should one feel about invasive species that threaten the integrity of local ecosystems? These higher-level concerns are what motivated Leopold to articulate his "land ethic," with its classic statement of the "ecocentric" perspective: "A thing is right when it tends to preserve the integrity, stability, and beauty of the biotic community. It is wrong when it tends otherwise."[37]

Many of these specific arguments are bolstered by a rhetorical strategy that points to various discredited hierarchies of the past, such as inherited aristocratic privilege, or the system of racial inequality that was used to defend slavery.[38] In each case, moral progress was a straightforward good-versus-evil battle between those who wanted to expand the circle and those who sought to limit it. The good guys, it is noted, have always been those who were in favor of expansion. Although the demands of abolitionists or feminists were initially condemned as "radical" or "absurd," they proved right in the end. So what can we learn from this? As an *inductive* argument, pointing to the history of expansionists being right clearly has no merit. After all, there must be *some* point at which the circle stops expanding, where it genuinely is too radical or absurd to go further. How do we know that we have not reached that point? The history of moral progress provides no resources for answering this question. And yet the historical analogies show up again and again, suggesting that they provide a powerful motivation for the expansionist view. After all, who wants to risk being on the wrong side of history? Who wants to be looked back upon and judged, the way that we judge apologists for slavery? Indeed, the historical argument for expansionism has some similarity to Pascal's wager, where the costs of overexpanding the circle appear trivial compared to the eternal opprobrium one might attract from fighting to keep it narrow.

And yet the expansionist argument runs the obvious risk of generating a reductio ad absurdum—something that authors in this tradition have not always been as concerned about as they should be.[39] Sue Donaldson and Will Kymlicka, for instance, offer as their primary defense of animal rights a Singer-style argument, claiming that there is no way of drawing a princi-pled distinction between humans and nonhuman animals.[40] And yet, if the boundary is not to be drawn at the species boundary, then it must be drawn somewhere else, such as the sentience boundary. But is there a principled way of drawing a distinction between the sentient and the non-sentient? When it comes to discharging this burden of proof, Donaldson and Kymlicka offer no more than hand-waving assurances that it can be done.[41] Thus there is a dramatic asymmetry in the burden of proof that they impose on the "spe-ciesist" and what they are willing to shoulder themselves. In arguing against a species boundary, they point to marginal cases, such as severely disabled children, to show the impossibility of providing criteria that include humans but exclude animals. But difficult cases like this can be used to undermine *any* boundary. Defining sentience in terms of the pursuit or possession of "interests," for instance, is subject to the obvious objection that plants also have "interests." Indeed, there is a sense in which any self-organizing, or autopoetic, system has "interests," and it is not clear why reflective aware-ness of those interests should be important.[42] This is precisely what drives the push toward a "life ethic," or even a "land ethic." There is some point, however, at which the circle becomes implausibly large. Fox's typology of moral expansionism goes beyond just ecosystem ethics, pausing to consider "earth system" ethics, before culminating in what he calls "cosmic purpose" ethics, or a general worship of negative entropy.[43] Assuming that we find this endpoint absurd, and yet can find no principled point at which to fix the boundary of moral "considerability" that stops short of this, then it follows that we will have to fix it somewhere arbitrarily, or in a way that is somewhat unintuitive (along the lines of a precisification of a vague predicate). Since any specification will require an intuitively unsatisfactory treatment of some marginal cases, it seems that we are free to pick and choose which ones we would like to impose this upon. If so, then what is to stop us from putting it back at the boundary of the species?

This analysis suggests that there is something wrong with the expansionist argumentation strategy. The "argument from marginal cases" is, in a sense, too effective, since it can be used to criticize *any* proposed limitation on the domain of moral considerability. The argument is commonly used to attack one particular way of drawing the boundary, when in fact it undermines every way of drawing the boundary. This makes it more like a skeptical argument than a substantive philosophical claim. At very least, it generates a regress problem that has a lot in common with certain traditional forms of skepticism. (Think, for example, of the difficulty one might have if given a room full of people and asked to separate those who were "bald" from those who were "not-bald." Every cutoff that one adopts will be vulnerable to the exact same objection that undermined one's previous attempt to impose a cutoff.)[44]

And yet if one drops the expansionism in favor of a more radically non-anthropocentric perspective, one that does not take as its point of departure any accepted criterion of human value, it gives rise to a different, and yet equally troublesome, set of skeptical problems. Suppose we stifle our initial reaction to the wolves eating the caribou, and instead of engaging in moralizing condemnation, grant that it may actually be "good" for the wolf, or for the ecosystem. We can perhaps follow Paul Taylor's recommendation, and adopt the "point of view" of other organisms. ("*Conceiving of it as a centre of life, one is able to look at the world from its perspective.*")[45] Think of this as "moral saltationism," where instead of expanding the human circle, we instead make a mental leap to adopt the perspective of different natural systems. (This is a bit more difficult to do in the case of an ecosystem, rather than a life form, but suppose that we are able to work something out.) This still leaves us with the question of what we are supposed to do with the judgments made from these perspectives. Most obviously, they need not be consistent with our own or each other—what is "good" for an organism might be "bad" for an ecosystem and vice versa. Furthermore, as Hayward notes, there is a question about why or whether the non-anthropocentric values should be action-guiding for us, insofar as they do not genuinely speak to us, or to our values. Should we defer to the wolf's good? Why not let the wolf see things from the wolf's perspective, while we continue to look at things from our own, human perspective? (Does that not seem more "natural"?) If we decide to give some weight to the wolf's values, it must be because there is some reason, grounded in *our* values, to accord such deference. But then we are back to the anthropocentric perspective. So again, our attempts to escape from the anthropocentric perspective seem to generate a raft of skeptical problems (many of which are familiar from discussions of moral relativism).[46]

There is another problem with the attempt to formulate a non-anthropocentric ethic, which affects not only the expansionist strategy, but also any attempt to develop a more radically non-anthropocentric ethics. The problem is that proponents of various non-anthropocentric schemes disagree with one another—in many cases profoundly and vociferously—about what the non-anthropocentric values are and where they lie. For instance, consequentialist animal rights theorists are often inclined toward the view that, if animal suffering is a great evil, then humans have an obligation not only to refrain from inflicting it themselves, but also to minimize the amount of suffering that occurs in nature.[47] In some cases, this can motivate highly interventionist projects for a "cruelty-free nature," which involve primarily limiting the activities of carnivores. For instance, there is talk of using genetic engineering as a way of transitioning carnivores to a vegetarian diet (and in the case of hopeless species, such as the *felidae*, mass euthanization). Environmentalists, obviously, are horrified by these ideas, which are interventionist beyond the wildest dreams of suburban developers and oil company executives. Fox describes these projects as "ecological lunacy" and as an extension of "the modern project of totally domesticating the non-human world."[48]

More generally, there is a tendency among ecologists to consider the concern over individual animals, or over animal suffering, to be mere sentimentality.[49] At very least, these views seem to be grounded in a fundamentally unnatural attitude toward nature. As Mark Sagoff observes:

> The principle of natural selection is not obviously a humanitarian principle; the predatory-prey relation does not depend on moral empathy. Nature ruthlessly limits animal populations by doing violence to virtually every individual before it reaches maturity; these conditions respect animal equality only in the darkest sense. Yet these are precisely the ecological relationships which Leopold admires; they are the conditions which he would not interfere with, but protect.[50]

It is this respect for the integrity of natural systems that leads environmentalists to support initiatives like the reintroduction of wolves into national parks, precisely on the grounds that they will restore greater balance to ecosystems from which apex predators have been eliminated. In this context, it is worth noting as well that Leopold was an avid hunter, and Callicott a principled omnivore—in both cases, commitments that they justified by adopting an "ecosystem" or "biome" perspective, rather than focusing on individual animals. (Both men regarded the preoccupation with individual members of a species as a bias inherited from traditional anthropocentric moralities.) Many environmentalists also see no use for various domestic species, and would be happy to see them phased out through mass sterilization or euthanization. (This is particularly true of housecats, which are ecologically catastrophic. One recent study blames cats for the killing of as many as four billion birds and 22.3 billion small mammals per year, in the United States alone. Domestic cats are also thought to be responsible, globally, for the extinction of at least 33 species.[51] This has given rise to particularly acrimonious debates, perhaps because the concern for animal welfare appears to be felt most acutely by individuals whose interactions with animals have been dominated by relations with household pets.)[52]

Setting aside animal rights theorists, one can find equally profound divisions among environmental ethicists. Many of the disagreements revolve around the question of how far one should go in trying to escape from the chauvinism of traditional morality and its preoccupation with characteristically human concerns. This line of thinking quickly raises some very difficult issues. Our concern about species extinction, for instance, seems heavily skewed toward vertebrates, to the relative neglect of invertebrates, much less microscopic or single-celled organisms. More generally, both speciation and extinction are natural processes, and while the rate of extinction currently being caused by human interference is alarming, a sufficiently robust ecosystem has the capacity to produce new species. As far as ecosystems are concerned, we tend to favor those with aesthetic qualities that we value. Thus we worry a great deal about forests, and much less about grasslands, and even less about microenvironments. Our concepts of integrity, or stability, also seem to be very much tailored to human sensibilities. Human life spans are extremely short on the timescale of planetary life. So when we worry about the fragility or stability of ecosystems, we are typically thinking of their dynamics on a timescale of years, perhaps decades. But why not "think like a mountain" and treat them as systems that persist over centuries, millennia, or even eons? (Climate change, for instance, will produce a large number of species extinctions in the short run, but temperature increase will favor increased biodiversity in the long run.[53] Which is the appropriate timescale to evaluate its effects?)

Furthermore, there is the danger that, in striving to escape from our narrow human prejudices, a non-anthropocentric morality will distance itself *too much* from human concerns, and wind up being misanthropic. Caring too much about nature can easily lead one to care too little about people. Callicott, for instance, earned the accusation of being an "ecofascist," particularly through his expressed desire to limit human population in order to minimize disruption of natural systems.[54] (His argument against vegetarianism, for instance, is that it would increase the amount of food available for humans, which would in turn promote ecologically damaging population growth. The implication would appear to be that a certain amount of human starvation is desirable, for essentially Malthusian reasons.) His response to the accusation was to insist that his theory of value was pluralistic, so that the concern about ecosystems did not rule out a concern for animal suffering or human welfare. But then of course the question of how to mediate conflicts between these different values becomes even more pressing (as Bryan Norton has observed).[55] To make things worse, many writers have insisted that the values in question are incommensurable, such that they cannot even meaningfully be compared, much less traded off against one another.[56]

Complicating the issue further is the fact that the various forms of intrinsic value posited by environmental ethicists do not automatically trump instrumental values.[57] Most people recognize that if it is possible to bring about some state of affairs with very large instrumental value, at the expense of some small sacrifice of intrinsic value, the trade-off must be contemplated. Environmental ethicists have often tried to finesse this point by insisting that the two forms of value are also incommensurable, and that deliberation over environmental questions should focus only on intrinsic value. This is an indefensible position with respect to environmental policy, since it tacitly assigns lexical priority to intrinsic values, no matter how trivial. But if anything less than priority is given to intrinsic value, then the two forms of value will in practice be traded off against one another, which means that they will be rendered commensurable. The only remaining question will be whether they are being commensurated in a way that is consistent across time and across decisions, something that is unlikely to occur if the weighting system is not formally articulated.

Turning to the specific problem of climate change, it is not difficult to see how environmental philosophy in the traditional style has had difficulty formulating a coherent response. Climate change is of course a very complex process, which will result in a vast number of different effects. The damage it will cause, however, can be analyzed under four major headings: (1) agricultural impacts, (2) extreme weather events, (3) sea-level rise, and (4) species extinctions.[58] What is striking about these effects is that, with the exception of the last, they are all primarily effects on human systems and human welfare. As a result, proponents of a non-anthropocentric ethic are likely to be somewhat more ambivalent about these effects than, say, members of the general public. We humans have carved out an ecological niche for ourselves, one that we think of as very robust, but in fact depends rather heavily on late-Holocene climatic conditions. It includes, for instance, reliance upon intensive agriculture, and by extension domesticated animals, for almost all of our nutritional needs. Much of this has been possible because of the unusual climatic stability that has prevailed for the past 10,000 years—which despite being no more than the blink of an eye in the history of the planet, represents the entire period in which complex human societies, based on agriculture and sedentary population centers, have existed.[59] Climate change threatens this niche, something that is a problem for us, but is not obviously a major concern for nature.

Indeed, agriculture is extremely unnatural and in many respects destructive, and so climate change offers nature an opportunity to reclaim some of what has been lost. For instance, large-scale collapse of agricultural productivity in India, due to failure of the monsoon, would be a *human* catastrophe. And yet human agriculture is widely regarded as being already an ecological catastrophe. Thus what we regard as an agricultural collapse could very well be regarded, from a non-anthropocentric perspective, as land being "rewilded." Similarly, the creation of millions of climate refugees from the global south, which would put enormous strain upon human social and political systems, might be largely a matter of indifference from the perspective of nature. Thus someone who takes a non-anthropocentric perspective seriously is likely to regard the effects of climate change on agricultural productivity as a mixed blessing.

Similarly, sea-level change is something that has happened quite regularly in the history of the planet, and indeed, has even fluctuated quite dramatically during the period in which humans have inhabited the earth. The oceans have risen by around 100 meters in the past 13,000 years, through entirely natural processes associated with the end of the last ice age. The problem posed by climate change is that, right now, 10 percent of the human population lives in coastal regions that are less than 10 meters above sea level.[60] This creates a problem of significant importance for people, while at the same time, one struggles to think of any reason why it would be a matter of less than total indifference to nature. The flooding of Venice would be a great loss to humanity, but Venice has only been a human population center for around 1,500 years. Before that it was a lagoon, and if it returns to being a lagoon, its brief metamorphosis into a center of human learning and culture will be but a small footnote in the much longer, natural history of the place.

Finally, it is worth observing that the basic mechanism through which climate change is occurring is not unnatural, in the way that many other human interventions in natural ecosystems are. The creation of plastics, or of synthetic chemicals such as dioxins, is much more unnatural, in the sense that these compounds do not exist in nature and cannot be broken down by microbes or other organic processes. Release of carbon dioxide into the atmosphere, by contrast, is something that occurs naturally, and indeed, by burning fossil fuels, humans are simply returning to the atmosphere carbon that was previously sequestered by plants. Looking at the history of climatic fluctuations (such as the period known as the Paleocene-Eocene Thermal Maximum [PETM], when the entire planet was ice-free), one can see that a stable climate is not the natural baseline.[61] (It is, in fact, a rather peculiar feature of the late Holocene.) The major problem with anthropogenic climate change is the abruptness of the changes (the current rate of carbon injection into the atmosphere is about 20–50 times greater than during the run-up to the PETM). If what we were doing was occurring more slowly, it would be difficult to formulate any ecological objection to it.

Apart from the issue of species extinction, it is almost misleading to characterize climate change as an "environmental" problem at all. It is a set of problems caused by changes to the environment, but the problems themselves are almost all ones that affect primarily people and that threaten the "built environment."[62] Climate change is, first and foremost, a threat to complex human civilization, particularly through its impacts on agriculture and infrastructure. While humans have destroyed the ecological niche occupied by an untold number of different species, what distinguishes climate change is the fact that it represents the first large-scale, credible threat to the ecological niche that *we* inhabit. Environmental philosophy, however, has encouraged a certain measure of ambivalence about the value of human civilization, and indeed, has often criticized it on the basis of its ecological impacts. As a result, that same philosophy tends to encourage ambivalence about climate change. At the very least, it offers no way of balancing all the different considerations that play a role in determining an appropriate response to the challenge of climate change.

This is why traditional environmental ethics has played no constructive role in the current debates over climate change policy. Several decades of reflection on intrinsic value and non-anthropocentric ethics have produced little more than a bewildering proliferation of candidate values, with no clear way of adjudicating among them.[63] Speciesism, sentientism, biocentrism, and ecocentrism are each, in their own way, plausible, and enjoy a loyal following. There is no sign of imminent convergence toward one or another view. The reason, I have suggested, is that the underlying moral expansionism is being driven by an argument that is a disguised form of skepticism, and so there is no principled resolution to the question. Beyond this, environmental philosophy has raised additional skeptical problems, about what a genuinely non-anthropocentric ethic would look like. If there is value in nature, how would we know what it is, and why should we care about it? And how should we balance or adjudicate among these values, in cases of conflict? As a result, it is not difficult to see why this particular stream of philosophical reflection has not had a great impact on policy. It is all so controversial, and so contested, that it is practically non-actionable.

1.2. Liberal Environmentalism

Those who approach these questions from a background in political philosophy will recognize the situation that I am describing. The easiest way to sum up the current state of environmental ethics is to say that the *fact of pluralism* prevails in this domain. Different people hold different values (or "conceptions of the good"), and yet their disagreements are reasonable, in the sense that no one is making any obvious errors, and it seems unlikely that further dialogue will produce any significant reduction in the level of disagreement. There is an element of *judgment* in all of these positions, as different individuals weigh the various considerations in different ways, often in the light of their own personal experiences, with no clear standard or index to guide them. In particular, the regress problem generated by the moral expansionism strategy shows that *any* way of fixing the boundary of moral concern is going to be fraught with difficulty, and thus will involve an exercise of judgment. This suggests that reasonable people may reasonably disagree about where the boundary should be drawn. The saltationist strategy leads to a similar difficulty, since it postulates a variety of "points of view," or centers of value, while at the same time undermining the authority of any perspective that could be used to select one of them as the basis for action. In the private sphere, this indeterminacy is not such a problem, but it does create a problem for policy, because the latter involves the use of state power. The fact that the law is coercively imposed on all citizens means that basing it on a controversial conception of the good has the potential to exacerbate the underlying conflict.

It is a natural assumption to make in these situations that value pluralism represents a significant impediment to the development of state policy, and so the only way to proceed is by determining who is right and who is wrong. The emergence of liberalism, in the early modern era, is based on a rejection of this assumption. In the *Leviathan*, Thomas Hobbes broke with the received tradition of Western political philosophy by dismissing questions of the good at the outset, arguing that they are essentially irresolvable ("whatsoever is the object of any mans Appetite or Desire; that is it, which he for his part calleth Good").[64] Hobbes observed that despite these disagreements, individuals may still benefit from entering into cooperative relations with one another (or in John Locke's later formulation, from respecting each other's rights).[65] Thus Hobbes laid down a set of principles for the organization of civil society that would not presuppose the correctness of any one individual's conception of the good. The goal was to develop principles that could guide state policy *despite* underlying disagreements about questions of the good.

At the time, the conflicts that Hobbes sought to avoid were all religious in nature, and liberalism arose as a response to the problem of religious warfare in Europe. And yet the basic blueprint for the state that later liberals laid out proved to be of much greater applicability. The idea of state neutrality with respect to controversial questions of the good has proven to be successful at managing a wide range of social and political conflicts. When extended from religion to ethnicity, for instance, it provides the basis for state multiculturalism and anti-discrimination policies, which established the basic resources for managing immigrant integration. State neutrality also provides the conceptual foundations for the doctrine of limited government, along with respect for private contract, which in turn played an important role in the emergence of the market economy. Under the banner of "consumer sovereignty," the concept of liberal neutrality has informed the view that markets should be responsive only to the intensity of individual preferences, rather than favoring one or another preference based on its content.

There are two central ideas in the liberal conception of the state, which historically were articulated in the language of social contract theory. The first is that individuals leave the state of nature and enter the "civil condition" only because it is mutually advantageous to do so. In other words, they are willing to subordinate themselves to the power of the state because this actually makes each individual better off, from his or her own perspective.

Thus the state, as well as the system of laws that it imposes, can usefully be thought of as creating a system of cooperation. The normative principles that govern the basic institutional structure of society are ones that are able to bring about agreement on such a system of cooperation. The second major idea is that individuals, upon entry into civil society, do not surrender all of their powers to the state, but only those that are necessary to bring about the relevant system of cooperation. The powers that are not surrendered constitute the domain of individual rights, which in turn serve as a constraint on the scope of state power.

Liberalism has over time developed into a complex family of views. What I have provided here is only a vague indication of the major animating ideas. These points are sufficient, however, to identify the defining features of what we may refer to, non-pejoratively, as "liberal environmentalism."[66] The centerpiece of this approach is the commitment to neutrality, or the desire to avoid taking sides on controversial conceptions of the good. With respect to the environment, this manifests itself primarily as an unwillingness on the part of liberals simply to declare environmental values superior to all others, such that they may dictate public policy. To caricature somewhat, if we think of society as divided into two groups, "tree lovers" and "tree haters," traditional environmental philosophy (and much of traditional environmental activism) sets itself the goal of converting everyone into a "tree lover," and thus seeks to make policy that directly enforces the love of trees. Liberal environmentalism, by contrast, starts out from the assumption that the conflict between the two groups may be irresolvable, and therefore seeks to moderate the inevitable conflicts that arise in a way that will be fair to all parties. Translated into policy terms, this means that when it comes to the conflict between conservationism and resource extraction, or between pollution and economic development, liberal environmentalists refrain from declaring one set of values superior to the other, but insist rather on treating them all as equally legitimate *from a policy standpoint*. Thus liberal environmentalists have attempted to apply the conceptual apparatus of contemporary liberal theories of justice—theories that are crafted with the aim of respecting neutrality—to the problem of global climate change.

This insistence on neutrality is the first, and in some ways the most important, stumbling block to the acceptance of the standard policy framework by traditional environmental ethicists. Most of us find it very easy to accept the constraints of liberal neutrality when these serve to limit the public influence, or access to state power, of groups espousing values that we *reject*. The majority of political philosophers, for instance, do not share the values of conservative Christians, and so are predisposed to accept a political philosophy that prevents these same Christians from using state power to enforce their views on, say, sexuality upon the population. Given that the private moral views of political liberals tend to be liberal as well (e.g. viewing consent as the primary criterion of permissibility in sexual behavior), their commitment to political liberalism does not require any sacrifice with respect to their private values. In the domain of the environment, by contrast, liberals are much more inclined to hold perfectionist values (such as a belief in the intrinsic goodness of ecosystems). As a result, the commitment to political liberalism requires, for the first time, that they restrain themselves, by refraining from using state power to impose their own values. This creates a tension between their political and their private commitments. At this point, many turn out to be "liberals of convenience"—willing to appeal to the doctrine of liberal neutrality to keep other people's values out of the domain of public policy, but unwilling to apply it to their own.

Some environmental philosophers have seen the difficulty here, and have responded by developing, with varying degrees of ingenuity, arguments aimed at showing that the environment is somehow exceptional, such that policy in this domain must be based on first-order value commitments.[67] Others philosophers have recognized that, no matter how sincerely held or strongly espoused, their personal value commitments remain controversial, and *as such* cannot serve as a basis for policy. Furthermore, when it comes to the problem of climate change, the urgency of the problem simply militates against any effort to secure value-consensus as a precursor to policy action. This explains the disconnect between the work done by second-wave environmental philosophers such as Henry Shue, Stephen Gardiner, Simon Caney, Catriona McKinnon, Lukas Meyer, and John Broome, and the older tradition of environmental philosophy (a disconnect that manifests itself, most obviously, in the fact that the latter groups of theorists are almost never mentioned or cited by the former).[68] Coming from political philosophy, second-wave theorists all accept the basic premise of liberal neutrality, along with its normative language of efficiency and optimality, of equality and distributive justice, and of human rights. As a result, they are not interested in the metaphysical questions that preoccupied the earlier generation of environmental philosophers—indeed, they regard these questions as precisely the sort that must be bracketed when addressing matters of public policy. (They are, in this respect, cleaving to Rawls's claim that justice is "political, not metaphysical.")[69] Whatever their other doctrinal disagreements, members of this second generation of environmental philosophers have been united in the view that the normative language of liberalism is adequate to the problem of climate change, and so there is no need to wade into the complex debates over anthropocentrism and intrinsic value.[70]

To take just one example, the concept of a "negative externality" provides a characteristically liberal way of formulating many environmental problems, because it looks at the specific harms that one individual's actions impose upon another. This boundary-crossing may be prohibited on various grounds consistent with liberal principles, e.g. as a harm, a rights violation, or a source of inefficiency. In particular, negative externalities, when unchecked, generate collective action problems (or "market failures"), which are just another way of describing failures of cooperation. Failures of cooperation make everyone worse off, each from his or her own perspective, and so there is no need to secure any deep value consensus in order to recommend their elimination. Thus it has struck many people that there is a great deal to be gained from formulating environmental problems in these terms, especially when the problem is urgent and requires a near-term political response. It allows one to develop solutions to environmental problems without having to appeal to the correctness of any specifically environmental *values* in their justification.

At the same time, these points about neutrality are quite subtle, and must be defended with a great deal of care. In the following chapters, I will get into some detail on several of these points. For the moment, however, I would like to observe that beyond the general point of agreement regarding the sufficiency of a liberal theory of justice there are many internal disagreements among political liberals. Although almost every liberal view combines a small set of normative principles in some configuration, there are differences in emphasis, which give rise to three major schools of thought:

1. *Welfarist liberalism.* Perhaps the least controversial principle of modern liberalism is the Pareto-efficiency principle, which states that an outcome is better, from the standpoint of justice (or of "society"), if it makes at least one person better off without making anyone worse off. The commitment to neutrality is reflected in the fact that what counts as "better off" for a particular individual is determined by that individual's own preferences. This is often signaled by a formulation of the principles of justice in terms of individual utility, or welfare, which

is defined in terms of preferences. In many ways, the commitment to efficiency in this sense is just a way of articulating the core idea of cooperation, which is that it constitutes an interaction that makes everyone better off. Welfarist forms of liberalism are ones that assign significant, and sometimes overriding, importance to Pareto efficiency. The most well-known versions of this are associated with the discipline of welfare economics, which for most of the 20th century was basically a school of applied Paretianism.

2. *Liberal egalitarianism.* The Pareto-efficiency principle is silent on the question of how the benefits of cooperation are to be distributed among the participants in the cooperative scheme. Thus its natural complement is a principle of equality, which specifies that the division should be one that, to the extent possible, equalizes these benefits. Again, this equality principle is one that can be given a neutral formulation, if benefits are specified in welfarist terms, or if the value of a bundle of resources is calculated using a subjective metric of value.[71] Ronald Dworkin's conception of equality as envy-freeness is perhaps the best-known example of the latter approach.[72] Regardless of the different goals and projects that individuals may be inclined to pursue (in accordance with their private conceptions of the good), there is a sense in which they can be given equal opportunities (or resources, capabilities, etc.) to achieve these goals and projects. This form of liberalism, which puts particular emphasis on the equality principle, has become perhaps the most widely held view in normative political philosophy.[73]

3. *Libertarianism, or "human rights" liberalism.* Older versions of liberalism were distinguished by a greater emphasis on individual rights. Classical liberalism tended to construe these rights as negative liberties, whereas more contemporary forms often work with a broader set of social rights. Either way, the doctrine can be given a neutral formulation, insofar as rights are seen as providing, or protecting, a certain set of opportunities individuals have to pursue their own projects. The term "libertarianism" is normally associated with views that put major emphasis on the commitment to liberty or autonomy, and see individual rights as the primary vehicle through which this commitment is institutionalized. More left-wing theorists often use the language of "human rights," which they interpret expansively, to include various aspects of economic well-being.

Most liberal views, it should be noted, assign some weight to all three of these basic normative principles, and so the question of which camp they belong to is determined primarily by the way that they prioritize them, when it comes to situations of conflict. The normative framework presupposed by many economists, for instance, is one that assigns lexical priority to the efficiency principle, then brings in the equality principle as a way of selecting among outcomes on the "Pareto frontier." Rawls's difference principle, by contrast, is essentially a formula for trading off equality concerns against efficiency gains.[74] It assigns priority to equality, but then permits deviations when doing so generates a benefit for the worst-off representative individual.

When it comes to the specific issue of climate change, policy discussions have been dominated by welfarist liberalism. On this view of things, the most fundamental problem with climate change is that it is inefficient. Greenhouse gas emissions are a textbook example of a negative externality—a byproduct of economic activity that has a negative impact on the welfare of others, but which is unpriced, due to incompleteness in the system of property rights. The presence of negative externalities generates a collective action problem (or a "tragedy of the commons"), in which individuals, as a group, make themselves worse off through their actions, but where no one has any incentive to stop, because the benefits of each individual's actions accrue to that individual, while the costs are born primarily by others.[75] To put it in more precise language, the situation is one in which the strategic equilibrium of the interaction is Pareto-suboptimal. In principle, therefore, it is possible to impose restrictions on individual actions that will make everyone better off, by moving the outcome in a Pareto-improving direction.

From this perspective, the fundamental problem is not that fossil fuel is being burned and greenhouse gases produced, but that *too much* fossil fuel is being burned, relative to the benefits that are being produced by its burning (similarly, too much concrete is being laid, too many forests are being cleared, and so on), because some fraction of the costs are externalized. *Some* greenhouse gas production is desirable, even if it results in overall warming of the planet, because the value of the economic activity that generates those gases as a byproduct is much greater than the damages caused. It would, for example, not be possible to feed all of the people alive today without significant consumption of fossil fuel. It is not even possible to make the transition to renewable energy without consuming fossil fuel, because both solar panels and wind turbines contain components that can only be manufactured using extreme heat that, for now at least, can only be produced by burning fossil fuel. The problem is that not only are these high-priority economic activities occurring, along with their associated carbon emissions, but that a great deal of discretionary activity is occurring as well. The primary policy question therefore involves determining how *much* fossil fuel we should be burning, and then how to bring actual emissions into line with those targets.

When the question is posed in this way, it naturally frames the issue in terms that are extremely familiar to economists, and amenable to modeling using the economist's traditional toolkit. Reducing greenhouse gas emissions (which we can refer to, for convenience, as "carbon abatement")[76] is subject to increasing marginal costs—the more of it we do, the more costly it becomes to do more—as well as decreasing marginal benefits—the more abatement we do, the less benefit we get from doing more (as shown in Figure 1.1). To put it somewhat roughly, but more tangibly: reducing expected warming from 3°C to 2°C will produce much greater benefits than reducing it from 2°C to 1°C, while achieving the first 20 percent reduction in global emissions will be significantly less painful than achieving the next 20 percent reduction.[77]

The primary question in climate change policy is how much carbon abatement should be undertaken. The representation of the problem in Figure 1.1 strongly invites the suggestion that carbon abatement should be undertaken up until the point at which the marginal cost (MC) of further abatement is equal to the marginal benefit (MB)—at point *r*. This transforms the question into a fairly standard optimization problem. Solving the problem is, of

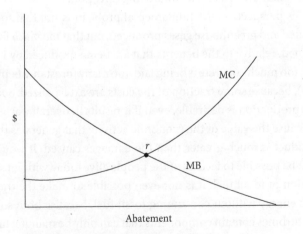

Figure 1.1 Costs and benefits of carbon abatement

course, fearsomely complex, but the *normative* question is relatively simple on this view.

Many philosophers, however, have resisted this way of posing the problem. This is true not only of traditional environmental ethicists, but also of liberal political philosophers. Some of this is just a hypertrophied aversion to "economism."[78] Many also feel that, when it comes to the environment, we should be atoning for our sins rather than calculating an optimal growth trajectory. In other cases, there is a principled objection to the way that the welfarist approach is willing to contemplate trade-offs of various kinds. After all, the procedure that one must follow, in order to determine the intersection of the two curves shown in Figure 1.1, is basically a cost-benefit analysis. Cost-benefit analysis, in turn, is often thought to presuppose a commitment to consequentialism, or utilitarianism, as a comprehensive moral view. Thus many liberal philosophers have sought to develop a more deontological approach to the problem, based either on an egalitarian or a rights-based theory. Some of the more moderate versions of this will be discussed over the course of the next two chapters. For the moment, however, I would like to consider two attempts that have been made to reject the entire welfare-liberal approach to the problem. The first of these essentially rejects the characterization of climate change as a collective action problem, instead choosing to treat it as a "cutting the cake"-style distribution problem.[79] Normatively, this suggests that the most important principle that must be brought to bear upon it is equality, rather than efficiency. The second approach tries to avoid dealing with trade-offs by framing the issue, not as an optimization problem, but instead as a question of rights, claiming that our greenhouse gas emissions are impermissible because they violate the rights of future people. Both of these approaches are, I believe, dead ends, and so I would like to discuss them primarily in order to set them aside, so that subsequent chapters can focus on what I consider to be more productive ways of thinking about the problem.

1.2.1. Climate Justice

To an economist, trained to think about problems in terms of efficiency, there may be a tendency to believe that everything is an efficiency problem until proven otherwise. For the post-Rawlsian political philosopher, trained to think about problems in terms of distributive justice, the tendency will be rather to construe everything as a distributive justice problem. Thus it is not surprising that many philosophers and political theorists have approached the climate change problem using this framework.[80] This is often signaled through use of the phrase "climate justice" to describe the normative problem. Consider, for example, the way that Steve Vanderheiden characterizes the central normative issue raised by anthropogenic climate change, in his book *Atmospheric Justice*:

> All of the world's nations and peoples need and want to emit GHGs; they need to emit survival emissions if their members are to meet basic human needs, and they want to emit luxury emissions so that those members can enjoy aspects of the good life that go beyond mere survival. These needs and wants can be thought of as claims to shares of atmospheric absorptive capacity, and through their current emissions levels all nations now make de facto claims on this capacity. Such claims cannot be infinite, given the atmosphere's finite capacity to absorb GHGs and satisfy those claims, and its capacity to sustainably do so is already being exceeded. These overall claims on the atmosphere must now be reduced if we are to avoid catastrophic climate change in the future, so the task for justice is to articulate and justify a distributive principle by which we may limit overall atmospheric claims in a manner that is fair to all nations and peoples, and to allocate those emissions shares in a manner that is likewise fair to all.[81]

On this view of things, the atmosphere is like a pasture that is in the process of being overgrazed, which must now be divided up and assigned to the nations of the world. The question of how *much* should be emitted is basically set aside, often based on the assumption that some warming target, such as 2°C, can be taken as given.[82] The question that gets foregrounded is the distributive one, of how the remaining emissions room should be allocated.[83]

This way of thinking about the problem, it should be noted, received considerable support from the United Nations Framework Convention on Climate Change (UNFCCC), particularly in the run-up to the Kyoto Accord. The principle of "common but differentiated responsibilities" put a great deal of emphasis on the question of emission allocations, resulting in an acrimonious debate over what national quotas should be. The international community essentially set itself the goal of determining, collectively, how much carbon abatement should be undertaken in each different nation or region of the world. This generated a great deal of controversy, particularly with the decision to exempt all developing countries from any emissions reduction obligations in the accord. More generally, it reinforced the tendency to see the problem as a "cutting the cake"-style distribution problem, with the central question being whether the emissions reduction burdens were being distributed in a way that was fair or just to all nations. This in turn gave rise to a number of difficult problems, such as the importance of historical emissions, adaptive capacity, land use changes, population growth, and so on.

Construed somewhat more abstractly, the most natural way of formulating the distributive justice approach has been to imagine that emission *permits* are equivalent to property rights, and so the central normative issue in debates over climate change involves developing a set of principles for the allocation of these permits. From an egalitarian perspective, it seems natural to suppose that an equal per capita allocation of these permits forms something of a baseline—with national quotas then being a function of population. For example, Dale Jamieson argued that "the most plausible distributive principle is one that simply asserts that every person has a right to the same level of GHG emissions as every other person. It is hard to see why being American or Australian gives someone a right to more emissions, or why being Brazilian or Chinese gives someone less of a right."[84] This gave rise to an enormous literature about the desirability of an equal per capita permit allocation scheme.[85] Despite the intuitive appeal of this baseline, the proposal immediately attracted a number of objections:

1. *What about historical emissions?* Most of the excess carbon dioxide that is currently in the atmosphere was put there by a very small number of countries, during the process of industrialization that resulted in their citizens now enjoying a very high standard of living. Should this not result in some reduction in the entitlement of current generations living in these countries to further emissions? Of course, most of the individuals who did the emitting are long dead, but their economic activities generated benefits that have been transmitted to subsequent generations. How should these be taken into account when determining emissions quotas?[86]

2. *What about circumstances?* A very prominent stream of egalitarian thinking suggests that individuals should not be held responsible for unchosen circumstances. But an equal per capita emissions scheme takes no account of circumstances. For instance, what about individuals living in northern climates, who have no choice but to consume

more fossil fuel in order to heat their homes? Should they not receive a larger allocation of emissions permits?[87]

3. *What about emissions reduction capacity?* Emissions reductions are easier for some nations to achieve than others. For instance, some are blessed with a geography that includes waterways that can be dammed to produce hydroelectric power, while others are not. Iceland has abundant geothermal power that makes it easy for its residents to achieve low emissions, despite its cold temperatures. Economic development in general means that abatement will require less sacrifice. Should this not result in a lower per capita emissions allocation?

If one is willing to accept that one or more of these issues should affect individual entitlements, it is easy to see that the distributive justice question quickly becomes extremely complex. To make things worse, many theorists have questioned the suggestion that the distribution of emissions rights should be resolved in a manner that is independent of the other distributive justice issues that exist in the world today.[88] Many academics have a moral sensibility that is exquisitely attuned to the detection of even the faintest injustice, and are highly resistant to the suggestion that any one problem can be resolved without facing "deeper" or "structural" issues, such as the legacy of colonial violence, or the inequalities of the neoliberal trade system.[89] Politically, this analysis lends support to the familiar strategy of establishing "linkage" between issues. Dominic Roser and Christian Seidel, for instance, have argued that

> climate ethics should be addressed in combination with other questions of global, intragenerational distributive justice—in particular, questions concerning the distribution of the costs of combating hunger and poverty, questions concerning the distribution of contributions to economic development, and questions concerning the distribution of water, food, medicines, technologies, seeds, and patents. Instead of inquiring only into the just distribution of emissions or climate-change-related advantages and disadvantages in isolation, we should instead address the question of the distribution of all of these goods together.[90]

Apart from adding further complexity to the problem, this approach generates serious practical difficulties. While the linkage strategy sounds virtuous, its practical consequences would be paralyzing. Roser and Seidel have basically assembled a laundry list of intractable global problems, then suggested that climate change should be tied to them, in such a way that it will no longer be possible to make progress on one without addressing them all.[91] In many ways this belies the urgency of the situation. If one is at all skeptical about the capacity of the international community to make serious inroads on the problem of global poverty or unequal development, then a conceptual strategy that ties action on climate change to the resolution of these problems seems like little more than a recipe for deferred action on climate change.[92]

Even if we take the problem of emissions allocation in its isolated form, the egalitarian agenda still gives rise to a chorus of realist objections, given the enormous differences that exist in the world today in emissions levels. Annual carbon dioxide (or equivalent) emissions in the United States average 20 carbon-tons (tCs) per person. Annual emissions in India are just slightly over 1 tC per person.[93] Suppose that the global sustainable level of annual carbon emissions, keeping current population constant, is approximately 2 tCs per person. It is difficult to imagine circumstances in which per capita emissions in the United States and India would converge on 2 tCs. Thus there is an obvious question that arises about whether principles of justice that impose such an obligation could ever serve as the basis of a global climate treaty—and if not, then how they should be modified in response.[94]

This entire discussion, however, is predicated on the assumption that these permits are going to function as hard caps—that individuals, or nations, will be assigned certain emissions rights, and that once they reach the limit of their assigned rights, they will simply have to stop emitting. Apart from being political unrealistic, this is not how any actual emissions control system functions. Domestically, emissions rights are typically allocated within a "cap and trade" system, where the "trade" component signals the fact that anyone's initial allocation of permits can always be expanded by purchasing additional ones. Internationally, the Kyoto system of national emissions quotas was accompanied by a "clean development mechanism," which allowed Annex I countries (those required to reduce emissions) to meet their targets in part by purchasing "certified emissions reduction" units from developing countries, in return for making emissions-reducing investments in those countries. In other words, emissions permits are always *tradable*.

There is a good reason for this, which is that the marginal cost of abatement differs enormously from one country to another, and since a given quantity of carbon abatement produces the same benefit regardless of where it occurs, it does not make any sense to insist that each country reduce its own emissions to the same per capita level. A country like France, which relies on nuclear power for over 75 percent of its domestic electricity production, has relatively little room to undertake further carbon abatement—as a result of which emissions cuts are extremely expensive. Countries that still produce electricity in coal-fired generation facilities, by contrast, can often achieve emissions reductions quite inexpensively. South Africa, for instance, produces about the same total emissions as the UK despite having an economy only one-eighth the size. As a result, any investment made in carbon abatement in France or the UK could instead be used to achieve several times more carbon abatement in another country. Given the seriousness of the climate change problem, no credible policy proposal can forgo these efficiency gains. Thus any permit system will necessarily be accompanied by a trading scheme.

If there is going to be permit trading, however, it raises doubts about the importance of the initial allocation of permits.[95] Suppose, for instance, that permits are allocated with no consideration for geographic latitude. This does not mean that people who live in cold countries are going to be left shivering in the dark when their permits run out. It just means that they are going to have to buy a certain number of permits from people in warmer countries, who have less need of them. Thus the allocation of permits does not determine where the abatement will actually occur. The fact that the initial allocation is relatively unimportant explains why the permit allocation process in domestic cap-and-trade systems is typically quite haphazard, and often involves significant giveaways to large emitters. What matters is not possession of the permit, but rather the *opportunity cost* of holding it, which is created by the trading mechanism. Even if a permit has been given away, the decision to use it is costly once there is a market in which that permit could be resold. Thus the important feature of a cap-and-trade system is the *carbon price* that emerges from the trading mechanism, and the incentive to reduce emissions that it creates. The way that the atmospheric "commons" gets divided will then be determined by who is willing to pay the price to emit carbon, not by any principles of distributive justice.

As a result, it is quite erroneous to think of the division of the atmosphere into a set of discrete emission rights as analogous to a Lockean "initial appropriation" phase, in which individuals enclose private property from the commons. With a trading system in place, the initial allocation of permits will have no effect at all on where the carbon abatement takes place; all it will do is generate a flow of payments from regions or individuals with high marginal costs of abatement to those with low. If we decide to give people who live in cold countries extra permits, in light of their circumstances, we are not really giving them extra shares of the atmospheric carbon sink; we are in effect just giving them money (or, more precisely, releasing them from the obligation to make certain monetary payments to people living in warmer climates). This does, of course, raise a question of distributive justice (is it fair to be giving them this money?), but this question is not what anyone has in mind when discussing "the problem of global climate change." Furthermore, there is no reason for philosophers to be worrying about whether extra permits should be assigned to some people in light of the fact that they live in a cold climate; the decision will be made by the actual people affected, who if they need the extra permits will simply purchase them.

Another way of seeing the point is to recall that a cap-and-trade scheme is considered, from a policy perspective, essentially equivalent to a carbon tax.[96] The former fixes the quantity of emissions, then allows the market to determine the price of carbon, while the latter fixes the price of carbon, then allows the market to determine the quantity of emissions. And yet it is easy to see that a carbon tax does not generate any permit allocation problem, because there are no permits to allocate. People who emit greenhouse gases must pay the tax, end of story. Furthermore, one need not address the issue of historical responsibility for emissions in order to determine where cuts must be made. The only question of distributive justice that is raised by a carbon tax concerns the way that governments choose to spend the revenue raised. Upon reflection, one can see that the permit allocation question raised under a cap-and-trade scheme is exactly the same. Thus the question of "climate justice" is, as Simon Caney (somewhat belatedly) observed, essentially a question about how the proceeds of a permit auction should be distributed.[97] This is a distinctly secondary question. Furthermore, it distracts attention from the difficult problem, which involves determining how high the carbon price should be (or equivalently, how many permits should be issued). And yet this is a question that principles of distributive justice provide no purchase upon, because it is fundamentally an efficiency question.

Thus the entire "climate justice" literature, which tries to frame climate change as a problem of distributive justice, rests on a fundamental misunderstanding of the issue. The confusion stems in part from a failure to maintain a clear distinction between collective action problems and distributive justice problems. The former are, first and foremost, about efficiency, although they raise a distributive justice question (secondarily, as it were), because any co-operative scheme must decide how the burdens and benefits are to be distrib-uted. This is different, however, from a distributive justice problem, in which there is some injustice that is being perpetrated and the only question is what sort of redistribution or redress should occur. Figure 1.2 illustrates the dis-tinction. Any movement to the northeast, from the status quo, represents a Pareto improvement, because it makes both players better off. Thus a move to point *a* would be an efficiency gain, with distributive implications, since the move generates greater benefits for player 2 than player 1. Movement to the northwest, by contrast, is redistributive, because it generates a ben-efit for player 2 and a *loss* for player 1. The line U represents the utilitarian indifference curve, so any point above that line constitutes a gain in aggre-gate utility over the status quo. Thus the move to *b* represents a "utilitarian" redistribution—it is a win-lose transformation, although the winner in this case gains more than the loser loses. Point *c*, by contrast, is a "pure" redis-tribution, in that player 2 gains and player 1 loses. Since the magnitude of the loss exceeds that of the gain, the only argument for it must be a distri-butive one, that player 2 is for some reason entitled to more, or player 1 less, than at the status quo. Both points *b* and *c* are Pareto-noncomparable to the status quo, and so the norm of efficiency has nothing to say about these redistributions.

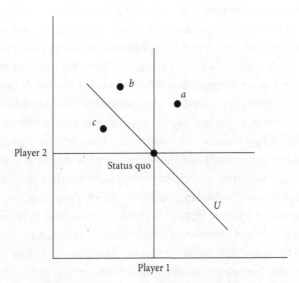

Figure 1.2 Distributive justice versus collective action problems

Many discussions of "climate justice" unfortunately conceive of climate change as a distributive justice problem—as though it were primarily a problem of wealthy countries enriching themselves at the expense of poor countries, or of an unfair division of a natural resource (atmospheric absorptive capacity).[98] There is, of course, no doubt that climate change will have some adverse distributive impacts. The fact that warmer countries are, on average, poorer than colder ones means that the damage from climate change will occur disproportionately in poorer countries. But this is not the primary issue. The primary issue is that climate change is producing damages that are both unnecessary and unjustifiable, and therefore outcomes that in fact no one wants to see occur. Solving the problem of climate change involves movement into the space to the northeast of the status quo (to a point like a). Here the norm of efficiency is relevant to selection of the outcome, and so efficiency concerns are not irrelevant in the way that they are with a movement to points like b or c. Thus the characterization of the climate change problem as one of "climate justice" is actually quite misleading.

1.2.2.   Climate Rights

Many people are happy to endorse the "polluter pays" principle when read left to right ("If you pollute, then you must pay"), but want to reject it when read right to left ("If you pay, then you may pollute"). The latter, it is felt, fails to capture the deontic significance of the principle—that because pollution harms other people, we should refrain from doing it. The fact that we are willing to pay should not give us prerogative to do so; indeed, even framing the issue in this way is likely to contribute to the problem. On this way of viewing things, environmental regulation should impose fines rather than taxes. (This is often combined, somewhat confusingly, with a fallacious distributive justice intuition that the "polluter pays" principle is problematic because it allows the rich to continue polluting, while preventing only the poor from doing so.)[99] For a long time many environmental organizations opposed carbon pricing on these grounds.[100]

The idea that climate change is primarily about harm, and harm-prevention, persuaded some theorists that an appropriate normative framework for responding to it would be one that preserves the deontic significance of the prohibition on harm (rather than dissolving it into the milky consequentialist bathwater of welfarist liberalism). One natural way of articulating this is in terms of individual rights, and the prohibition on rights violations. Perhaps the most influential exponent of this view has been Simon Caney. As Caney describes it, the central attraction of the rights framework is that "human rights generally take priority over moral values, such as increasing efficiency or promoting happiness. They constrain the pursuit of other moral and political ideals, and if there is a clash between not violating human rights on the one hand and promoting welfare on the other, then the former should take priority."[101] Thus the rights framework appears to cut the Gordian knot, offering a blanket condemnation of global warming, without getting into the messy business of analyzing the costs and benefits. Indeed, Caney considered it one of the merits of his view that "a human rights approach rejects the tradeoff between burdens and benefits that other approaches endorse."[102]

This is Caney's early formulation of the argument, which is very much influenced by Ronald Dworkin's characterization of rights as "trumps." (As Dworkin put it, "Individual rights are political trumps held by the individuals. Individuals have rights when, for some reason, a collective goal is not a sufficient justification for denying them what they wish, as individuals, to have or to do, or not a sufficient justification for imposing some loss or injury upon them.")[103] Caney goes on to specify three rights, all of which are threatened by climate change:

> HR1, *the human right to life.* Every person has a human right not to be "arbitrarily deprived of his life."
>
> HR2, *the human right to health.* All persons have a human right that other people do not act so as to create serious threats to their health.
>
> HR3, *the human right to subsistence.* All persons have a human right that other people do not act so as to deprive them of the means of subsistence.[104]

Caney argues that the violation of these rights is impermissible, and so "if emitting greenhouse gases results in rights violations, it should stop, and the fact that it is expensive does not tell against that claim. A human rights approach thus requires us to reframe the issues surrounding the costs of mitigation and adaptation."[105]

From a policy perspective, the idea that greenhouse gas emissions must simply stop, without any attention paid to the cost of doing so, is so extreme as to disqualify itself from being taken seriously. While no reasonable person doubts that, in the long run, greenhouse gas levels in the atmosphere must be stabilized, the problem is that some ways of achieving this would be catastrophically costly, while others would be considerably less so. (Caney's use of the term "expensive" is not a particularly good way of describing an outcome in which millions of people starve to death, which is what would happen if we were to immediately halt production of synthetic fertilizers through the Haber-Bosch process.) Thus a normative framework that features a principled refusal to consider costs has essentially nothing to contribute to any discussion over climate change policy.

In addition to this pragmatic concern, there are more subtle philosophical problems with Caney's analysis, which can be seen by considering each of the rights in turn:

1. *Moral gridlock.* Caney observes that climate change will result in a number of deaths (indeed, millions of deaths) that otherwise would not have occurred, through extreme weather events, crop failure, flooding, heatwaves, and so on. This violates HR1, the human right to life. And yet he ignores the fact that any carbon abatement policy adopted will also result in deaths that otherwise would not have occurred. People die in heatwaves, but they also freeze to death. More importantly, there is a strong correlation in developing countries between economic growth and declining infant mortality.[106] If climate change mitigation were to depress the rate of growth in developing countries (as is likely), then this would violate HR1, as millions of infants would die who otherwise would not have. One might want to argue that *more* people would die under the business-as-usual scenario than under a climate change mitigation policy, but this is precisely the sort of trade-off that Caney's framework is intended to prohibit. The consequence is that his system of rights generates moral gridlock, where all actions, including inaction, are prohibited, because every option violates someone's rights.

2. *Rights are not trumps.* Dworkin's conception of "rights as trumps" has attracted its share of criticism from those who argue that, while rights may function as weighty moral considerations, they do not really allow individuals to veto any collective plan that impinges upon them. The rights of one individual are routinely weighed against the rights and interests of others. Even free speech has limits (such as the restrictions on libel). Whatever the global merits of this criticism, the rights enumerated by Caney, such as his HR2 right to health, are obviously unsuited to function as trumps.[107] Caney claims that "acting to expose

others to dangerous diseases manifests a lack of respect for their status as free and equal persons."[108] And yet, if this were true, we would be much more aggressive at quarantining people with contagious diseases, so that they do not expose others. One could see in the response to the COVID-19 epidemic that countries committed to respecting human rights all engaged in significant balancing between the liberty interests of the ill against the "right to health" of others. It has also been pointed out that everyday acts, such as driving a car, can pose serious threats to the health of others. The risk is of course rather slight, but strict deontological frameworks are not well suited to handling probabilities of rights violations.[109]

3. *Dealing with adaptation.* Finally, Caney recognizes that climate change policy is going to contain some mixture of investment in mitigation (preventing climate change from occurring) and adaptation (reducing the damage done by the climate change that does occur). In many cases, it is possible to avoid a rights violation through either strategy. For instance, Caney's HR3 right to subsistence specifies that individuals should not be deprived of the "means of subsistence." He gives the example of Bangladeshi peasants who may lose their farms due to flooding. But if these peasants are offered compensation, even in the form of food aid, then their right has not been violated, since they still have access to the means of subsistence. The question therefore becomes whether we should save their farms by limiting sea-level rise, or by building dikes, or let the farms flood, relocating the peasants to higher ground, or providing them with food aid. From the standpoint of HR3, these are all equivalent strategies, and so the rights framework offers no basis for choosing between them. The most plausible way to make the choice is to consider which of the various mitigation and adaptation strategies is the most cost-effective. But this means that the question of whether we must stop emitting greenhouse gases winds up depending upon the calculation of trade-offs that Caney was trying to escape.

Henry Shue has advanced an argument similar to Caney's, claiming that "a person's right to physical security against, among other things, anthropogenic malignancies is non-marketable and can, therefore, not be purchased for any amount of returns from alternative investments."[110] Since climate change jeopardizes the physical security of future people, it is impermissible, even if they are offered compensating benefits. This suggests that the harms caused by climate change cannot be traded off against the benefits of fossil fuel consumption. Shue develops an analogy between greenhouse gas emissions and a person who, as a hobby, likes to construct land mines that will detonate in several decades. Suppose that this person also endows an annuity, which matures around the same time that the land mine detonates, and provides financial compensation to the victims of the explosion. This does not make the hobby permissible, Shue observes, because the right to physical security serves as a strict deontic constraint, and so no amount of compensation could justify its infringement.[111] Climate change, he suggests, is the same—we are violating the physical security of future generations, and so no amount of indemnification can render this permissible.

The success of Shue's argument, it should be noted, hinges upon the deontic significance of two closely related distinctions, the first between doing and allowing, and the second between intended effects and byproducts. While we often are not permitted to do things that jeopardize the physical security of others, these dangers may be permissible when they arise as a byproduct of our activities (such as driving a car), or if we merely fail to prevent them. With the land mine scenario, the injury arises as an intended consequence of the bomb-maker's action. If it arose as an unintended effect of an otherwise valuable action (such as a nation protecting its borders) and full compensation was offered, then it is not so obvious that the arrangement is unacceptable. Climate change, however, is obviously a byproduct effect, not the intended result of our actions. As to whether it should count as a "doing" or an "allowing," this seems to be largely a matter of framing. The "business as usual" (BAU) scenario is routinely described as one in which we "do nothing" to prevent climate change. Furthermore there are very few scenarios under which carbon abatement occurs *merely* through people stopping doing something that they were doing (i.e. burning fossil fuel), rather than through switching to something else (e.g. making investments in renewable energy). Thus both climate change and climate change mitigation can both plausibly be described as "doings." Once this is granted, however, the three objections to Caney's argument apply with equal force to Shue's: (1) a strict prohibition on damaging anyone's physical security is gridlocking, since carbon abatement will have the same effects, (2) physical security rights do not function as trumps, since individuals are subjected to such risks, for the convenience of others, on a routine basis, and (3) sufficient investment in adaptation can make it so that certain harms do not occur, and so the case for mitigation is underdetermined.

This last point is perhaps worth dwelling upon for a moment. Shue's example of the bomb-maker providing victims with cash compensation evokes a familiar trope from the literature on climate change. Those who worry about the trade-off between economic growth and climate change mitigation are often portrayed as wanting to do something similar—by privileging economic growth, it is suggested, they are destroying the planet, then offering to write future generations a check to cover the damages. This is further nourished by the stereotype of the economist as one who knows the "price of everything, but the value of nothing."

Despite its popularity, this entire line of argumentation involves a clear misunderstanding of the relevant issues. The compensation that future generations would be offered for climate change would obviously not take the form of money, but would generally take the form of goods that people want and need. Consider, for example, Shue's example of physical security. People are subject to a wide range of threats to their physical security, many of which can be avoided at some cost. This is why rates of death from accident and injury tend to be much lower in developed countries—buildings are constructed to a higher standard, automobiles and roads are safer, healthcare is superior, and so on. If one were to prioritize economic growth, to the relative neglect of climate change, certain risks to physical security would increase (e.g. death from flooding), but other risks would decrease (e.g. death from earthquakes).[112] It is not clear in what sense the "rights" of future generations have been violated if we increase their risk of dying from one cause but decrease their risk of dying from some other. What the economist is trying to figure out, by looking at the trade-offs and contemplating the costs and benefits of a carbon abatement policy, is which option provides the *best* outcome for future generations. It is this calculation that Shue, Caney, and other deontologists seem intent on avoiding. And yet when one considers the matter clearly, it is difficult to see how this opposition could rest on anything other than confusion about the set of choices we are confronting.

In later work, Caney recognizes that a strict deontological framework is not able to address the important questions that arise in the climate change policy debate. This has become particularly apparent now that China has overtaken the United States as the world's largest emitter of greenhouse gases. In a wealthy country like the United States, it is easy to imagine that further economic growth will do nothing to improve the human condition, and so the country is producing little more than luxury emissions that can be eliminated without serious cost.[113] In China, by contrast, it is impossible to ignore the massive gains in human well-being that have occurred over the past two decades. This makes it more difficult to pretend that cutting back on energy consumption would not have significant human costs. At the same time, cutting back would produce some benefits, in the form of reduced global warming. The need to compare these costs to the benefits therefore seems inevitable. Indeed, the nature of the climate change problem seems to leave us no choice but to consider the trade-offs involved in each of the available policy options. Any normative framework that is incapable of contemplating these trade-offs excludes itself of serious consideration ab initio.

Caney's response to this line of argument has been to move away from a strictly deontological view of rights toward a "Razian" conception.[114] According to Joseph Raz's influential formulation, rights are best understood as just a protected class of interests, where the only difference from everyday interests is that they are sufficiently weighty that they impose duties on others: "X has a right if X can have rights, and, other things being equal, an aspect of X's well-being (his interest) is a sufficient reason for holding some other person(s) to be under a duty."[115] Thus the reason we have a right to health, on Raz's view, is that our interest in being healthy is sufficiently important that we take others to have a duty to refrain from endangering our health. The duty, however, need not be strict. For instance, it might mean that others have an obligation to avoid needlessly endangering the health of others, or that they must take reasonable precautions against doing so, but it does not mean that they are prohibited from doing *anything* that endangers anyone else's health. The question may come down to how important the relevant interest is, compared to the interest motivating the other person's action. In any case, because rights, on this view, are just a protected class of interests, there is nothing to prevent balancing, consideration of trade-offs, or indeed, full-scale cost-benefit analysis. Thus Caney's rights framework—which instructs us to consider the effects of our policies on health, life expectancy, and economic well-being—winds up becoming practically indistinguishable from most other forms of welfare liberalism, with the only important difference being that it relies upon a somewhat restricted concept of welfare. In the context of climate change policy this is not really an important restriction, given that the latter is primarily concerned with large-scale damage to important human interests.

Thus what began as a striking contrast between the "economic" approach to the problem of climate change and more strict "deontological" approaches coming from the liberal tradition of political philosophy has become significantly eroded over time. Much of this stems from the recognition that climate policy necessarily involves major trade-offs, and so any normative framework that seeks to move beyond stating generalities about the badness of global warming will have to provide some way of evaluating these trade-offs, as well as judging the relative merits of second- and third-best scenarios. Consequentialism, of course, excels in these contexts, and given its natural affinity with economic modes of reasoning, has been well represented in the literature. Deontologists, by contrast, have had to back away from the stricter formulations of their views, when pressed to offer policy-relevant normative guidance.

Among those who continue to resist the economistic approach, the fallback position has been to adopt some form of "sufficientarianism."[116] This normative framework is one that can be arrived at, quite easily, through relaxation of either the egalitarian view or the rights framework. An egalitarian, for instance, recognizing that perfect equality is unlikely to be achieved, and that serious measurement problems arise when it comes to comparing the situation of individuals in vastly different circumstances, might choose to weaken the commitment to equality by turning it into the demand that everyone have "enough," or that no individual fall beneath a certain threshold of "sufficiency." Similarly, a rights theorist, recognizing that economic and social rights cannot be applied as trumps, but must be balanced against one another and "realized progressively," may instead choose to interpret them in terms of thresholds, or as the specification of a basic minimum. Both views are sufficientarian, in that they reject the economist's emphasis on maximization, arguing instead that our obligations consist in ensuring that no one falls below a certain threshold of sufficiency, with respect to satisfaction of their basic interests. This approach has its own difficulties, which I will analyze in the following chapter. My broader task, which will occupy the remainder of the book, will be to show how a more plausible, non-sufficientarian deontological framework can be developed that can serve as an alternative to the consequentialist framing of the problem.

1.3. Conclusion

Ever since Socrates wandered the streets of Athens, asking people questions, then showing them how little their beliefs could survive critical scrutiny, philosophers have prided themselves on their unwillingness to accept at face value the commonsense views of the day. No doubt because of this, philosophy has evolved into a discipline that prizes the ability to *create* problems much more than the ability to *solve* problems. In contemporary academic philosophy, discovering a new, more difficult problem has become the surest route to reputation and fame (think of Derek Parfit's "nonidentity" problem, Nelson Goodman's problem of induction, Edmund Gettier's problem regarding knowledge, etc.).[117] As a result, it is perhaps too much to expect philosophers to have made, or to make, a productive contribution to solving the problem of climate change. Indeed, their contributions so far have been almost entirely in the problem-causing vein. For the most part, they have refused to accept even the standard definition of what the problem of climate change is, much less the prevailing understanding of the principles that should guide our approach to finding a solution.[118]

This focus on problematization is fairly evident when it comes to traditional environmental ethics. Although this movement in philosophy began with the laudable goal of encouraging deeper reflection upon our relationship to the natural world, in a sense the very deepness of the questions posed wound up fragmenting the responses, so that the discipline quickly became divided into different schools of thought. The attempt to make them all compatible, or at least non-antagonistic to one another, by adopting some form of value pluralism only exacerbated the underlying problem. The result has been a set of views that cannot easily be translated into policy guidance. Indeed, if the goal had been to paralyze decision-making, it is not clear how environmental ethicists would have proceeded much differently. (Daniel Farber describes environmental ethicists as having pursued the following approach to deciding policy issues: "*Step 1*: Settle the question originally raised by Plato by providing an indisputable definition of the nature of 'the good'. *Step 2*: Apply the results of step 1 to the particular problem of environmental quality.")[119]

This value pluralism, I suggested, has been one of the factors driving the rise of liberal environmentalism, and the fact that current discussions of climate change in philosophy are now dominated by theorists whose background training, as well as basic approach to the problem, is grounded in political philosophy, and in particular, post-Rawlsian theories of justice.[120] And yet these second-wave theorists have by no means escaped from the same problem, of promoting philosophical views that encourage policy paralysis. Much of this arises from an exaggerated aversion to anything that smacks of economism, and thus a powerful desire to avoid any serious contemplation of the difficult trade-offs involved in climate change policy. Perhaps the dominant impulse has been to preempt such concerns by adopting a deontological framework that will generate univocal results. As I have tried to show, these attempts to think about climate change primarily as an issue of rights, or as a question of distributive justice, have wound up generating more heat than light. Climate change is a collective action problem, which means that it is, first and foremost, an issue of efficiency and optimization. The "polluter pays" principle, for instance, was never intended to be a principle of distributive justice; it is an efficiency principle, designed to achieve the optimal pollution level. The fact that these concepts may be initially unintuitive does not provide legitimate grounds for resisting this framing of the issue. The seriousness of the problem simply does not permit us the luxury of indulging these doctrinal and disciplinary prejudices.

2

Climate Change and Growth

The discussion so far has proceeded without mention of what many regard as the most important feature of the debate over climate change, which is the temporal dimension of the problem. This has two aspects. First, among the greenhouse gases that are being released into the atmosphere, carbon dioxide is remarkably persistent. Some fraction of it can be expected to remain airborne, and to influence the earth's climate, for thousands of years. Thus the decisions that we take now have consequences that will be felt for a very long time. Second, because of "inertia" in the climatic system, there is a fairly long delay between our actions and their effects. Carbon emissions cause the average global temperature to increase, but with a delay of approximately 40 years between the time of the emissions and the point at which 50% of the consequent temperature increase has occurred. Thus most of the climate change that will occur over the next several decades has already been locked in. The current policy debate is primarily concerned with how we should be shaping the climate that will prevail in the second half of the 21st century and beyond.[1]

Because of this, many theorists have been inclined to classify climate change as a problem of "intergenerational justice," or of what we owe to future generations. The latter is a topic that predates the climate change discussion, among both economists and philosophers, and so the most common response has been to integrate the climate change issue within the established normative frameworks developed to deal with this question. Historically, the discussion of intergenerational justice has been focused primarily on the issue of economic growth. Of course, economic growth is only one of the ways that we produce benefits for people in the future. Under ordinary conditions, future generations will also stand to inherit whatever progress we make in science and technology, medicine, art, literature, philosophy, and law. But these benefits are ones that they receive "for free," as it were, or as byproducts of our ordinary activities. It does not take much effort on our part to make them available to future generations, largely due to the low costs of reproduction. Medical innovation, for instance, is primarily motivated by the desire to assist those who are

Philosophical Foundations of Climate Change Policy. Joseph Heath, Oxford University Press. © Oxford University Press 2021. DOI: 10.1093/oso/9780197567982.003.0003

suffering from various diseases and ailments right now. But once an invention occurs, it becomes available to all subsequent generations, and usually winds up producing a much greater sum of benefits for those who were not even alive at the point of invention. It is this cumulative quality of human cultural production that has been, as Joseph Henrich puts it, "the secret of our success" as a species.[2]

Economic growth is slightly different, however, because its rate depends heavily upon the level of productive investment. It is essential that, instead of consuming the entire social product, individuals invest time, energy, and resources in the production of capital goods, which amplify the effects of human labor, making it easier to produce the same social product, including the capital goods, in subsequent years. This investment is financed through saving, which represents the difference between each individual's production and consumption (and thus constitutes the economic "slack" that can be invested in the production of capital goods). Because consumption and savings are mutually exclusive, economic growth is often thought to pose a question of justice. Consumption generates immediate benefits, while investment makes possible a greater level of consumption in the future. Thus a normative issue arises over what the appropriate level of investment or savings should be. This is often referred to as the question of "just savings."

It should be noted, at the outset, that climate change mitigation looks like an investment in this sense, since it involves forgoing consumption in the present in order to achieve benefits that accrue gradually over time. There are many difficulties with this characterization, which I will discuss in greater detail as the occasion merits. For now I would just like to observe that the integration of the climate change problem into the just savings framework produces an obvious difficulty for many philosophers, because of the somewhat casual, and in some cases positively dismissive, attitude that many have had toward economic growth. Economists, of course, are disposed toward thinking that economic growth should be maximized. Philosophers, partly in reaction against this, have been more inclined to adopt sufficientarian views, according to which present generations have an obligation only to ensure that future generations meet a certain threshold level of well-being, and that any benefits beyond that are entirely discretionary.[3] While it is typically acknowledged that growth produces benefits in developing countries, and so living standards there should be raised, it is often suggested that additional growth in developed countries is absorbed entirely by consumerism, and thus produces no increase in average well-being.[4]

The problem with this sufficientarian position is that under all of the most probable scenarios, the benefits that we could be providing to future generations through ongoing economic growth are enormous, relative to the costs that will be imposed upon them by climate change. As a result, if we are under no obligation to maximize growth—indeed, if we are permitted to pass along to future generations an economy that will permit them to achieve a standard of living no greater than what we enjoy now—it would seem that the least costly course of action for us is to let climate change occur, then help future generations minimize the impact (or even reverse the process) by making resources available to them to cover the costs of adaptation. The central objection to the "business as usual" (BAU) scenario, in which we do nothing to slow or prevent climate change, is that it is non-maximizing. But if we are under no obligation to maximize the welfare of future generations, then following this suboptimal path does not appear to violate any duties of justice.

There is one obvious way to escape this tension, which is to adopt some form of "strong sustainability" commitment.[5] This involves the claim that, when we consider the "bequest" that we are obliged to make to future generations— which will consist in a complex bundle of resources, goods, and amenities— we are not allowed to substitute goods from one category for those of another. "Weak sustainability," by contrast, permits substitution. Appealing to a strong version of sustainability offers one way out of the climate-growth dilemma. The basic problem confronting those who would like to take action to combat climate change is that the benefits of economic growth are enormous, so that if we are not obliged to produce them for their own sake, then they become available to compensate future generations for the costs of adapting to climate change, which in turn erodes the case for mitigation. The natural way of avoiding the problem, therefore, is to say that we owe future generations a "list of stuff," and that one of the items on that list is a global climate that averages 14°C, so that if we allow it to rise to 16°C, or 17°C, we cannot make up for it merely by increasing other resources available to them.[6]

The structure of the chapter is as follows. I begin by articulating the core dilemma for the sufficientarian somewhat more perspicuously, beginning with a discussion of economic growth, and why many theorists believe that present generations are not obliged to maximize benefits to future generations. I will mention also "limits to growth" views, which consider continuous growth to be practically unobtainable, even if it were normatively desirable, and explain why I think that these views do not foreclose the discussion.[7] From this, it is easy to see the dilemma that emerges. Climate change policy necessarily involves a balance between investments in mitigation in the near term and adaptation in the longer term. If we take a steady-state economy as the normative baseline, then all the fruits of the (now discretionary) growth rate become available to fund adaptation efforts, which in turn undermine the case for strenuous mitigation efforts. For example, on the sufficientarian view, it is difficult to see what is wrong with an arrangement under which, 50 years from now, more than half of the global economy is dedicated to carbon scrubbing, sea wall construction, and other forms of climate change adaptation.[8] This brings the issue of weak versus strong sustainability to the forefront. I will argue that the strong sustainability thesis, although attractive when thinking about some environmental issues, is much less compelling when it comes to the problem of climate change. Finally, I will discuss low-probability catastrophe scenarios, and suggest that these constitute largely a distraction from the normative issue being discussed.

My reason for pressing this issue, I should note, is not to extol the virtues of economic growth, but rather to encourage more serious reflection upon the fundamental philosophical question of what we owe to future generations, and how we should conceive of our relationship to those who are yet unborn. The current philosophical discussion is dominated by two views that are, in my view, implausibly extreme. The first is the utilitarian view, which says that we are obliged to maximize the welfare of all future generations. I will defer detailed discussion of this position until the next chapter. For now I would like to focus on the second, "steady-state" or sufficientarian view, which says that we are not obliged to increase their welfare at all, beyond what is required to maintain something like current levels of well-being. The extremism of the second view has not been as apparent, and so I hope to demonstrate it by drawing out one of its implications, which is that we might not be obliged to care about climate change. This does not mean that we must immediately switch to the utilitarian position. It does, however, suggest that there is room for the development of a more moderate position, which says that we have some obligation to limit the effects of climate change, even if we are not committed to maximizing the welfare of future people.

2.1. The Undemandingness Problem

A commitment to the maximization of economic growth is one that follows most naturally from utilitarianism. If the goal is to produce as much happiness as possible, without any particular concern for who receives that happiness, then making productive investments in the economy, so that future generations are able to meet their needs with less effort than must presently be expended, is an extremely effective way of pursuing that end. Since some type of utilitarianism (or welfare consequentialism) is very much the received normative view among economists, the commitment to the maximization of growth is widespread among members of that profession. This commitment is amplified by an awareness of the extraordinary efficacy of economic growth at addressing many of the persistent ailments of the human condition.[9] This is largely due to the fact that economic growth is exponential. At a 5 percent rate of growth, it takes just slightly over 14 years to double the size of an economy. As a result, if there is a choice between consuming some resources now or using them to make a productive investment that will contribute to growth, it is very difficult to compete with the investment. Because of compounding, investment generates an open-ended stream of future benefits. Investment not only makes it easier to produce a surplus in the year following, but that in turn makes it easier to generate the investments that will make it easier to produce an even larger surplus in the year following that, and so on.

Indeed, it is an awareness of the extraordinary power of economic growth to improve human welfare that accounts, at least in part, for the tendency among economists to disregard or downplay questions of distribution.[10] One can see this, for instance, in the widely cited remark by Robert Lucas that "of the vast increase in the well-being of hundreds of millions of people that has occurred in the 200-year course of the industrial revolution to date, virtually none of it can be attributed to the direct redistribution of resources from rich to poor. The potential for improving the lives of poor people by finding different ways of distributing current production is *nothing* compared to the apparently limitless potential of increasing production."[11]

This may strike some as being an economist's hyperbole. And yet nonspecialists often ignore the importance of growth, simply because exponential functions are unintuitive, and so it is easy to underestimate the magnitude of the gains that can be achieved in this way. For instance, $1,000 invested at a 5 percent rate of return yields, at the end of a 100-year term, an impressive $131,501. Consider then a slight modification of the case, with the $1,000 being invested at a rate of 8 percent for the same period of time. When asked for a ballpark estimate of how much that yields, most people guess an outcome that is far lower than the actual sum of $2,199,761. Thus a relatively small difference in the rate of growth can make an enormous difference over time—often enough to completely eclipse whatever is being done on the distribution front. (For instance, if the sum generated by this investment was divided up between two people, one could give 97 percent of the product of the 8 percent growth rate to one person, and yet still ensure that the *other* one received more, in absolute terms, than he would have under an egalitarian division of the outcome produced at 5 percent.) Thus it is no surprise that economists have put so much emphasis on economic growth—it follows almost directly from the concern with welfare-maximization.

There are, of course, well-known limitations of gross domestic product (GDP) as a measure of well-being, or of social progress: it is subject to declining returns in utility, it does not keep track of the stock of wealth, it does not correctly measure the value of public services, it ignores leisure and household production, etc. For this reason, some economists speak of GDP+ as the appropriate measure of progress, where the "plus" indicates the other major variables associated with well-being that can be enhanced through the division of labor.[12] Most of the argument in this chapter is unaffected by whether one thinks of GDP in the narrower, or in the GDP+, sense, because access to the goods designated by the "plus" are all strongly and positively correlated with GDP growth, and in many cases obtained through growth, either directly (e.g. health) or indirectly (e.g. leisure).[13] Thus in the context of discussions of intergenerational justice, it is convenient to use the term "growth" to refer to the very broad class of benefits that can be achieved through productive investment.

In any case, the widespread disregard for growth in contemporary political philosophy is not due to such cavils. It is based primarily on a rejection of the commitment to welfare-maximization. This is particularly true of views that assign significant weight to the principle of equality. Indeed, one of the most unwelcome consequences of the principle of equality, or of any strongly egalitarian conception of justice, is that when applied to questions of intergenerational justice, it seems to assign too much weight to the interests of *present* generations. Precisely because economic growth is such a powerful means of improving the human condition, those of us alive today are the "worst off" generation in almost any likely scenario. Unless handled carefully, therefore, egalitarian theories can produce genuinely perverse outcomes when applied to intergenerational questions. This is something that Rawls realized when he considered applying his "original position" construct to the question of determining a "just savings rate."[14] If one were to imagine individuals choosing the rate of savings behind a veil of ignorance, not knowing which generation they belonged to, and one applied Rawls's recommended form of "maximin" reasoning, then it follows that the savings rate should be one that generates zero growth. For the same reason that the difference principle assigns lexical priority to the worst-off individual, here one would find lexical priority being assigned to the worst-off generation, which is, given the presence of any productive investment opportunities, the present one. This suggests that those who are currently alive should be maximizing their own consumption, subject only to the constraint that they not make future generations worse off than they themselves are.[15] This obviously has the capacity to lock society into "eternal poverty," by essentially prohibiting any savings or growth. Since any improvement in the human condition makes later generations better off, it winds up being unfair to earlier generations.[16] On this basis, certain philosophers have concluded that savings are not only non-obligatory, but actually impermissible.[17]

Rawls, to his credit, recognized this as an untoward consequence, and so simply suspended application of the original position to questions of intergenerational justice.[18] Nevertheless, he formulates a principle of intergenerational justice that puts very little weight on the interests of future generations, essentially recommending growth until a just basic structure is established, followed by a steady state. ("Once just institutions are firmly established, the net accumulation required falls to zero. At this point a society meets its duty of justice by maintaining just institutions and preserving their material base.")[19] Similarly, Brian Barry concludes on egalitarian grounds that our only obligation toward future generations is to leave them, in some sense, no worse off than we are. Thus he endorses what he calls a principle of "equal opportunity across generations,"[20] which suggests that "the requirement is to provide future generations with the opportunity to live good lives according to their conception of what constitutes a good life."[21] (His only real concern is that our respect for liberal neutrality makes it difficult to know exactly what opportunities will be valued by future generations.) When he goes on to discuss specific implications of this view, Barry assumes that a "good life" is approximately what people in the industrialized world enjoyed at the time of writing, and therefore interprets the principle as requiring the transmission of a stock of productive assets roughly equal to the present (i.e. at the time of writing) per capita endowment. Thus his view is that present generations have an obligation to refrain from doing anything that will foreseeably worsen the net circumstances of future generations, but that there is no positive obligation to improve it. (It is perhaps worth noting that Barry wrote this in 1978, at a time when real per capita GDP in the United States was half of what it is now.)

This view, which we might refer to as the steady-state principle, is both obviously and self-consciously non-utilitarian. Nevertheless, it is worth pausing to note just how strongly anti-utilitarian it is. Consider, for instance, something like a minimal principle of positive beneficence, stating that if one is in a position to perform an action that will bring a very great benefit for a large number of people at very little cost to oneself, then one has an obligation to do it.[22] This is a minimal principle in the sense that it does not require an equal weighting of one's own welfare with that of others, the way classical utilitarianism does, but permits a significant degree of egoism. Nevertheless, the steady-state savings principle clearly violates it, since maintaining a constant stock of productive resources means forgoing opportunities for investment that could produce far greater benefits for future generations. If the rate of return on productive investments is 5 percent, then one person consuming $1,000 worth of goods in the present means depriving more than 130 people of $1,000 worth of consumption a century hence.

Furthermore, the steady-state principle is extremely undemanding. In the absence of technological progress, it is possible to maintain a constant welfare rate with zero net savings. But if there is any technological progress, the principle permits dissaving. For example, over the second half of the 20th century, total factor productivity in US agriculture (which is to say, the growth in productivity that is accounted for by factors *other* than an increase in inputs, such as land, capital, or labor) grew by over 250 percent.[23] This means that if the obligation were merely to allow future generations the opportunity to produce the same quantity of food, then Americans in 1950 could have begun to deplete the stock of agricultural resources (e.g. destroying farmland, failing to replace equipment, etc.), so that there would have been, by the end of the century, only 40 percent of the level of inputs available that had been available 50 years earlier. Put more dramatically, technological innovation has been such a powerful force for improvement in agriculture that if one were concerned only with outputs, it would have been possible for one or two generations to destroy more than half of the farmland in America and yet still leave future generations just as well off, by increasing the yields that they are able to get from the remaining land. (This is what generates the so-called environmentalist's paradox, which is that, despite the degradation that has occurred in most ecosystem services, along with increased human demand for precisely those services, average well-being has nevertheless continuously increased.)[24]

This example illustrates a very general point, which is that the bequest made to future generations involves much more than just a stock of productive capital. The latter has attracted a great deal of attention, because making investments in productive capital (such as buildings, machinery, computers, communication and transportation infrastructure, etc.) requires forgoing consumption in the present, and so there is an interesting question that arises about how the two are to be balanced. But many other benefits that are provided to future generations involve byproduct effects that are costless in the present. Scientific and technical discovery, for instance, is typically motivated by present concerns, but a particular bit of knowledge, once discovered, becomes available to future generations at practically no cost. The same can be said for various forms of cultural production, such as literature, music composition, and, of course, philosophy.[25] To see this, we need only consider the inestimable value of the bequest that we ourselves inherited from our ancestors—the great works of art, the scientific knowledge, the techniques of production. Consider, for instance, the improvements in medical technology that have occurred in the past 150 years, including the development of antibiotics, anesthesia, vaccination, radiology, and chemotherapy. Faced with a choice between a 19th-century material standard of living combined with a 21st-century medical technology, and a 21st-century material standard of living combined with a 19th-century level of medical technology, how many people would choose the material standard of living over the medical technology? My suspicion is very few (and after a few anesthesia-free tooth extractions, none at all). This suggests, however, that most people value the increase in knowledge more highly than the economic growth that occurred over the course of the 20th century. It follows that, if previous generations had in fact engaged in significant dissaving, such that growth rates were negative, they could still have made it up to us simply by making this technological knowledge available.[26]

So if one adopts a broad view of the bequest that each generation makes to the next—expanding it beyond just economic and environmental goods, to include the benefits that are culturally transmitted as well—it is difficult to imagine many circumstances short of nuclear war in which one generation would *fail* to improve the circumstances of the next. Thus the steady-state principle seems a bit too easy to satisfy. Nevertheless, it has been embraced in some form by many political philosophers and environmentalists, although of course not without some controversy.[27] Perhaps the most well-known statement of it, in the environmental context, is the so-called Brundtland conception of sustainable development, which defines it as "development that meets the needs of the present without compromising the ability of future generations to meet their own needs."[28] If the obligation of present generations is to use resources "sustainably" in this sense, and the needs of future generations are roughly the same as ours, then technological progress alone makes this standard extremely easy to satisfy, and indeed, is compatible with significant resource depletion and environmental degradation.[29]

Thus sufficientarianism suffers from what might be described as an "undemandingness" problem when applied to questions of intergenerational justice. Because future generations can expect to derive such enormous benefits from our activities—just as we have derived untold benefits from the labor, not to mention the genius, of our ancestors—it seems that we can also inflict serious harms on our descendants, without actually impairing their ability to ensure that no one falls below the level of sufficiency. In particular, if economic growth is not obligatory, then the growth that does occur creates something of a perverse "harm budget," in that it specifies the amount of harm that we can inflict upon future generations without actually worsening their condition.

2.2. Limits to Growth

There is, of course, a reason that many environmentalists have been drawn toward such weak standards of intergenerational obligation. It is not due to a lack of concern for future generations, it follows rather from an application of the "ought implies can" principle. Given that we are already overtaxing the earth's resources, they believe that we are close to reaching the limits to growth, and since continuation of the growth trajectory that our economies have seen over the past two centuries is impossible, there cannot be an obligation to keep it going. (The most common version of this, I should note, is not that economic growth will be impossible per se, but that further increments in growth will be accompanied by such deleterious environmental harms that the *net* effect on welfare will be either zero or negative. On this view, we may be able to maintain the illusion of growth, but only because environmental externalities are left out of the accounting.) If this argument were correct, then there would be no difference between the utilitarian and the steady-state view—both would recommend a zero-growth policy.

This argument is one that must be approached with a great deal of caution, because there are elements of truth to it, but in many instances it rests upon a confusion, including in some cases a shifting definition of what economic growth consists in. This is reflected in numerous presentations of the argument, which sound as though they are articulating a deep insight about the human condition, but in fact are using key terms in a very misleading way. Consider, for instance, the claim by Tim Jackson, in his widely praised book, *Prosperity without Growth:* "The idea of a non-growing economy may be an anathema to an economist. But the idea of a continually growing economy is an anathema to an ecologist. No subsystem of a finite system can grow indefinitely, in physical terms. Economists have to be able to answer the question of how a continually growing economic system can fit within a finite ecological system."[30] His answer, of course, is that it cannot, and so we should attempt to make the transition to a zero-growth economy.

Peter Victor says very much the same thing in *Managing without Growth.* Against the mainstream economic view, he argues that "a different conception of an economy that provides a better starting point for a book on growth is as an 'open system' with biophysical dimensions. An open system is any complex arrangement that maintains itself through an inflow and outflow of energy and material from and to its environment."[31] On this basis, he represents economies as essentially throughput systems that transform resources into waste: "What goes into an economy must come out eventually. Such is the nature of economies as open systems. The vast quantities of wastes that our economies produce must go somewhere: on land, into water or into the air."[32] We are, however, reaching the limit of expansion of this system. "Historically these material and energy flows have increased with economic growth. Now the flows are so large that there are concerns over future supplies of resources such as oil, concerns over the impacts that waste energy and material are having on the environment, and concerns that life-support and amenity services provided by the environment are being damaged beyond repair."[33]

What is noteworthy about both of these arguments is the way that they use "systems" vocabulary to suggest that economists are making some kind of conceptual or logical error in thinking that economic growth could be open-ended or unlimited.[34] This suggests, in turn, that the details of what economists have to say are not that important, because their entire way of thinking about the question is incoherent. If one pauses to reflect upon it, however, one can see what an extraordinary claim this is. Economists may be many things, but stupid is generally not one of them. Nor has the entire profession become enslaved by ideology. The more likely explanation for these wildly divergent claims about growth is not that one side is committing an elementary error, but that there is some misunderstanding or miscommunication underlying the disagreement.[35]

In order to sort things out, there are several issues that must be disentangled. The first point is that popular, and sometimes even technical discussions of "growth" often do not distinguish *aggregate* from *per capita* growth. A certain fraction of the growth in many countries merely reflects an increase in population. For instance, at the time of writing, the economy of Nigeria is growing at a rate of 2.35 percent, while the population is increasing at a rate of 2.8 percent. The economy of China, by contrast is growing at a rate of 6.9 percent, while the population is increasing at a rate of only 0.5 percent. Thus in per capita terms, Nigerians are actually getting slightly poorer, despite the positive growth rate, while the Chinese are getting much richer. High rates of population growth are extremely undesirable, from an environmental perspective. Furthermore, the claim that population growth is subject to limits is obviously correct—we cannot just keep adding people to the earth. However, it is also generally granted that once people are born, it is important that they be able to satisfy their needs, and thus there is nothing objectionable about the economic growth that tracks these efforts.[36] Here, however, the environmental issue is not really about limits to growth but rather limits to population, and there is widespread agreement that we should be doing what we can, within the limits of respect for individual autonomy, to limit population growth, and perhaps even draw it down.[37] It is, however, needlessly confusing to talk about growth in this context, when this growth is just an indirect consequence of population increase. It is better to separate the two issues, and restrict the discussion of growth to growth of *per capita* income. The question is whether it is possible for us to make future generations richer in *individual* terms.

With this clarification in place, we can turn to the major issue, which is how Jackson and Victor represent growth. The central problem is that both of them consistently lapse into a mode of expression whereby they present growth as though it represented a *material* quantity of production. And yet the GDP measure, from which growth statistics are derived, is not actually a measure of material production in the economy. It is a measure of the monetary value of goods and services exchanged among individuals in the society (primarily through the market, but also through taxation and public goods provision). In this respect, it is quite different from something like the NMP (net material product) statistics compiled in the former Soviet Union, or the planning objectives that specified outcomes in material terms such as the number of tons of steel, kilograms of pork, or bushels of wheat that were to be produced. GDP statistics measure the level of economic activity, providing an aggregate measure of the value of economic transactions to the participants involved. This means that there is an intangible aspect to GDP, since it is possible to increase the intensity of transactions, or the value of those transactions, without actually increasing the level of material production. To take just one example, a thousand pairs of shoes, produced at a time when shoes are in great demand (and thus prices are high), will make a much larger contribution to GDP than a thousand pairs of shoes produced at a time when there is less demand. GDP also measures the value of services exchanged, so haircuts and massages count just as much as steel and wheat. Thus a person who starts a part-time babysitting service, and spends the money on yoga classes, is increasing GDP, even if there is essentially no material dimension to the transactions (or no net increase, relative to what the individuals involved would otherwise have been doing with their time).

In order to see the significance of this, consider a completely autarkic economy, in which private individuals provide for themselves all the necessities of life. There will be a certain level of material production in this society, although the GDP will be zero, because nothing is being traded. Now suppose that these individuals begin to exchange goods and services with one another. Even though the level of material production remains constant, GDP will increase. And as more goods and services are brought into the realm of exchange, or the "cash nexus," the economy will register growth—even though, in material terms, nothing is happening. Production is not increasing, what the growth registers is merely an improvement in the allocation of goods, the desire for which is driving the transactions.

This example is in one sense artificial, because as allocative efficiency increases, labor will tend to become more specialized, which will increase productive efficiency. Assuming these efficiency gains are not all consumed in the form of leisure, this means that the level of material production in the economy is likely to increase, which will also register as an increase in the GDP. In other words, when people are able to meet their needs more easily, they will begin to produce more stuff, which means using more energy, raw materials, and natural resources. The important point, however, is that GDP statistics measure this increase in material production only indirectly, via the value of the transactions that it generates. As a result, there is a great deal of noise that stands between the GDP measure and any measure of material production, and there are many circumstances in which growth could occur without there being any increase in "resource flows" into the economy— either because individuals are exchanging services that have no resource implications, or because they are improving the allocative efficiency of the economy. It is also possible to decrease resource flows into the economy, while maintaining constant GDP, simply through technological innovations that reduce the resource content of a commodity, while maintaining its market value (e.g. producing lighter cars that contain less metal, etc.).

Despite these complexities, Jackson has a habit of speaking of GDP as if it were a direct measure of the level of material consumption in a society. In the passage previously quoted, in which he sets up the tension between a finite ecosystem and an infinitely growing economy, he makes the crucial qualification that "no subsystem of a finite system can grow indefinitely, *in physical terms.*"[38] He is, of course, aware that an economy might be growing, but not in physical terms. And yet he constantly reverts to this mode of expression, describing the economy as involving a "material flow," or as involving "material consumption" or "material commodities."[39] This is a grave distortion, since most Western societies have economies that consist primarily of services being exchanged (for example, all agricultural production combined currently contributes less than 1 percent of US GDP).[40] Victor observes that "material and energy flows have increased with economic growth," but grants that this is not a necessary connection, only a historical regularity. And yet the systems language makes it seem as though the relationship is a structural feature, such that the larger the system grows, the more it must take in and dispose of.

What is peculiar about this conflation of economic and physical growth is that steady-state theorists have themselves developed the vocabulary required to express their view more precisely. Herman Daly, for instance, in his ur-text *Steady-State Economics*, draws a distinction between quantitative and qualitative growth, where the former represents an increase in the absolute quantity of goods produced and consumed, while the latter involves merely making better use of what we have, either with respect to production or consumption. What steady-state theorists are really forecasting (or advocating) is a halt to quantitative growth. Thus what they should be advocating, with respect to economic growth policy, is a strategy of "decoupling," or "dematerialization," which would involve a reduction in the material content of production and consumption, to the point where qualitative growth can occur that does not increase material flows, beyond the sustainable capacity.[41]

Seen from this perspective, the limits to growth thesis do not follow from first principles. When Jackson says that "no subsystem of a finite system can grow indefinitely, in physical terms," he is tacitly acknowledging that, if the economy were not growing in physical terms, then it should not be "an anathema" to the ecologist to contemplate the possibility of a "continually growing economy." Thus what Jackson's and Victor's thesis comes down to is skepticism about the possibility of decoupling. (Indeed, Jackson has a chapter criticizing what he calls the "myth of decoupling.") This constitutes a tacit concession that economists are not making a conceptual error when forecasting unlimited growth. The issue is actually an empirical one, about the possibility of decoupling economic from physical growth. And when it comes to assessing this empirical debate, it is difficult to avoid noticing that both Jackson and Victor tend to overstate how much decoupling must actually occur in order for growth to continue.

Victor's "open system" model, for instance, represents the economy as a fairly simple conversion mechanism that takes "resources" in on one side and produces "waste" out the other. If resources are finite and nonrenewable, then it follows that there will still be hard limits to growth, absent total decoupling. When put this way, the argument is reminiscent of the 1970s Club of Rome analysis, *The Limits to Growth*, which forecast an end to growth within the next hundred years due to resource constraints ("the most probable result will be a rather sudden and uncontrollable decline in both population and industrial capacity").[42] Despite the complexity of the model, its fundamental structure was quite simple. If one assumes that production always involves some use of nonrenewable resources, and that these resources are finite, then it follows inevitably that at some point in time further increases in production will be impossible and that, in the long run, production will drop to zero.

However, the idea that the economy takes resources as input and reduces them to waste is also misleading. The earth may be a finite system, but it is also a (relatively) closed system. This means that the "nonrenewable" resources are not literally consumed or destroyed; they are typically just reduced to a low-energy state.[43] For instance, metals are often classified as "nonrenewable" even though they are typically not destroyed through human use. Steel and iron, for instance, may become dispersed in the form of rust, but none of the metal is actually lost. It is merely transformed into a state in which it is very difficult to make use of. "Difficult," in this context, means "energy-intensive."[44] Similarly, although fossil fuels are often described as nonrenewable, this should not be interpreted literally. The hydrogen and carbon in hydrocarbon fuels is not consumed by burning; it is merely transformed into (predominantly) carbon dioxide and water. We currently have the technological capacity to capture atmospheric carbon dioxide, combine it with water, and turn it into diesel fuel.[45] The chemistry is quite simple. The problem, of course, is that the process uses a vast amount of energy (at very least, the same amount that is released by burning it). Technological improvement (the process is solar-powered) could change this, which would in turn make diesel fuel a renewable resource. The point generalizes—if one is willing to expend enough energy, any type of "waste" can be converted back to a "resource." Atmospheric carbon, in particular, can be captured and turned into carbon fiber, carbon nanotubes, carbon-based fuels, etc. As a result, no resource is truly nonrenewable; it all depends upon the state of energy technology.

Thus when thinking about the limits to growth, we should be formulating the issue in terms of energy, not resources. As Vaclav Smil puts it, "From a fundamental biological perspective, both prehistoric human evolution and the course of history can be seen as the quest for controlling greater stores and flows of more concentrated and more versatile forms of energy and converting them, in more affordable ways at lower costs and with higher efficiencies, into heat, light, and motion."[46] This is a point that was clearly understood by early pioneers in the field, such as Daly. Echoing a point made by Nicholas Georgescu-Roegen in 1971, Daly observes that the issue is really about entropy.[47] The crucial distinction is not between "resource" and "waste," but rather between low-entropy (or negentropic) and high-entropy structures. As Georgescu-Roegen put it, "Our whole economic life feeds on low entropy."[48] The question, therefore, is not whether we are going to run out of "resources," but whether we are likely to run out of negentropic structures. These are, however, absurdly plentiful. The largest source of low-entropy structures that we use involves, in one way or another, solar energy (the others are geothermal energy and nuclear fission). Even the energy in hydrocarbon fuels, which we rely upon for the majority of our energy needs, is just solar radiation converted by plants to high-energy chemical bonds through the (extremely inefficient) photosynthesis reaction.[49] The fact that we are relying upon solar energy captured by plants eons ago, instead of drawing it directly from the sun (and wind, which is an indirect effect of solar radiation causing differential heating of the planetary surface), is not due to any paucity of solar radiation; it is entirely due to the limitations of existing technology.

Steady-state economists sometimes draw attention to the fact that solar radiation is limited by the intensity of the sun and the surface area of the planet, as though this supported the limits to growth thesis.[50] And yet this is more of a theoretical limit, since the energy budget made available to us from the sun vastly exceeds human needs. Total human energy consumption in the world today is just under 18 terawatts (projected to rise to 43 TW by the end of the century). The total amount of solar energy striking the planet at any given time is 180,000 TW.[51] Factoring out albedo, atmospheric absorption, cloud cover, and night, the average amount of solar energy striking the planet surface at any given time is approximately 90,000 TW.[52] Of this, plant photosynthesis manages to capture only about 130 TW (which is why fossil fuels represent such an important resource). Thus the earth is literally awash with surplus energy. A typical lightning strike peaks at 1 TW. The problem is just that we are incredibly limited in our capacity to convert this energy into usable form.

Thus the only hard constraint on growth is the solar energy budget, which at this point so greatly exceeds human usage that it is difficult to imagine circumstances in which humans could use even a perceptible fraction of it.[53] The real constraint is just the practical one, of finding ways of converting this energy into usable form. (There is also the additional wrinkle that solar electricity generation, the most promising form of renewable technology, produces energy at zero marginal cost, which creates the possibility of a counterintuitive scenario in which there is no longer anything to be gained from curtailing energy demand.)[54] Thus the view that, over the next one or two centuries, economic growth will be subject to "limits" essentially involves making a bet against technological progress. "Resources" are not really finite, and "waste" is not really waste, and therefore neither imposes any hard constraints on growth. As a result, complete decoupling or dematerialization is not even required to preserve growth. The question comes down to whether or not we can harness energy to avoid depleting negentropic structures. Thus the claim that there are limits to growth is basically just a bet against basic human ingenuity. Furthermore, it is a bet that the future will not resemble the past—that the pace of technological innovation and scientific discovery cannot continue as it has for the past two hundred years.

There are, it should be noted, some who have made this argument explicitly. Thomas Homer-Dixon's *The Ingenuity Gap* is one instance.[55] And yet the claim is essentially speculative. When it comes to human energy systems, it is important to recognize that the problem (capturing ambient energy and transforming it into a usable form) is not intrinsically that difficult. Producing renewable energy is not like vaccine development, or genetic engineering, which involves intervention in hypercomplex systems. The core of our energy systems involves recombining the chemical bonds between hydrogen, carbon, and oxygen atoms. Furthermore, whether we can expect progress in this domain is not simply a question of optimism versus pessimism, as though technological innovation were an exogenous process. An important objective of environmental policy is to stimulate investment in research and development aimed at improving sustainability. One can, of course, debate the prospects of fundamental innovation in this domain. This is, however, a discussion worlds removed from the systems-theoretic vocabulary used to formulate the limits-to-growth thesis, which makes it sound as though there were some scientific principle at work, imposing a hard constraint on economic growth. This is not the case (which is one of the reasons why the models used by the IPCC assume ongoing economic growth). Indeed, the claim that economic growth will continue is essentially the consensus view among economists, in the same way that the phenomenon of global warming itself is the consensus view among climate scientists.

So despite its widespread prevalence, the limits-to-growth view has little to recommend it as an empirical or scientific claim. It is no accident that the date at which growth is supposed to stop keeps getting pushed back further and further, while proponents of this view have learned to be increasingly vague about when they expect the effect to materialize. It is useful therefore to distinguish *normative* from *empirical* steady-state theorists. The former say only that growth is not obligatory, and perhaps not even permissible or desirable. The latter say that growth is not possible. I have tried to show, in this section, that the claims made by the empirical steady-state theorists are unproven and essentially speculative. Thus I will be setting aside these views in the discussion that follows, and focus my attention exclusively on normative steady-state theorists.

2.3. Impacts of Climate Change

Anthropogenic global warming has been described as "the mother of all collective action problems." Greenhouse gas emissions, largely from consumption of fossil fuels, are slowly changing the composition of the earth's atmosphere, in a way that increases the amount of energy that is retained from the solar radiation that strikes the planetary surface. This will have a number of negative consequences, some of which are being seen already: increase in average temperature, increase in extreme weather events, sea-level rise, and ocean acidification. The problem is difficult to solve, unfortunately, for three reasons. First, individuals derive significant benefits from fossil fuel consumption, whereas the costs are extremely diffuse and almost entirely born by others. Thus it is in practically no one's personal interest to stop producing emissions, despite the deleterious consequences—this is what gives rise to the core collective action problem. Second, because the consequences of actions being taken now will not begin to be felt for decades, those who will bear these costs are not a powerful constituency, present to press their objections. This makes political mobilization around the issue difficult, because the most important interest group is absent. And finally, the atmosphere is a global commons, which does not discriminate between emissions from different sources, and so there is little to be gained from any nation taking unilateral action, if all this does is give other nations the opportunity to delay or avoid taking action of their own. This creates a significant compliance problem.[56] The confluence of these three factors makes the free-rider problem particularly acute, and thus raises real questions about whether our species even has the capacity to cooperate at the level required to respond to this challenge.[57]

Perhaps because of the difficulty of eliciting cooperative behavior in response to the problem, there has been an unfortunate tendency to ratchet up the rhetoric, and in some cases to overstate the seriousness of the likely effects of climate change. Steve Vanderheiden, for instance, in *Atmospheric Justice*, suggests that, "as a result of choosing a high-growth, high-consumption, and high-pollution path, the planet's future capacity to fulfill human wants and needs will likely be significantly diminished by environmental degradation and climatic instability, worsening conditions for those inhabiting the future world."[58] Tim Mulgan has set about the task of reconceptualizing ethics for a "broken world," "a place where resources are insufficient to meet everyone's basic needs, where a chaotic climate makes life precarious and where each generation is worse off than the last."[59] Naomi Oreskes and Erik Conway have suggested that climate change may produce "the collapse of Western civilization."[60] Catriona McKinnon pushes it one step further, opening her book on *Climate Change and Future Justice* with the suggestion that, "in worst-case scenarios of runaway climate change, *Homo sapiens* could go extinct."[61] This echoes a theme from Clive Hamilton, who wrote a book on climate change called *Requiem for a Species*.[62] And these are just academics; a number of popular writers—not to mention Hollywood movies—have been even more alarmist.

Luckily for our descendants, there is no plausible scenario in which climate change results in the extinction of our species, and no probable scenario in which it brings about the end of civilization. Humans possessed of only stone-age technology survived 4°C warming at the end of the last ice age. Thus the description of climate change as posing an "existential threat" to humanity should not be taken literally. Even the suggestion that it will bring about an absolute decline in living standards is considered an outlying possibility, subject to considerable uncertainty. Of course, if it were the case that climate change stood poised to bring about one of these catastrophic outcomes, then the normative problem would be much simpler. A deontic prohibition on harming others, along the lines initially sketched out by Simon Caney, would be sufficient to resolve the question of what we should do. This sort of simplification, however, obscures the difficulty of the problem. It is precisely because climate change is *not* likely to produce an absolute reduction in living standards—and is almost certain not to within the next century—that very difficult normative judgments must be made about the competing priorities of economic development and climate change mitigation.[63] (Needless to say, the temptation to inflate one's empirical estimate of the dangerousness of climate change, in order to salvage one's normative theory, or to avoid having to contemplate difficult trade-offs, is one that should be resisted.)

It is, however, not difficult to see how many people could have gotten the wrong impression about the likely effects of climate change. Consider, for instance, the *Stern Review on the Economics of Climate Change*, which was commissioned by the UK government to provide a synoptic overview of the anticipated costs of climate change, in order to guide policy. The review garnered headlines throughout the world when it was released, in part because it recommended much stronger action, and much more aggressive carbon abatement targets, than mainstream policy analysts had been recommending. Stern took care to emphasize, in painstaking detail, the incontrovertible fact that climate change will be enormously destructive and disruptive and, therefore, expensive. Depopulation of coastal regions and equatorial zones, along with mass migration of hundreds of millions of people, are not outcomes to be contemplated with equanimity. However, it sometimes escaped the attention of nonspecialists—non-economists in particular—that Stern followed the IPCC in assuming that economic growth rates will remain positive, even under the BAU scenario, under which nothing is done to slow the progress of climate change. Indeed, one of the central objections to Stern's analysis, among economists, was that the primary benefits produced by the aggressive mitigation efforts he was recommending would be enjoyed by future generations who, according to his own model, would be vastly better off than we are now.[64]

This fact was, unfortunately, often obscured by Stern's mode of expression. For example, when calculating the potential loss of GDP from climate change, he sets up a range of scenarios, based on different estimates of climate sensitivity (how much the temperature will change in response to a given increase in atmospheric carbon dioxide), as well as "catastrophe" scenarios (caused by feedback loops, in which global warming itself sets off processes that result in additional warming). He then calculates that in the worst-case scenario, with both high sensitivity and risk of catastrophe, "the mean loss in global per-capita GDP is 0.4% in 2060. In 2100, it rises to 2.7%, but by 2200 it rises to 12.9%. Adding non-market impacts, the mean loss is 1.3% by 2060, 5.9% in 2100 and 24.4% in 2200."[65]

This conclusion is illustrated by various graphs, which appear to show a decline in GDP per capita over time.[66] He adopts a similar mode of expression when he suggests that "in the baseline-climate scenario with all three categories of economic impact, the mean cost to India and South-East Asia is around 6% of regional GDP by 2100, compared to a global average of 2.6%."[67] The casual reader could be forgiven for thinking that the reference, when he speaks of "loss in GDP per capita," is to *present* GDP. What he is talking about, however, is actually the loss of a certain percentage of expected *future* GDP. In some cases, he states this more clearly: "The cost of climate change in India and South East Asia could be as high as 9–13% loss in GDP by 2100 compared with what could have been achieved in a world without climate change."[68] The last clause is, of course, crucial—under this scenario, GDP will not be 9–13 percent lower than it is *right now*, but rather lower than it *might have been*, in 2100, had there not been any climate change.[69] Similarly, what Stern's graphs represent is the loss of GDP per capita, as a fraction of what it is expected to be in the future. In other words, what Stern is saying is that climate change stands poised to depress the rate of growth.

This type of ambiguity has unfortunately become common in the literature. An important paper in *Nature* by Marshall Burke, Solomon M. Hsiang, and Edward Miguel, estimating the anticipated costs of climate change, presents its conclusions in the same misleading way.[70] The abstract of the paper states that "unmitigated climate change is expected to reshape the global economy by reducing average global incomes by roughly 23% by 2100." The paper itself, however, states the finding in a slightly different way: "Climate change reduces projected global output by 23% in 2100 . . . relative to a world without climate change."[71] Again, that last qualifying clause is crucial, yet it was the unqualified version of the claim found in the abstract that made its way into the headlines when the study was published. Thus many people came away from it with the impression that climate change would actually make people 23 percent poorer in 2100 than they are now.

It should be emphasized that these losses of potential GDP are enormous, and they call for a strong policy response in the present. At the same time, what these economists are describing is not a "broken world," in which "each generation is worse off than the last." On the contrary, they are describing a world in which the average person is vastly better off than the average person is now—just not as well off as he or she might have been had we been less careless in our greenhouse gas emissions. For example, the annual rate of real per capita GDP growth in India, at the time of writing, is 6.3 percent, and so what Stern describes, in his analysis of impacts on South Asia, is equivalent to the loss of approximately two years' worth of growth.[72] At the present rate of growth, living standards of the average person in India are doubling every 12 years. There are fluctuations from year to year, but the mean expectation of several studies, calculated by William Nordhaus, suggests that the GDP of India will be about 40 times larger in 2100 than it was in the year 2000 (which implies an average real growth rate of 3.8 percent).[73] The 9–13 percent loss, due to climate change, is calculated against the 40-times-larger 2100 GDP, not the present one. Thus Figure 2.1 shows a conventional way of representing the "loss in GDP per capita" that Stern calculates for the worst-case scenario (13 percent).

Of course, in order to get a scientifically accurate projection, one would have to use an "integrated assessment model" (IAM) in order to account for the feedback relationship between the rate of growth and the extent of climate change. Commentators sometimes forget that a significant fraction of the climate change that is expected to occur beyond the 50-year horizon will be a consequence of the economic growth that will occur during the interim period—so that if the growth does not occur, then neither will some of the climate change. Figure 2.1 is intended just to show what a 13 percent loss of potential GDP looks like, when expressed as a fraction of the anticipated

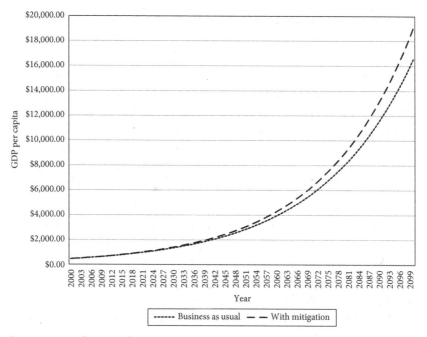

Figure 2.1 India growth trajectory, with and without climate change

GDP of India in 2100. (From a GDP of $457 per capita in 2000, GDP is expected to rise to $18,342. A 13 percent loss of potential GDP would result in it rising only to $15,957. Either way, GDP in 2100 is expected to be much larger than it is currently.)

It is not difficult to see what the problem with the BAU scenario is from a utilitarian perspective—the loss of $2,385 GDP per capita, in a country that will likely have over 1.5 billion inhabitants, is an enormous cost. Indeed, the expected annual losses caused by climate change in the year 2100 are more than five times larger than the entire GDP of India in the year 2000. The fact that this loss could be avoided by a much smaller investment of resources in carbon abatement, in the present, reveals a dramatic failure to maximize well-being. Inaction on climate change is like allowing water damage throughout one's house, in order to avoid the cost of fixing a leak in the roof. And yet for the normative steady-state theorist, the scenario creates something of an embarrassment. According to this view, there is no obligation to maximize future GDP, and so there is no injustice committed against future generations by a zero-growth policy, which would impose enormous losses upon them—losses that easily dwarf those associated with climate change.

And yet if the zero-growth policy violates no obligations, it is difficult to see how the BAU scenario, which offers a combination of growth and climate change with positive net value, could be prohibited.

Consider a specific illustration of the problem. Catriona McKinnon follows Rawls in claiming that we have an obligation to maintain the rate of savings necessary to provide our descendants with the productive resources required to sustain just institutions. Thus societies will undergo an "accumulation phase" in which net savings are positive, before entering into a "steady state," in which they fall to zero. Indeed, McKinnon appears sympathetic to the view, defended by Axel Gosseries, that once the steady state is reached, positive savings are not just non-obligatory, but actually impermissible, because any positive surplus should be directed to helping the worst-off, who are the poorest among those alive right now. Climate change, however, creates an exception to this, because of its negative impact on future generations. Thus our obligation, she claims, is to maintain positive savings sufficient to compensate them for those harms: "We ought to save now in order to provide future generations with funds and resources adequate to compensate them for the harm at which we are possibly (with respect to climate change catastrophes) and probably (with respect to non-catastrophic but still harmful climate change) putting them at risk."[74] Absent these harms, there is no obligation to save beyond capital replacement.[75]

McKinnon therefore recommends the creation of an "Intergenerational Climate Change Compensation Fund" for this purpose.[76] There are some questions about how she views the economics of this, since she discusses the fund as though it were a money balance in a bank account, e.g., "The current generation should discharge the obligations created by this liability through contribution to an intergenerational fund from which any future generation suffering harm can draw adequate compensation."[77] Similarly, she talks about the savings being "held" until a particular point in time.[78] Expressed in this way, the view does not make sense—it is like imagining that we could compensate future generations for the burdens of climate change by writing them a check that they could cash sometime long after we are dead. In order for there to be genuine compensation, consumption must be forgone now, in order to make real resources available to them later. The idea that we might set aside a stock of goods that future generations could tap into once the harms begin to materialize is impractical. If a compensation scheme were to be organized, it would be by having present generations make investments and produce capital goods, with the benefits "paid forward" in the form of increased economic growth, which will provide the resources required either to adapt to climate change or to provide compensating benefits for the harms suffered.

In other words, what McKinnon's proposal for climate compensation boils down to, once the economics are sorted out, is that each generation is obliged to maintain a rate of economic growth sufficient to produce a steady state, factoring in all costs of climate change.[79] No special principle of compensation is required. Because climate change stands poised to generate significant costs, this will require a positive savings and growth rate in order to avoid having future generations fall below the level needed to sustain just institutions. In effect, since the stock of "natural capital" (in the form of an atmospheric good) is being depleted, each generation will be obliged to increase the stock of economic capital, so that future generations come out even. (Note that whatever reservations one might have about GDP growth as a measure of increased welfare are not germane in this context, since we are considering GDP as a measure of resources available for climate change adaptation and compensation.)

Stated in this way, the view may sound like a plausible requirement of justice, and yet it amounts to the claim that, given the current growth trajectory of the world economy, we are entitled to completely ignore climate change. A few modifications of the data presented in Figure 2.1 should make this clear. First, steady-state theorists seldom specify in dollar terms what would constitute a reasonable level of GDP per capita, sufficient to maintain just institutions, although most seem to assume that it will be higher than many countries, such as India, are at now.[80] (Indeed, both Jackson and Victor grant that underdeveloped countries are entitled to grow a significant amount, and so the limits to growth injunction/thesis apply only to rich countries.) Based on the statistics that Jackson presents, regarding education, health, and infant mortality, let us suppose that accumulation should continue until a relatively generous $5,000 GDP per capita is reached.[81] Figure 2.2 shows the steady-state (SS) growth path as a third alternative for India. The fourth data series added shows how much GDP per capita would need to increase, along the steady-state path, in order to fully compensate individuals for the effects of unchecked climate change, based on Stern's estimates.

This graph, I should note, is merely for purposes of illustration, in order to provide a sense of the magnitude of the costs and benefits involved. In fact, the adoption of a steady-state growth trajectory would slow the rate of increase in global warming, and so less compensation would be required than is

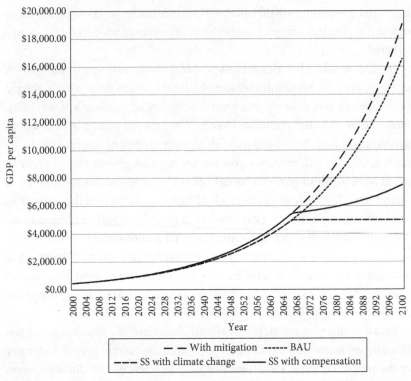

Figure 2.2 Steady state trajectories for India, with and without compensation

shown (calculation of the exact trajectory would require an IAM). The point, however, is that the BAU scenario—one that extends the current growth trajectory while ignoring climate change—easily generates sufficient benefits to fully compensate the entire population for the effects of climate change. By 2100 individuals under the BAU scenario enjoy a standard of living more than twice that of what they receive on the "steady state with compensation" path, and are receiving "compensation" almost five times greater than what they are owed. As a result, even using very conservative growth estimates and fairly pessimistic climate change damage estimates, as I have done here, our obligations to future generations, as specified by the steady-state view, are easily discharged under the BAU scenario, even though we do nothing to mitigate climate change.

The lesson to be learned from this is not that we should be ignoring climate change, but rather that philosophers such as McKinnon should be taking economic growth much more seriously. She is hardly alone in this. Caney, for instance, after shifting to a Razian theory of rights, interprets our obligations to future generations in terms of a set of "moral thresholds." Rights, on this view, specify a minimum level of key-interest satisfaction below which no one should be permitted to fall. And yet, if one considers the triumvirate of rights that Caney enumerates (to life, health, and subsistence), all of them permit substitution and compensation. For instance, while climate change may lower life expectancy, increasing per capita GDP by an order of magnitude is likely to raise it by a great deal more, effectively ensuring that no one's rights will be violated (or need be violated) under the BAU scenario. Thus Caney's normative framework is one that, when translated into real numbers and expectations, winds up also permitting inaction in the face of climate change. If our concern is merely to ensure that no one falls below a certain threshold—a threshold that, it should be noted, at least a billion people are currently below—then "BAU plus redistribution" is a much faster, less costly way of achieving that outcome than phasing out consumption of fossil fuels.

More generally, it is strange to think that we might be perpetrating a great injustice against future generations by adopting a BAU climate change policy that will lead to losses equivalent to 13 percent of potential GDP, or even 23 percent of potential GDP, but that it is perfectly acceptable for us to adopt a steady-state growth policy that will lead to a loss of 70 percent of potential GDP. It is also worth noting that these growth outcomes are obtainable by the end of this century, the period before truly dangerous climate change is likely to occur. It is difficult to make confident predictions beyond that period, given the numerous uncertainties that cloud the picture (involving climate, but also involving technological progress). Thus the policy problem is that, in the relatively near term (i.e. on the order of decades), we face a choice between rapid economic growth and a warming process that is slow, but gathering momentum. Given the trade-offs between the two, trying to find the right level of carbon abatement is a nontrivial problem. Furthermore, given the unintuitiveness of the exponential growth process, the answer is going to rely on some complex calculations, not brute moral intuitions or crude deontic constraints.

Finally, it should be noted that support for the steady-state view is sometimes motivated by the suggestion that, once GDP per capita passes a certain threshold, something like satiation is reached, and so further increases do not generate any improvement in human well-being. Thus it does not make sense to adopt policies that do significant damage to the environment, in the name of maximizing growth, because the growth does not actually make anyone better off.[82] This may be true, but if so, it also undermines the case for climate change mitigation, by suggesting that future generations should not mind having to redirect trillions of dollars' worth of resources toward adaptation efforts, or to fund a compensation scheme, or to engage in carbon scrubbing of the atmosphere to reverse climate change. Since these costs affect people in the future much less than they affect us, it seems to follow that they should be the ones to bear the burden.

This point applies with equal force to a number of other environmental problems. For example, there are many parts of the world in which both farmers and consumers are draining water aquifers at an unsustainable rate, leading to the expectation that fresh water will become increasingly scarce in many parts of the world. And yet fresh water is not literally becoming scarce, it runs off into lakes and oceans, where a certain fraction evaporates and is returned to land in the form of rainfall. So again, our "consumption" of fresh water is not really consumption, but rather just an increase in the entropy of freshwater stores. So even though responsible water use is an important component of sustainable agriculture, future generations could be compensated for depleted aquifers by being given alternate ways of collecting and transporting surface water (e.g. pipelines and canal systems), or even desalination plants to purify ocean water. These are all extremely capital intensive, and so not cost-effective alternatives in the present. However, if we do not owe future generations any increase in their material well-being, and if ordinary growth would have been absorbed into "an orgy of self-indulgent consumerism," then it does them no injury to have to redirect these resources to the transportation or production of fresh water.[83]

Thus there is something of a paradox that arises within the standard environmentalist position. On the one hand, there is a tendency to dismiss the importance of economic growth, on the grounds that it reflects a set of "materialistic" values, or that it does not serve important human needs.[84] On the other hand, there is enormous concern about the environmental problems that we are bequeathing to future generations. And yet many of these "problems" are only problems in the sense that they would be extremely costly to remedy (or work around, etc.). If future generations are going to be suffering an embarrassment of riches, then it follows that they should be able to fix these problems without much hardship (or with less hardship than we would experience), which in turn reduces the extent to which we should be worried about them. As a result, there is something inconsistent in a view that exhibits both a concern over environmental degradation and a lack of concern over economic growth. (This issue is one that recurs in the debate over the social discount rate, and the elasticity of marginal utility of consumption, which will be discussed in Chapter 6.)

2.4. Sustainability and Fungibility

It is unlikely to have escaped anyone's attention that the argument above involves monetizing the harms associated with climate change. This is not meant literally, of course. Stern's calculations include non-GDP impacts. The harms are merely being expressed in monetary form, in order to compare the magnitude of the damages to that of the benefits provided to future generations, in order to determine whether they are being left better off or not. Furthermore, *pace* Henry Shue and others, future generations are not being offered merely "financial compensation" for the damages caused by climate change. Under the standard compensation scenarios, they are being offered real resources, which will permit them to adapt to climate change in ways that are expected to make them better off on the whole. Money is being used merely as a *metric of value*. It is a general feature of theories of justice that, whenever individuals are being allocated a complex combination of goods and bads, some basis of comparison is required, in order to determine whether or not the allocation is just.[85] There are various well-known difficulties associated with using money as that metric, but it is important to recognize the importance of having *some* metric of comparison—thus the onus rests on the critic of monetary valuation to propose some feasible alternative.

At the same time, there is a substantive and controversial assumption underlying the discussion in the previous section, which is that there is "fungibility" in the goods (and bads) that we are leaving to future generations.[86] This is what makes it possible to talk about adaptation and compensation in lieu of mitigation. Following Bryan Norton, we can define our "bequest" as the total state of the world that we pass on (and thus, "the sum total of our impacts on subsequent generations").[87] This bequest can be analytically divided into a number of different dimensions or categories. It will contain, for instance, cultural artifacts (e.g. art and literature, scientific knowledge), technological expertise (e.g. manufacture of plastics, construction of aircraft), institutions (e.g. states, universities), durable consumer goods (e.g. housing stock, highway infrastructure, antiques), capital goods (e.g. "plant and equipment," machines, computers), other elements of the built environment (e.g.

dams, canals, monuments), natural resources (e.g. fossil fuel and mineral reserves, aquifers), ecosystems (e.g. forests, lakes, wetlands), and "dependent" natural resources (e.g. agricultural land, domesticated animals).[88]

In some of these categories, the stock that is being passed along will clearly be degraded in terms of quantity, quality, or both. This is obviously true in the "natural resource" and "ecosystem" categories, but also in some of the social categories, such as "transportation infrastructure" and "political institutions" in some countries. A key question, from the perspective of intergenerational justice, is whether we are obliged to pass along to subsequent generations as much and as good in each category, or whether a decline in one category can be made up for by an increase in some other. If we accept such substitutions, the most important question then becomes how we are going to measure value in each category, so that we can determine how much of an increase in X will be required in order to compensate for a decline in Y. The natural impulse here is to introduce, as the metric of comparison, something like welfare, or well-being, or needs, or "opportunities to lead a good life," or some other placeholder for the value that we expect future generations themselves to place upon the bequest. In this way, the question of whether the compensations we are offering are adequate will be answered by whether future generations consider themselves adequately compensated.

Again following Norton, we can then distinguish different normative views in terms of how *structured* they believe the bequest should be.[89] A completely unstructured bequest would be one that does not prohibit any substitution from one category to another. (Such a view was expressed canonically by Robert Solow in his "Sustainability: An Economists' Perspective": "Resources are, to use a favourite word of economists, fungible. . . . They can take the place of each other. That is extremely important because it suggests that we do not owe to the future any particular thing.")[90] This is known as "weak sustainability." Figure 2.3 shows a model of this sort, with circles of different sizes showing the various elements of the bequest, and the arrows between them showing the permissible substitution relations (each arrow could also be given a weight, showing the rate at which substitution is permitted between the categories). In the weak sustainability model, a bidirectional arrow links each component of the bequest to every other, in order to represent the fact that none of these substitutions are prohibited by the normative theory.

Since we know very little about the preferences of future generations, the best way to discharge our obligations is to make sure that a large fraction of the bequest is passed along in the most abstract form possible, e.g. in the

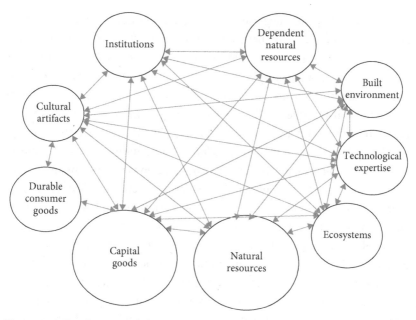

Figure 2.3 Weak sustainability model

form of a pool of savings or capital, which can be gradually adapted with the passage of time.[91] This is what gives rise to what Norton refers to as the "grand simplification," which treats the entire intergenerational justice problem as essentially a question of "just savings."[92] According to this view, there is no special issue of environmental sustainability, since any losses within the natural categories can, in principle, be compensated for by an increase in one or more of the social categories (in particular, the stock of capital goods).

Of course, the idea of a bequest in which anything can be freely substituted for anything else is not realistic, and is adopted by economists largely for modeling purposes. For policy purposes, any plausible development of the weak sustainability view will need to acknowledge certain limits to substitution. What makes these views still weak is that the limits *are not specified by the normative theory*, but follow rather from the combination of empirical features of the world and the anticipated valuations of future generations. Thus it may be ridiculously costly, and therefore not feasible, to replace certain natural resources, but as long as it is not normatively prohibited, then there is no incompatibility between weak substitutability and the practical impossibility of making substitutions in this dimension. Economists are well aware that such empirical limits on substitution exist, because many of them arise even within the categories. For instance, agriculture contributes only about 4 percent of total global GDP, but obviously if one generation were to set into motion events that would foreseeably destroy all agriculture on the planet, it could not make up for it by increasing production by 4 percent in other areas. Even if we are unable to foresee the specific preferences of future generations, it is a safe bet that they are going to need food. Thus it is perfectly consistent with a weak substitutability view to insist that the economy we leave to our descendants must include an agricultural sector. This means that even *within* the category of GDP, substitution is not, as a matter of fact, unlimited.[93] Similarly, proponents of the "critical natural capital" (CNC) approach to sustainability have observed that certain natural systems perform environmental functions that cannot be replaced by synthetic alternatives.[94] Although often described as a strong sustainability position, because of the limits it seeks to impose on substitution, this perspective is actually consistent with a weak sustainability position, insofar as we can predict that future generations will value those environmental functions. For instance, it is not difficult to predict that future generations will want there to be an ozone layer. Thus it is not the normative theory that prohibits substitution for its services, but rather practical features of the world and the limitations of human technology.

Finally, it is worth emphasizing that the weak sustainability position is not committed to the view that substitution between various forms of natural and synthetic capital is a practicable option, only that it is normatively permissible and possible in principle. It may also be punitively expensive. For instance, proponents of CNC point to examples such as the role that wetlands play in purifying water. This is a valuable ecosystem service, yet as we have seen, it is possible to replace this with water purification plants, or even desalination facilities near the ocean. Thus when CNC advocates say that certain forms of natural capital "cannot be substituted for" by manufactured capital, what they often mean is "cannot be substituted for at reasonable expense."[95] But this transforms the environmental issue into essentially an efficiency concern. In terms of maximizing the value of our bequest to future generations, it is at the moment much less costly to preserve wetlands than it is to provide them with a technological substitute for the services provided by these forms of natural capital. But that does nothing to impair the claim that, if we were to destroy these forms of natural capital, we would not be wronging them if we were to provide a substitute that left them just as well off.[96]

At the opposite extreme are views that require a *completely structured* bequest, in which the normative theory does not permit substitution across categories. Norton refers to these as "list of stuff" views, on the grounds that, when substitution is forbidden, the obligation to future generations winds up being an obligation to transmit a stock of specific goods, and so they can be enumerated as a list.[97] (Daly refers to this view, dismissively, as "absurdly strong sustainability," since it is difficult to think of any reason why *all* substitution between categories should be prohibited.)[98] Of course, there will always be substitution within categories, and, depending upon how the distinctions are drawn, the view may be more or less restrictive. Thus it makes more sense to characterize the remaining set of views as more or less structured, depending upon how broadly the categories are drawn, or what kinds of substitutions are prohibited. For instance, there is a very common "strong" view that distinguishes three very rough categories: "heritage" artifacts, economic "capital," and "nature," between which substitution is impermissible. (Thus, for instance, one is not allowed to destroy works of art, or old buildings, even if one promises to build new ones. Similarly, one is not allowed to log old-growth forests, even if the timber will be used to produce very durable furniture.) There are subcategories as well, since substitution is often forbidden within those broad categories. (For instance, it may be permissible to swap one tract of conservation land for another, but one might not be allowed to make up for the extinction of a species by increasing the amount of land under conservation.)[99]

Norton refers to structured views as requiring "strong sustainability" (although he tends to speak of this as just one view, as opposed to a spectrum of views, as I am presenting it here), because of the constraints on substitutability—or, more accurately, on compensability. As has been shown, the details of these views may become quite complicated. Figure 2.4, for instance, shows a view in which certain forms of substitution between categories are prohibited entirely or prohibited in one direction. In particular, most of the incoming arrows to the "natural resources" and "ecosystems" categories have been removed, suggesting that if these are eroded, the loss cannot be made up for by improvements in other domains. (The example is hypothetical, because proponents of strong sustainability have generally not articulated their view at this level of detail.)

It is also worth noting one other theoretical possibility, which is a view that does not prohibit certain substitutions, but where the normative theory imposes a very high rate of substitution (e.g. higher than would be imposed

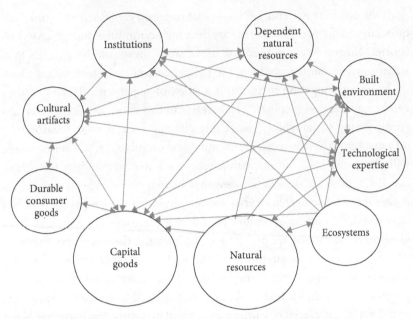

Figure 2.4 A strong sustainability view

by the mere preferences of future generations). For instance, it might permit the substitution between human and natural capital, but require that a great deal more capital be provided than just that which is required to keep future generations satisfied, or allow them to meet their needs. (Visually, one could make use of varying the width of the lines, in order to show different rates of substitutions between categories.)

The most important question that arises is whether compensation should be permitted from the artifactual categories to the natural ones. Since we are currently degrading and damaging the natural environment in myriad ways, the question is whether we can make up for it by providing future generations with compensation, in the form of various artifacts of human ingenuity, or even just labor and sacrifice, that will enhance their lives in other ways. Although the environmentalist impulse will be to say no to this question right away, there are many instances in which the answer must clearly be yes. When it comes to most questions of resource depletion, for instance, it is difficult to come up with any reason to care about the resources as such; the question is simply what can be done with them. Iron ore deposits, for instance, are extracted and used to make steel, which is used to make automobiles, buildings, household appliances, wind turbines, and a thousand other things. If technological innovation makes it possible to increase the tensile strength of steel, so that less material is needed in construction, or if new manufacturing processes allow sheets to be made thinner, so that automobiles require less metal, or if new plastics are developed that can replace steel in certain applications, all of this means that future generations will have less need for iron, and so the fact that we have depleted the most accessible deposits should not be of concern to anyone. Furthermore, as we have seen, the iron is not being consumed, but merely dispersed, and so even improvements in mining technology, which make it easier to access previously marginal deposits, or blasting, which allow us to extract the iron more efficiently from ore, should leave future generations perfectly indifferent with respect to the total stock of deposits we leave to them. The same can be said for most other so-called nonrenewable resources. As a result, *some* substitution from the artifactual to the natural category must be permissible.

The structure of this argument, with respect to future generations, is basically, "If they don't care, why should we care?" What is striking about this argument, of course, is the assumption that what *people* care about is all that matters. This suggests a theory of value that is entirely anthropocentric. Thus the standard strategy for defending strong sustainability, among environmental ethicists, is to reject this theory of value and make the claim, which was analyzed in the previous chapter, that certain elements of the natural environment have value, or are good, in ways that do not depend upon human valuation, and that these values are what should be driving policy. Thus we can speak of "natural goodness" as a form of value that nature supposedly possesses, and would possess, even if there were no human beings in existence to value it. This goodness might then be thought to be what we have an obligation to preserve.

This approach to defending the strong sustainability thesis suggests that even construing the problem as one of intergenerational justice, involving a bequest to future people, is misleading. The obligation to preserve certain forms of natural goodness is not an obligation to future persons, it is an obligation that flows directly from the goodness of those natural states, and is one that, *ex hypothesi*, we would have even if there were no future people. Thus the debate over sustainability actually runs together two quite distinct obligations. We have an obligation to future people to create conditions that will result in the realization of certain forms of anthropocentric value, or human goods (such as the satisfaction of needs). And we also have an obligation to refrain from destroying certain forms of natural goodness. The strong sustainability thesis then follows from the claim that, in addition to being different, these two types of goodness are incommensurable, so that we cannot make up for the destruction of natural goodness by increasing the stock of human goodness, and vice versa. On this view, then, obligations of ecosystem or species preservation might be non-compensable because they involve natural goodness, while resource depletion would permit compensation because it involves only human goodness.

There are, as we have seen, some well-known difficulties with this argument. Because of their differences in perspective, proponents of natural goodness have been unable to present anything like a unified front in the debates over environmental policy. Given this fact of pluralism, one way of proceeding would be to decide that, since we have no basis for declaring one view to be the correct one, policy should be determined, not by the forms of natural goodness as such, but rather by how much different individuals value various forms of natural goodness. Assuming that human forms of valuation are commensurable, this then undermines the strong sustainability thesis. For our purposes, however, there is no need to insist upon this point, because when it comes to climate change, there is actually not that much natural goodness at stake, relative to the human welfare impacts. Even if one were inclined toward a strong sustainability perspective, any plausible view is going to acknowledge that what we describe as "environmental problems" are going to involve a combination of natural and human impacts. At one extreme, we have an issue like ozone depletion, which is important because of its human impacts, but which (one can only assume) is a matter of relative indifference to nature. At the other extreme, we have the extinction of an undiscovered species of spider in Amazonia that has essentially no significance for humans but that represents a loss to nature. The question is whether climate change is more like the former or the latter. The answer, I would argue, is that most of the damages caused by climate change are due to its human impacts, not its effects on nature. Climate change, in other words, is going to do a lot more damage to stuff that people care about than it will do to anything that nature cares about.

The first and most fundamental point is that the changes we are bringing about in the earth's atmosphere, and as a consequence, its climate, are well within the range of natural variation. Indeed, the past 50,000 years have been a period of unusual climatic stability, and the last 10,000 years even more so. Instability has been much more the norm—instability that includes dramatic changes in both atmospheric carbon and temperature levels. Carbon concentrations recently surpassed 400 ppm, more than twice the preindustrial level. This may be alarming for humans, but from the perspective of nature, it barely registers. During the Cambrian period, carbon levels were more than 7,000 ppm. Furthermore, we are certainly not the first living organism to change the composition of the earth's atmosphere with its waste products. The oxygen that we breathe, it is worth recalling, is entirely a byproduct of the metabolic processes of plants. And indeed, we are not so much introducing carbon into the atmosphere as reintroducing it, since the carbon content of fossil fuels is all originally atmospheric carbon that was sequestered by plants, millions of years ago. Thus the changes we are bringing about are unnatural only if one takes a fairly narrow temporal perspective. There is nothing we are doing that has not also happened before, without our involvement. This is in stark contrast to many other things that we are doing, such as creating plastics, or producing nuclear waste, or releasing chlorofluorocarbons into the atmosphere, that are unprecedented, and thus more obviously unnatural.

Again, none of this is to suggest that climate change is not a very serious problem. It is to say that a non-anthropocentric perspective, or system of values, does not come anywhere close to capturing the true extent of the problem. If one looks at where the major damage from climate change comes from, the categories are agricultural impacts, sea-level rise, extreme weather events, and mass migration. These are all very much problems for people. Furthermore, it is important to recognize that one of the major forms of compensation we may be able to offer to future generations is the capacity to remove greenhouse gases from the atmosphere. Genetic re-engineering of plants to improve photosynthesis, for instance, is an innovation that is likely to occur regardless of our climate policies, because of its capacity to improve agricultural yields. But it also has the capacity to transform our energy systems and to increase carbon sequestration by plants. There are also a variety of direct carbon-scrubbing technologies available that work perfectly well, but that are currently far too expensive to scale up to the level at which they would have much impact on atmospheric carbon concentrations. But if future generations are significantly wealthier than us, one of the things they can do with the capital goods and technology that we bequeath to them is *reverse climate change*. This may sound like a farfetched possibility, until one considers the fact that steady-state theorists are contemplating a scenario in which the "compensation fund" being made available to future generations will be several times larger than the size of the current global economy. As a result, a world in which future generations are dedicating most of the economy to carbon scrubbing is one that, according to the steady-state theorist, need not be unjust. This form of adaptation is, of course, not a magic bullet, since it is constrained by the inertia of the climate system. Nevertheless, it seems impossible to deny the possibility of certain forms of substitution between natural and artifactual capital in cases where the latter permits restoration of the former. While certain forms of damage due to climate change are not reversible, climate change itself is—given sufficient resources and technology.

Of course, not all negative impacts of climate change are human impacts. There are two major categories of climate impact that could plausibly be construed as losses of natural goodness, and not just problems for people. These are ocean acidification and species extinction (two effects that are, I should note, not entirely distinct, since ocean acidification will be the cause of many extinctions). If one abstracts radically from the anthropocentric perspective, the fact that we are currently experiencing the sixth great extinction since the beginnings of life on this planet suggests that periodic die-offs are not unnatural either, since the previous five had nothing to do with us. Nor did they do anything to halt the long-run trajectory of increased planetary biodiversity. Adopting a more commonsense view of the problem, however, it is easy to see that there is something profoundly disturbing about the present rate of species extinction, and it seems clear that there is something problematic about the idea that we might absolve ourselves of the consequences of our actions by providing future *human* generations with compensating benefits.

Even if one acknowledges this, however, by partially structuring our bequest to separate out the issue of species preservation, it must be acknowledged that climate change is only *one* of the factors driving the current rate of extinction. The introduction of invasive species into new ecosystems, or the destruction of forests through logging or land use changes, are both important forces driving mass extinction.[100] And since it is inconceivable that we would make the changes necessary to bring extinctions as a whole to a halt, the appropriate normative framework for thinking about the issue is probably a "harm reduction" one. This implies, however, that we should be choosing the extinction-reduction options that have the greatest impact, relative to their cost. If this is the framework, then it is not obvious that having everyone on the planet reduce their consumption of fossil fuels is a particularly effective way of reducing species extinctions. (Stern, for instance, recommended an immediate expenditure of 1 percent of global GDP on carbon mitigation. With that sort of a budget—approximately $770 billion—one can imagine a large number of other initiatives that might have much greater impact on biodiversity preservation.) So while the species preservation issue stands out as one in which the weak sustainability position seems inadequate, the magnitude of the carbon mitigation efforts that are being contemplated, within mainstream policy discussions, is clearly being driven by the anticipated *human* cost of climate change, not just the damage that is being done to nature.

2.5. Catastrophe

There is another objection to the way of thinking I have been recommending, one that raises a host of different issues. Stern's high-end estimate of the anticipated damages from climate change included a factor that accounted for "risk of catastrophe." However, the way that he accounted for it was to treat it in the ordinary cost-benefit style, which was to take the magnitude of the damages that might occur under each different catastrophic climate-change scenario, multiply it by the best estimate of the probability of that scenario coming to pass, and then add these all up. As a result, even though many of these scenarios would push GDP growth below zero, the *expectation* of them does not, because each is quite low-probability. This is why expected growth remains strongly positive in Stern's analysis. This is, in turn, what produces the embarrassment for normative steady-state theorists, of recommending zero-growth policies while nevertheless condemning climate change, despite the fact that the former is likely to be far more costly to future generations than the latter.

One apparent way of avoiding this embarrassment would be to reject the cost-benefit framework as an adequate way of dealing with the risk of catastrophe, and thus to separate it from the whole issue of growth and compensation. On this way of thinking, climate change policies are to be adopted, not to diminish the pedestrian damages to the global economy envisioned under the most likely set of scenarios, but rather to protect against the "long tail" events, or the catastrophic damages that would occur if some runaway process were to be unleashed.[101] One could then argue that, while we have no obligation to raise the living standards of future generations, we do have an obligation to avoid acting in ways that create the possibility of a "broken world," no matter how small the risks (based perhaps on some version of the precautionary principle). Thus climate change mitigation would fall into the same category as nuclear disarmament, asteroid detection, pandemic readiness, and other policies aimed at eliminating the risk of improbable but extremely destructive events.

Indeed, one of the most prominent criticisms of the Stern Review, by the economist Martin Weitzman, focused precisely on the way that it deals with catastrophic damages.[102] Weitzman criticized Stern for, in effect, trying to shoehorn two quite separate issues into a single framework. On the one hand, there is the most likely scenario, in which carbon emissions generate a steady increase in global temperatures, with some amplification due to feedback effects (such as the methane released by the melting of the permafrost in the arctic tundra). Here damages can be well represented with a continuous function. On the other hand, there are the less likely but nevertheless possible scenarios, in which sudden, dramatic events occur, such as an abrupt reversal of the North Atlantic current, cessation of the Indian monsoon, or the collapse of a major ice sheet into the ocean. Here damages would be very large and discontinuous. Stern was concerned about both, but because he insisted on a unified analytic framework, he wound up recommending a policy that was, in Weitzman's view, adequate to neither—too strict for the pedestrian damages, but not focused on eliminating the catastrophic risks.

There is an important point here, although unfortunately it is often expressed in a highly misleading way in the literature. It is often suggested that, instead of adopting a cost-benefit perspective, and undertaking mitigation efforts up to the point where the benefits they produce become lower than their costs, one should instead think of carbon abatement as akin to purchasing an "insurance policy" against certain climate risks. The suggestion is that, given a significant risk of catastrophic loss, one should be willing to pay something upfront in order to eliminate that risk. This is, it is claimed, like insurance.[103] For instance, even though the risk of one's house burning down in any given year is close to zero, the financial loss is enormous, and so the average person is willing to pay the premium on an insurance policy in order to eliminate the expected loss. The cost of carbon abatement, on this view, is like the payment for an insurance policy. One is not expecting to make money on it (as a cost-benefit perspective would suggest); one is instead trying to eliminate a risk—and the amount that one is willing to pay should be dictated by the severity of the risk and how difficult it is to eliminate.

Yet while this analysis has struck many people as plausible, it rests on a rather significant mischaracterization of how insurance systems function. Most importantly, buying insurance against a catastrophic event does not actually reduce the risk of it happening. Insurance systems arise when a group of individuals, facing similar risks, agree to pool their resources, in order to indemnify those who suffer the loss. If a person has a 1 percent chance of someday needing a medical treatment that will cost $1 million, setting aside $10,000 to cover that eventuality is useless—either she will need the treatment, in which case the $10,000 is not enough to pay for it, or she will not need the treatment, in which case the money has been set aside to no purpose. If, however, she can find 99 other people facing the same risk, and they each set aside $10,000, then their pooled premiums are just enough to pay for the one person who is likely to actually require the treatment. But the probability of any one person needing the treatment is unaffected by the insurance scheme.

Turning to the climate change case, if we could find other inhabited planets, each with an industrial civilization slowly changing the composition of its atmosphere, creating a small risk of catastrophic environment damage, then we might be able to enter into some sort of insurance arrangement, whereby the planets that suffer catastrophe get bailed out by those that do not. But global climate change is obviously a one-off event, and there is no one with whom we can pool the risks. Thus any talk of an insurance approach is simply confusing. Catastrophic climate change damage is simply not an insurable event, and "saving" for catastrophe—if that is what we were doing—would be useless, since we are guaranteed to be saving either too little or too much.

What proponents of the insurance argument are really doing is just describing the phenomenon of risk aversion. When confronted with a particular stochastic process, a risk-averse individual is one who is willing to accept a substitute gamble that has a lower mean outcome, but also lower variance. This second gamble has a lower expected value, because of the reduced mean, but is also less "risky," in that the lower variance results in a reduced chance of winding up with one of the "extreme," or tail-end outcomes. Figure 2.5, for instance, could be used to represent the returns to two different crops over many seasons. One has a higher average yield, along with greater "upside," while the other offers a lower average return with greater certainty. The former, however, has a "fat tail" at both ends, in the sense that it has a much greater chance of producing far-from-average returns. If having returns fall below a certain threshold results in catastrophe (e.g. starvation), then one might quite reasonably prefer the option that provides the lower average return, on the grounds that it provides much greater confidence that returns will not be catastrophically low. This is risk aversion.

The willingness of individuals to purchase insurance contracts is one *manifestation* of risk aversion, but there are many others, including the various

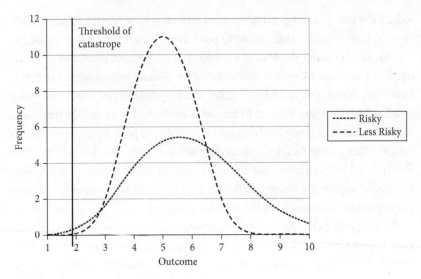

Figure 2.5 Eliminating tail-end risk

investments we make in safety equipment and regulations. Wearing safety glasses while cutting with a power saw, for instance, entails a cost, but at the same time, essentially eliminates a small risk of very serious harm. This is not "risk pooling," of the sort that an insurance arrangement achieves, but is actually "risk management," which involves making investments that change the profile of risks one is confronting. What proponents of climate change mitigation have in mind, when they describe it as an "insurance policy" against catastrophic risk, is actually something more like a risk management approach. They are claiming that we should be willing to accept certain costs now in order to reduce or eliminate tail-end risks of climate catastrophe. Normatively, this is equivalent to saying that we should be more risk-averse (compared to the typical cost-benefit analysis).

There is an interesting argument to be had over whether we should be more or less risk-averse in our attitude toward future generations. The issue is one that intersects in complicated ways with the debate over time-discounting (to be discussed in Chapter 6), since risk aversion, when dealing with a series of gambles, is in the long run equivalent to time-discounting (although, strangely, most proponents of risk aversion in the face of climate change are also opposed to time-discounting when assessing the costs and benefits). Luckily, neither issue needs to be resolved in order to see that the risk management perspective does very little to rescue the steady-state theorist from the embarrassment described in the previous sections. Recall that, on the risk management view, we should be willing to invest in carbon abatement, not up until the point at which the benefits produced are no larger than the costs, but up until the point at which the risk of certain catastrophe scenarios, and thus losses, drops below some acceptable level (possibly to zero). Proponents of this view sometimes fail to grasp how much this shift in normative perspective changes the conversation about carbon abatement policy. After all, what does it mean to say that policy should be driven, not by the desire to avoid the pedestrian damages expected in the most-likely scenarios—that is to say, the damages in the range of 10–20 percent of future GDP by the end of the century that we are expecting—but rather by the desire to eliminate certain tail-end risks of catastrophic damage? The central problem of climate change is not that we are afflicted by uncertainty about whether certain scenarios will come to pass. The problem is that we are extremely confident that the current growth trajectory, with essentially unrestricted greenhouse gas emissions, will bring about sea-level rise, widespread crop failure, an increase in extreme weather events, etc.—and this is what we are seeking to head off.

To illustrate the difference between the two approaches, suppose that we could eliminate the tail-end risks without actually doing much to avoid the pedestrian damages. Would this constitute an acceptable policy response? What if it were possible to eliminate the catastrophic risks without actually engaging in broad-based carbon abatement? Under these circumstances, the steady-state catastrophe theorist would seem to be committed to the view that we should allow BAU climate change to occur, because the loss of potential GDP is easily made up for by the intervening economic growth, which is, according to this view, supererogatory.

This thought experiment is not entirely hypothetical. Every low-probability climate change catastrophe scenario relies upon a dramatic amplification of one or more of the feedback processes that are expected in the high-probability scenarios. What if we were to develop a geoengineering strategy that, while not generally desirable, was nevertheless capable of interrupting these feedback effects (such as dispersing sulphate aerosols in the lower stratosphere, or seeding the ocean in order to encourage algae blooms)? The intervention would not do anything to counteract ordinary climate change, but could be deployed in an emergency to prevent catastrophic scenarios from unfolding. (This is hardly implausible. One of the problems with the more apocalyptic climate change scenarios is the implicit assumption that people would just stand around and do nothing while temperatures spiked by seven or eight degrees. This is highly unlikely. Very rapid warming of two or three degrees would undoubtedly result in a geoengineering response, such as the "sulfur nightmare" proposal made, with cheerful insouciance, by Steven Levitt and Stephen Dubner.)[104] The effect of such interventions would therefore be to eliminate tail-end risk, without doing anything to reduce pedestrian damages. On the steady-state "catastrophe avoidance" view, the development of such a strategy would fully discharge our obligations to future generations, leaving us with no obligation to engage in broad-based carbon abatement. This strikes me as quite implausible.

Several conclusions can be drawn from this. First, it suggests that the focus on catastrophic damage is something of a distraction, used to turn the conversation away from the fundamental perversity of the steady-state view, which is that it cannot explain why present generations should care about ordinary, business-as-usual climate change. Second, it suggests that there are actually two issues involved in the climate change debate, the first having to do with the high-probability, expected effects of the atmospheric externality, and the second having to do with the long-tail, or low-probability, high-damage events. The two should probably be dealt with separately, while resisting the temptation to reduce one to the other. What I have been calling the "standard policy debate" is basically about the first problem—the high-probability, easily foreseeable, pedestrian damages. Here the best way to proceed, from a policy perspective, would be to handle this in the manner of standard cost-benefit analysis, multiplying the damages under each scenario by our best estimate of the probability of it occurring. Once a policy response has been formulated, one can then turn to the question of how this affects the long-tail probabilities of catastrophic damage, with an eye to determining whether any *additional* policy response is required. After all, it is possible that the level of carbon abatement recommended in the standard policy framework will be sufficient to eliminate the catastrophic risks. If not, however, one might make the case for supplementing those initiatives. Broad-based carbon abatement, however, is only one possible response to these residual long-tail risks. It must compete with other options, including geoengineering. It is of course absurd to be discussing geoengineering in a context in which we have made practically no effort to curb carbon emissions, and the growth trajectory we are on is deeply suboptimal. It is, in other words, irresponsible to suggest geoengineering as an *alternative* to carbon pricing. On the other hand, if we had in place a fully responsible, comprehensive carbon-pricing regime, which was on track to reduce the major damages associated with climate change, it does not seem so unreasonable to contemplate geoengineering as an emergency measure, in the event that some low-probability, "black swan" event were to occur. This is, however, not the debate that this book is intended to address.

2.6. Conclusion

I stated at the outset that my goal was not to make a case for the importance of economic growth, but merely to expose an inconsistency in the views held by many philosophers who have espoused steady-state or sufficientarian normative theories. Part of my reason for doing this is to narrow the gap between the discussion about climate change that occurs in philosophical circles and the one that is occurring in policy circles, concerning the appropriate public response to the crisis. One of the major differences is that the policy debate is conducted under the assumption of ongoing economic growth, as well as an appreciation of the importance of growth for raising living standards in underdeveloped countries. The philosophical discussion, on the other hand, is dominated by the view that ongoing economic growth is either impossible or unimportant, leading to widespread acceptance of steady-state normative theories. These steady-state views are, however, a complete nonstarter as far as the policy debate is concerned, because the normative constraints are too easily satisfied. As a result, their widespread acceptance among philosophers (and environmentalists) has led to further self-marginalization. Furthermore, it creates a danger of motivated reasoning about the expected damages of climate change. Since the sufficientarian normative position is only plausible under the expectation of catastrophic damage, there is a temptation to insist that climate change will be catastrophic (e.g. an "existential threat," the creation of a "broken world," etc.), not because the science says so, but merely as a way of insulating the normative theory from the threatened reductio, that it recommends inaction in the face of climate change. This generates widespread rejection of IPCC damage projections, or overemphasis on low-probability scenarios, which further isolates philosophers from the mainstream policy discourse.[105] This state of affairs has left utilitarianism (or welfarist consequentialism) as the only game in town, as far as policy-relevant normative theory is concerned.

With respect to environmental pessimism, it is important to recognize that the limits-to-growth thesis is not based on any deep scientific or conceptual insight into the structure of human societies. It is based rather on pessimism about the possibility of technological progress in human energy systems. Whether one views this pessimism as warranted depends, in large part, on how one interprets the fact that the 20th century was a period of relative stagnation in this domain. Indeed, while we have seen phenomenal development in computing, information transmission, pharmaceuticals, flight, and many other areas, most of our energy systems continue to rely upon 19th-century technology. The only genuine invention of the 20th century was nuclear power, and even there, the mechanism that actually produces the electricity in a nuclear reactor is a steam-powered generator. Internal combustion engines, batteries, electrical motors, and generators are all 19th-century technology, and solar, wind and hydro power are much older still. The question is whether the relative paucity of new inventions is due to the intrinsic difficulty of the task, or whether it was because *fossil fuels were so cheap*, which deprived people of the incentive to research alternative energy technologies. My inclination is to think that the stagnation was caused by the glut that followed the discovery of fossil fuel, which suddenly made energy available to us on a scale that had barely been contemplated, and so simply eliminated the need to use it efficiently, much less to investigate other sources. Absent these circumstances, there is no reason to think that innovation is unlikely or impossible. Much of this assessment is based upon the recognition that the basic task is not intrinsically that complicated—it is to capture some of the ambient solar energy, which is both ubiquitous and available in astonishingly large quantities, and to transform it into a form that can be controlled, such as the high-energy chemical bond in the glucose molecule produced through photosynthesis. I cannot see any reason to think that this problem should prove resistant to human ingenuity, given the enormously more complicated problems that people have been able to resolve when appropriately incentivized. As a result, there is simply no reason for environmentalists to be opposed to, or pessimistic about the prospects for, economic growth. They should merely insist that it be "clean" growth, based on full internalization of all costs.

3

Intergenerational Justice

The casual reader could be forgiven for thinking that the discussion so far has been an extended brief in support of consequentialism, as the only plausible normative framework for thinking about climate change policy. It is true that I have argued against all of the major alternative approaches. First, I suggested that traditional environmental ethics has generated a set of conflicting views whose very rivalry serves as a perfect illustration of why approaches of this sort are an inappropriate basis for public policy. This motivated the turn toward liberal theories of justice. I went on to argue, however, that attempts to apply a strict deontological framework to the problem, whether it be rights-based or egalitarian, have proven to be a conceptual dead end— they are structurally incapable of providing answers to the most pressing questions that arise in deliberations over policy. Finally, I showed how the orthodox Rawlsian approach to intergenerational justice, which generates a sufficientarian or steady-state view of savings, under a plausible set of empirical assumptions, fails to motivate any concern about climate change at all (or is, at very least, much more problematic than its proponents have realized). What does this leave, other than the sort of welfare consequentialism that economists have been urging (and that, among philosophers, John Broome has defended)?

To a certain extent, the pressure toward a more consequentialist style of thinking is imposed by the climate change problem itself, which is one that involves a very complex set of trade-offs. The challenge of weaning our societies from fossil fuel is like trying to change the direction of a supertanker traveling at full speed. Indeed, when one begins to look at the full scope of the necessary changes, the task in many ways seems overwhelming.[1] Furthermore, the problem is not just that fossil fuel consumption is such an embedded feature of our energy systems, building methods, transportation networks, agricultural practices, settlement patterns, and of course, leisure activities, it is that our current consumption generates more than just ephemeral benefits, it lies at the heart of a system of production that is expected to produce nontrivial benefits for future generations. This makes the

Philosophical Foundations of Climate Change Policy. Joseph Heath, Oxford University Press. © Oxford University Press 2021. DOI: 10.1093/oso/9780197567982.003.0004

question of how much carbon abatement should be undertaken extremely difficult to grapple with. Everything involves trade-offs, in some cases very difficult ones.

Consider just one example, Canada currently has 291 "remote" communities—defined as permanent settlements with more than 10 dwellings not connected to the power grid or a natural gas pipeline—170 of which are Indigenous.[2] Of these, only 11 have renewable power, in all cases hydro, since solar is nonviable in northern climates. The rest get their electricity from local fossil fuel-powered generation facilities, almost all diesel-based (fuel that is, in many cases, flown in by bush plane). All vehicular transport is fossil fuel-based as well, and since lithium batteries are severely degraded in temperatures below −20°, there are minimal prospects of moving away from fossil fuel sources with current technology. In other words, without diesel fuel almost every one of these communities becomes completely nonviable. For this reason, and given the poverty of northern communities in general, Indigenous communities in particular, fuel consumption is heavily subsidized by the government. What exactly does "climate change" policy look like in this context, given the competing priorities (e.g. combating climate change, promoting northern development, avoiding the "cultural genocide" of Indigenous nations, etc.)?

The example here is particularly vexing, but it is generally the case that policy is about making trade-offs, in some cases very difficult ones. Whenever there are competing priorities and a budget constraint, the only rational way to approach decision-making is to figure out what one is hoping to achieve, what one's priorities are, and then try to do the best one can, relative to those goals and priorities. This necessarily involves looking at the likely consequences of different options, and finding some way of comparing their relative advantages and disadvantages. Whether one chooses to articulate this in terms of "pros" and "cons," or "costs" and "benefits," is largely a stylistic question. Every plausible approach to policy must enter into these sorts of calculations, and therefore every plausible normative theory must be able to say something about how these considerations should be assessed.

One might think that this settles the question in favor of consequentialism, except that, as a matter of philosophical commitment, one need not be a consequentialist in order to be concerned about the consequences of one's choices. There are various reasons why one might be moved to care about the effects of one's actions. Consequentialism, as a moral theory, along with its influential variants, such as utilitarianism, is characterized by the view that certain states of affairs in the world possess intrinsic goodness, and that the imperative to bring about these states of affairs is what makes certain actions obligatory. On this view, the moral qualities of our actions are determined by, and only by, the consequences that they are expected to produce. Nonconsequentialists, by contrast, are committed to the negation, but not the obverse, of this claim. In other words, they do not deny that consequences matter, they simply deny that consequences are the *only* thing that matters. The most straightforward example of such a theory is contractualism, which says that the moral qualities of actions are determined by their conformity to principles that could be the object of unforced agreement. If one were to ask why individuals might agree or disagree with certain principles, the answer will of course refer to the expected consequences of those principles, and the actions that they license. But one must appeal to more than just the bare consequences in order to make the action obligatory. It is the expected agreement that confers normative authority upon the principles.

My goal in this chapter and the next will be to defend a contractualist approach to the problem of climate change, and to the formulation of climate change policy, that avoids the various pitfalls that other deontological views have fallen into. The position I will be defending is the minimally controversial contractualism outlined in the introduction. In terms of policy, the implications of this philosophical view are quite standard—I defend the imposition of a carbon-pricing regime, based on our best estimate of the social cost of carbon, using a moderately low social discount rate. What distinguishes my view is rather the philosophical motivation for this policy— I will try to show how this relatively standard set of policy prescriptions can be derived without requiring any commitment to maximizing the welfare of future generations, but merely from the requirement that our existing institutions satisfy certain basic requirements of fairness.

In the discussion that follows, I will be adopting an essentially Coasian view of environmental regulation. Among Ronald Coase's many influential contributions was his observation that, when it comes to resolving any particular externality problem, the initial assignment of rights to the parties involved does not determine the ultimate resolution.[3] This is because, in the absence of transaction costs, the parties themselves can be expected to negotiate a mutually advantageous solution, one that will possess the additional property of being Pareto-efficient. Given the attractiveness of these private solutions, cases in which the state is called upon to regulate will typically be ones in which individuals, largely because of transaction costs, are unable to negotiate an agreement. Under such circumstances, Coase maintained, the outcome that the state chooses to impose should be the one that the parties themselves would have contracted to had they not been prohibited by trans- action costs from doing so. The normative authority of this outcome can, of course, be interpreted in different ways. A consequentialist could look at it and say that the Pareto efficiency of the outcome is what makes it authorita- tive, while the contracting of the parties is simply a discovery mechanism, or a procedure that has probative value when it comes to identifying the efficient outcome. A contractualist, on the other hand, would say that the outcome is authoritative *because* it is the one that would have been contracted to by the parties. On this view, Pareto efficiency is attractive as a normative principle because it picks out a set of outcomes that people would be likely to agree upon. It is this latter interpretation that I will be adopting. So while I will be focusing on efficiency as the most important principle governing climate change policy, this should not be equated with a commitment to consequen- tialism, since the norm can just as easily be given contractualist foundations.

The discussion proceeds along a slightly circuitous route. I will begin by outlining, in a general way, what I think is wrong with consequentialism, and thus what the motivation is for adopting an alternative, contractualist approach to the problem of climate change. Before going on to develop the latter, however, I will pause to deal with a fundamental objection that has been made to the entire contractualist project. While I have, in the previous two chapters, presented some criticisms of specific Rawlsian approaches to thinking about climate change, consequentialists have developed a much more radical critique of the entire Rawlsian way of thinking about questions of justice. Climate change, they argue, is fundamentally an issue of intergen- erational justice. Contractualists, however, cannot have a coherent theory of intergenerational justice because it is impossible for non-contemporaneous generations to cooperate with one another: while we are able to affect future people, future people are unable to affect us in return. Critics have argued, on this basis, that the "contract" metaphor is inapplicable, and any theory of justice that relies upon it will be incoherent in this context. Some have gone even further, claiming that climate change cannot even be characterized as a collective action problem.[4]

This is a more radical critique than anything I have considered so far— if correct, it really would leave consequentialism as the only game in town, as far as climate change policy deliberations are concerned. It is, however, mistaken. Unfortunately, while the argument against contractualism is both rhetorically powerful and intuitively plausible, the counterargument is some-what unintuitive, and so an effective response requires careful analysis of how different systems of cooperation can be organized and how intergenerational cooperation is in fact structured in our society. The upshot of the discussion will be a demonstration that there is a structure of intergenerational coop-eration underlying the entire system of saving, investment, and growth in our society, and so there is nothing incoherent about contractualists having a theory of justice that constrains the way that individuals interact with one another over vast stretches of time. Careful attention to the structure of sav-ings, however, shows that climate change is not best thought of within a "just savings" framework at all, or even as an issue of intergenerational justice. In the following chapter, I will attempt to show that the best way of framing the climate change problem is as an ordinary collective action problem, caused by a negative externality, which should be approached using the ordinary conceptual apparatus used in thinking about most other forms of environ-mental regulation.

3.1. The Consequentialist Challenge

My attempt to defend a contractualist alternative to the welfare-consequentialist analysis of the problem of climate change is motivated in part by the standard philosophical objections to consequentialism, such as its failure to respect the distinctness of persons, its lack of concern over distribution, its inability to distinguish the moral significance of acts from omissions, etc. Some of these classic problems become more acute in the case of climate change. For instance, the question of why one should not free-ride, when there is a large-scale collective action problem, and any one individual's contribution is not essential to the realization of the cooperative outcome, becomes particularly pressing when it comes to climate change (as Walter Sinnott-Armstrong has argued).[5] Derek Parfit's "nonidentity" problem is also a constant worry for consequentialists, and clearly generates difficulties when it comes to thinking about climate change.[6] Finally, Parfit's "repugnant conclusion"—that consequentialists seem to be committed to maximizing human population, even if they can foresee a very low standard of living for those people—is also a looming threat, the significance of which has been somewhat underappreciated.[7] Human population growth is in many respects a more severe threat to the environment than climate change, and so a normative view that recommends maximizing population (even pro tanto) is distinctly "off message" in discussions over environmental sustainability.

But setting aside these more specific issues, much of my desire to develop an alternative approach to thinking about climate change is motivated by frank incredulity—shared by many—in the face of the consequentialist's sense of obligation to maximize the welfare of future generations. As we have seen, consequentialism can quite easily justify action to combat climate change, but only as a byproduct of the rather implausible view that we are committed, right here and now, to maximizing the welfare, not just of those who are alive today, but of all those who will ever live (and that the "wrongness" of climate change consists in a failure of maximization). The fact that we are able, through productive investment, to create enormous benefits for future generations, at very little inconvenience to ourselves, does lend support to the view that we are obliged to do at least something to increase their well-being. On the other hand, to suggest that we are obliged to do *everything feasible* to improve their circumstances seems far too extreme. Indeed, if one were to think through the implications of this carefully, the resulting set of obligations would bear practically no resemblance to everyday morality.

As a heuristic, it is perhaps useful to contemplate our own material circumstances, and the set of developments that led us to enjoy the standard of living that we enjoy. Looking over the broad sweep of history, most of us no doubt feel quite lucky to be alive today, enjoying the standard of living that we do. And yet, if there were an obligation to maximize welfare, then luck would have nothing to do with it—we are not just *entitled* to everything that we have, we were actually entitled to considerably more. If anything, we should feel righteous indignation and resentment. It is only because of the abysmal dereliction of duty, on the part of our ancestors, that we are not vastly richer. Throughout most of the history and prehistory of our species, humans lived in a "Malthusian trap"—infant and childhood mortality was so high that any increase in the social surplus was immediately absorbed through an expansion in population.[8] Thus, throughout most of human history, each generation handed down to its descendants a steady-state economy. The only real bequest was technological improvement and cultural production, and even then, the pace of change was glacial. As a result, from the earliest appearance of *Homo sapiens* until about 1800, humans managed to achieve only an approximate doubling of (average global) material living standards. Then between 1800 and 1870, living standards doubled again. And since 1870, they have increased tenfold.[9]

I am curious how many people feel anger and resentment toward our own ancestors, for their rather abject failure to make any effort to improve the living standards of their descendants—not their own children, but future generations generally. Even a tiny effort on their part (e.g. slightly less concern for monumental architecture, slightly more investment in productive technology) could easily have put us in a world with GDP per capita of $50,000 instead of just $10,000, as it is now. One could perhaps excuse earlier generations on the basis of their ignorance. For the past two or three centuries, however, the basic mechanism of accumulation has been well understood. Yet I doubt that many people feel the relevant Strawsonian moral emotions, with respect to the failure of earlier generations even to think about our well-being, much less try to improve it. And yet, if we do not think that earlier generations had an obligation to put us where we are today, what reason do we have to regard ourselves as being under an obligation to dramatically improve the circumstances of future generations? There seems to be a peculiar asymmetry, in which we hold ourselves to a very high standard of obligation to future generations, but do not fault past generations for their failure to adhere to anything like the same standard.

One possible explanation for these peculiarities in our moral attitude is that we are living in a highly anomalous time. As Bradford DeLong has observed:

> Perhaps the best indicator of the extraordinary level and rate of advance of material well-being and productive potential is that we take it for granted. If in the eighteenth century people began to think of the idea of progress, and in the nineteenth there actually began to *be* visible progress, in the twentieth century we *expected* and today we expect progress. We assume that each generation will live between half again and twice as well in material terms as its parent generation. We find it hard to imagine what it would be like to live in a society not experiencing rapid material progress.[10]

I would add that, not only do we "expect" progress, but we feel obliged to continue and extend it. Our concern over climate change seems to be an expression of this concern. After all, of the climate change scenarios that are routinely described as "catastrophic," few are so bad as to reduce human beings to the standard of living or the population level that prevailed in the 19th century, much less the 18th. Of course, such a change would no doubt be experienced as deeply traumatic by those who lived through the transition. But from a more abstract historical perspective, was the 18th century really a catastrophe for humanity, as far as the material standard of living was concerned? Far from it;

indeed, from the perspective of the time, the 18th century was the best on record. From a moral perspective, therefore, what the talk of "catastrophe" amounts to is the concern that future generations might receive a bequest that is only somewhat better than what almost all human beings have received, throughout almost all of human history. The thought that this involves an extreme dereliction of duty on the part of present generations is not self-evident. Are the billionaire's children *entitled* to a multimillion-dollar inheritance? Obviously there are arguments that can be made either way. My point is simply that the answer should not be regarded as self-evident, or beyond argument.

This is why, I believe, the consequentialist commitment to maximization fails to offer a compelling articulation of our sense of the wrongness associated with climate change. It gets us to the right conclusion—that we are obliged to undertake some climate change mitigation—but for the wrong reason. Many others, I suspect, share this sense, which is what accounts for the enduring popularity of the sufficientarian and rights-based views. The sense of wrongness lies rather in a "do no harm" intuition, of the sort that is evoked by many instances of pollution and environmental damage. The concern is that we are not entitled to impose these costs on others. Once one realizes, however, that the industrial processes that generate climate change as a byproduct are ones that stand poised to deliver enormous benefits to future generations, then a simple prohibition on harm fails to motivate concern over anything less than extreme climate change. The simpler idea, it seems to me, is to insist that when it comes to our level of greenhouse gas emissions, our actions are *unjustified,* or the principles we are acting upon are not ones that could be agreed to by all, because we are not being forced to take into account in our decisions the damages that are being imposed on others. We are, to put it simply, free riding. According to this view, climate change is a collective action problem, just like ozone depletion (caused by chlorofluorocarbon emissions) or acid rain (caused by sulfur dioxide emissions). Failure to do our part to resolve this problem is objectionable, not because it involves a failure to maximize welfare over time—although it does involve such a failure—but simply because we are not paying our way. We are externalizing costs in a way that is *unfair,* even if in the aggregate it may be outweighed by benefits we are producing.

Unfortunately, this straightforward analysis is also one that proves difficult to defend. Although the IPCC has stated that climate change is a collective action problem with "high confidence," many philosophers have seen fit to contest this claim.[11] Their reasoning is connected to the intergenerational structure of the problem. On the one hand, greenhouse gases are a fairly typical negative externality. Although emissions create costs, there is no mechanism to charge people for these costs, because the atmosphere is a global commons. As a result, no one has any incentive to refrain from producing these emissions, which results in overconsumption of fossil fuels, along with a price distortion that results in systematic overproduction of emissions-intensive goods and services. This makes everyone worse off than they otherwise might have been, since at the margin, the costs associated with these emissions is greater than the benefits.

So far this is all standard environmental economics, no different from the analysis that has been applied to chlorofluorocarbons and sulfur dioxide. The difference, in the case of carbon emissions, is that there is a significant delay between the time that the emissions are produced and the time at which the damages occur. As a result, many philosophers have felt that this cannot be a standard collective action problem, because it is not the case that "everyone benefits" from a solution, or that a carbon-abatement policy would be Pareto-improving. With sulfur dioxide, the benefits of emissions reductions were felt almost right away, and so those who shouldered the costs of increased energy prices also received the benefits of reduced acid rain. With climate change, however, the present generation is being asked to shoulder the costs of carbon abatement, yet the most significant benefits will not be felt for several generations. So one cannot say that everyone benefits from a solution, or that it constitutes a Pareto improvement. It looks more like a *transfer* to future generations.

On this view, the situation between generations, with respect to the atmosphere, is like that of people living along the banks of a river who pollute it in various ways, diminishing the value of the river for those living downstream. Because the water only flows one way, there is no possibility of cooperation— there is no point in one person saying, "I won't pollute the river if you don't." Reciprocity, which is fundamental to cooperation, appears to be precluded by the structure of the interaction. Each person is at the mercy of those who live upstream and is, in turn, in a position to unilaterally determine the circumstances of those who live downstream. There is room for altruism here, or benevolence toward those downstream, but no room for cooperation.[12] Since a collective action problem is, by definition, one that calls for cooperation, and since a cooperative solution to climate change is impossible, it follows that it cannot be a collective action problem.

This analysis can be generalized to provide the basis for a radical critique of contractualism, and to contractualist approaches to thinking about climate change. Since Rawls, it has been common to conceive of a contractualist theory of justice as a set of principles used to allocate the benefits of burdens of cooperation. If there is no such thing as intergenerational cooperation, then it follows that there is no place for a social contract to determine how the benefits and burdens of cooperation are to be assigned. Since the theory of justice is the set of principles used to effect this assignment, the implication seems to be that intergenerational relations are not governed by principles of justice. If this were true, then it would obviously be a very significant defect in that theory. It is this claim about the application of contractualist theories that will be my primary focus of evaluation in this chapter. My objective will be to show that the supposed problem for contractualism is not really a problem, since the "non-reciprocity" problem stems from the adoption of an overly narrow, direct conception of reciprocity. Cooperation, however, can also be sustained by systems of indirect reciprocity, where there is no requirement that the person *to whom* one supplies a benefit be the person *from whom* one receives a benefit. Thus I will show that there is no problem in principle, or in practice, with a system of intergenerational cooperation in which benefits flow only one way. Having set aside this issue, I will go on to argue that the delay involved in climate change mitigation does not change the nature of the problem in any fundamental way, because it is embedded in a much broader system of intergenerational cooperation, the norms of which prohibit environmental externalities of the sort that we are currently producing.

3.2. The Structure of Intergenerational Cooperation

Social contract theory, as we have seen, provides a distinctive approach to the derivation of moral principles, taking as its point of departure the observation that, in many interactions, not everyone is able to have his or her own way, and so in order to avoid the conflict and disorder that might arise from this goal incongruity, the parties adopt a set of rules that constrain each individual's actions, creating an arrangement under which each is able to pursue his or her own goals in ways that are maximally compatible with everyone else doing the same. This basic construct provides an answer to the fundamental question of what morality is (a set of shared rules of conduct), what problem it constitutes a response to (goal incongruity), and why we should be moral (because we are all better off, when we all act morally). It also provides guidance on the core normative question, of how we ought to act. This becomes a matter of determining the *terms* that individuals would accept when it comes to constraining their conduct. Since Rawls, it has been common to describe the abstract formulation of these terms as the principles of justice.

The suggestion that social contract theory might have difficulty accounting for obligations of intergenerational justice has been around for a long time, but has acquired increased salience in the context of debates over climate change. The basic criticism was made most forcefully and influentially by Brian Barry. Barry took issue with what he regarded as a "Humean" strain of thinking in Rawls's thought, according to which justice is an artificial virtue, which individuals cultivate in order to secure "mutual advantage."[13] According to Hume's line of thinking, the obligations of justice are not a structural feature of the human condition (e.g. something that we are subject to qua rational subjects); rather, they arise in response to certain quite particular "circumstances." Specifically, under conditions of moderate scarcity and limited generosity, individuals may find it mutually beneficial to institute a system of rules for the allocation of goods (such as a system of property rights).[14] Rawls endorsed this general idea, in claiming that the Humean "circumstances of justice" describe "the normal conditions under which human cooperation is both possible and necessary."[15] Justice then arises from the need to adopt principles to determine the allocation of the benefits and burdens of cooperation.

There is, it should be noted, some debate over how seriously these passages in Rawls's *A Theory of Justice* should be taken.[16] In the intervening years, it has become conventional in some quarters to introduce a terminological distinction between versions of social contract theory that assign central importance to the "mutual advantage" postulate and those that do not. Thus David Gauthier's approach, which sees justice as essentially an indirect strategy for maximizing individual utility, is often referred to as "contractarianism," while T. M. Scanlon's version, which focuses purely on agreement, while ignoring the circumstances that have traditionally been thought to elicit it, is referred to as "contractualism."[17] However, once it is recognized that whenever Rawls uses the term "cooperation," he is referring to interactions that are "mutually advantageous," and that he consistently defines the "basic structure" as a "system of cooperation," it seems as though he must also be classified as a "contractarian." This is, however, terminologically revisionist, since Rawls is conventionally described as a "contractualist."[18] Thus I will follow Elizabeth

Ashford and Tim Mulgan in distinguishing contractualism in the "narrow sense," which refers only to Scanlon's view, from contractualism the broad sense, which is used to describe any view that regards principles of morality or justice as based on contract.[19] When I use the term "contractualist" it will typically be in the broad sense.

It is the emphasis on cooperation or mutual advantage in Rawls's philosophy that gives rise to Barry's objection. The problem, as Barry sees it, is that cooperation is impossible between non-contemporaneous individuals, and therefore the circumstances of justice do not obtain between present generations and those who will be born sometime in the future. As he puts it:

> Whether or not the circumstances of justice obtain among nations is an empirical matter. They may or they may not. Whether or not they obtain between the generation of those currently alive at one time and their successors is a logical matter. They cannot. The directionality of time guarantees that, while those now alive can make their successors better or worse off, those successors cannot do anything to help or harm the current generation. "I would fain ask what posterity has ever done for us," as the wag put it: if justice equals mutual advantage then there can be no justice between generations.[20]

It follows from this analysis that, since the basic structure clearly does persist over time, and across generations, it cannot be a system of cooperation. It must instead have at least some element that is essentially altruistic—involving a one-way transfer from present to successive generations. Similarly, the argument implies that the way we think about obligations to future generations, with respect to climate change and other environmental issues, must be framed in terms of altruism (i.e. present sacrifice for future benefit). Thus Barry winds up recommending a more Kantian approach to thinking about intergenerational justice, which would not have the problem he attributes to Rawls, because the obligation to act in conformity with principles that would be acceptable to all would not be tied to the existence of any specifically cooperative interactions between individuals.

If this argument were correct, it would obviously provide a powerful motivation for thinking about these issues in consequentialist terms. As Stephen Gardiner observes, "If, in the intergenerational setting, we depart from the prisoner's dilemma model, then the traditional contract approach seems to be undercut, and in a particularly deep way: *its basic analysis of the problem appears not to apply.*"[21] Daniel Attas has echoed these sentiments, claiming that the "absence of mutuality between generations poses an insurmountable problem to any contractarian theory that aims to ground obligations on the idea of mutual advantage."[22] Gustaf Arrhenius has claimed, on this basis, that "intergenerational justice remains an embarrassment for contractarians and will, I surmise, continue to be so."[23] It is very seldom in the field of philosophical ethics that anyone has the opportunity to make a decisive objection to any widely held position. Philosophical views tend rather to become worn down over time, by an accumulation of difficulties, each rather minor in itself, but perhaps adding up to something more substantial. In the case of intergenerational justice, however, these critics of contractualism are claiming to have uncovered a fundamental incoherence in the foundations of the Rawlsian edifice. The basic structure, they claim, cannot be "a system of cooperation between generations over time" (as Rawls describes it)[24] because there is no such thing as intergenerational cooperation!

Luckily for Rawls, not to mention his many admirers, this claim is mistaken. The problem with the argument is that the conception of reciprocity that it relies upon is extremely narrow. It assumes that the system of reciprocity at the heart of any cooperative arrangement must be of the "You scratch my back, I'll scratch yours" variety—any act that benefits another person must be reciprocated directly, by that same person, on the model of a two-player prisoner's dilemma. There are however many systems of cooperation that do not have this structure. Consider the case of a market economy. This is obviously a system of reciprocity, in that goods are ultimately exchanged for other goods. In a barter economy, the reciprocity is direct—person a gives b something that b wants, in exchange for something that a wants. This direct reciprocity requirement is what generates the "coincidence of wants" problem, which is resolved through the introduction of money. Now person a can give something to b in return for money, which serves as an abstract store of value, representing the magnitude of the benefit that she has provided to some other person. She may hold it for a while before deciding to "cash it in" by using it to purchase goods that she desires from some person c. In this case, the person to whom she has provided the benefit (b) is not the same as the person from whom she receives a benefit (c). Structurally, therefore, it does not matter whether the person she helped has the power to help her in return, because that is not the person from whom she expects to be repaid. The entire system is still based on reciprocity, but the form that it takes is *indirect*.

Once this observation has been made, it becomes immediately clear that Barry's argument is insufficient to show the impossibility of intergenerational cooperation, because all that he does is point to the impossibility of direct reciprocity between generations far removed from one another in time. He makes no attempt to show that these generations cannot be participants in a beneficial system of indirect reciprocity. The most plausible explanation for this omission is that he did not realize that systems of reciprocity could have a more complex structure, and so did not consider it necessary to make an argument. Many philosophers have followed suit, assuming that the non-reciprocity problem is a serious challenge, while tacitly construing all reciprocity as direct.[25] In order to assess these claims, and whether a more sophisticated version of the non-reciprocity problem can be defended, we require a better understanding of indirect reciprocity, as well as cooperative systems more generally.

While the conviction among philosophers that there is a problem with intergenerational cooperation remains widespread, it is difficult to find any economist who shares this view. This is largely due to familiarity, on the part of economists, with a number of textbook models of the "folk theorem," which specify the conditions under which cooperation can be sustained in repeated games.[26] One of the well-known variants involves an "overlapping generations" model, which starts with indirect reciprocity and extends the system of cooperation out indefinitely into the future.[27] Since the mathematical formulation of these models no doubt constitutes an obstacle to a more general appreciation of their significance, I will begin with a non-technical presentation of the central ideas, using a typical public goods game as an example.

Cooperation, in the sense in which the term is being used here, arises in situations in which individuals are in a position to achieve mutual benefit, but where no one has an individual incentive to act in a way that will produce that benefit, and so the strategic equilibrium of the interaction is suboptimal. Thus individuals are called upon to exercise some restraint in their pursuit of self-interest, in order to achieve this mutually beneficial outcome. Acting on the basis of this restraint is referred to as "cooperation."[28] This is a narrower definition than the ordinary-English sense of the term, which is often used to describe situations of mere coordination as well—where individuals need to act together, in some sense, regardless of whether anyone has an incentive to deviate from the plan. The possibility of cooperation, in the narrower sense of the term that I am using here, arises only in situations in which the strategic equilibrium of an interaction is Pareto-suboptimal (which is to say, worse for at least one person and better for no one), and so individuals have an incentive to deviate from any cooperative plan (i.e. to free-ride).

In these situations, as Rawls noted, individuals have both a common interest (in maximizing the benefits of cooperation) and a conflict of interest (over how these benefits are to be distributed). A theory of justice, again in the technical sense of the term, constitutes a set of principles designed for the specific task of "choosing among the various social arrangements that determine this division of advantages."[29] It is the function of these principles to bring about the agreement required to induce individuals to act in a non-maximizing fashion (i.e. to refrain from defecting, or to respect the constraints required to achieve the cooperative outcome).

The paradigm case of a cooperative interaction is the prisoner's dilemma (PD), such as shown in panel A of Figure 3.1. This is a two-person interaction in which the Pareto-optimal outcome can be achieved only if both players select a strongly dominated strategy. For concreteness, imagine that the game involves the following exchange: each player has a token, which she may elect to keep or pass to her partner. The token is worth $2 in the hands of the person who was originally assigned it, but $5 in the hands of her partner. Each must decide what to do independently of the other, and only after the decision has been made are the results revealed. The best outcome is for each person to hand the token over to her partner, resulting in them both receiving a payment of $5. There is, however, a free-rider incentive, which is to keep the token. If one's partner passes his token, this results in the highest possible payoff of $7 (and if one's partner does not pass the

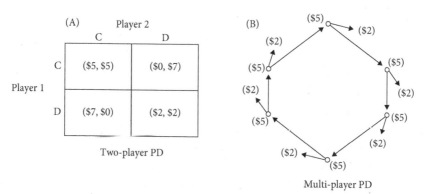

Figure 3.1 Two-player and multiplayer prisoner's dilemma (PD)

token, it provides the secure payoff of $2, instead of nothing). Because of this free-rider incentive, in a one-shot version of the game the only equilibrium outcome is for both players to keep the token, resulting in the suboptimal ($2, $2) outcome.

The most basic "folk theorem" result is based on a repeated version of a game such as this.[30] When the stage game is played more than once, players have the opportunity to develop conditional strategies, so that what each one does in round 2, round 3, etc., can be made conditional upon what the other did in the previous round(s). A strategy in this game is a set of actions, specifying an action for a player at each stage, for each possible history of play. This allows players to respond to each other's moves, by making the action at any stage depend upon the other player's actions at any previous stage. A set of strategies is in equilibrium if each player's action is (or would be) a utility-maximizing response to the action of the other player, at the time at which it is played.[31] It is easy to show that repeated play of actions that are in equilibrium in the stage game will also be in equilibrium in the repeated game. Since mutual defection (D,D) is the sole equilibrium of the one-shot PD, the strategy pair ([D,D,D, . . .], [D,D,D, . . .])—i.e. "both players defect forever"—will be an equilibrium of the repeated game. If one player is always defecting, then it will never be in the interest of the other player to start cooperating. Conversely, simple cooperation remains out of equilibrium, for the same reason that it is out of equilibrium in the stage game. If the expected strategy pair is ([C,C,C, . . .], [C,C,C, . . .]), then either player could benefit by defecting at any stage game. If, however, the players adopt strategies in which cooperation is conditional upon a history of successful cooperation, then it is possible to sustain a strategy profile of actions that would be out of equilibrium in any one stage game. In particular, if both players adopt what is known as a "trigger" strategy, "Play C, and continue to play C as long as the other player plays C; if the other plays D, then play D forever," then an equilibrium path in which both players choose C can be sustained as an equilibrium of the repeated game (as long as the discounted sum of future cooperative benefits outweighs the one-shot benefit of defection).[32] Or to put the same in plainer language, it is possible to sustain cooperation in the stage-game PD by having each player adopt a strategy such as the following: pass the token, and continue to pass the token, so long as your partner passes as well, but if your partner keeps the token, then keep it as well in all subsequent rounds. With a sufficiently low discount rate, and a sufficiently high probability each round that the game will continue, it becomes rational for each player to pass the token in every stage game PD, allowing the players to achieve the optimal outcome of $5 each per round.

In this game, each player cooperates because he knows that, if he fails to cooperate his partner will retaliate by keeping the token in future rounds. Thus they are both "helping" one another, because they know that their partner is in a position help or harm them in future rounds. This is, however, not an essential feature of the equilibrium. The cooperation that is established in this game must be based on direct reciprocity because there are only two players. If we add more players, however, additional options become available. Suppose that four more players are introduced to the game and are each given the same set of options: keep the token, in which case it is worth $2, or else give it to some other person, to whom it will be worth $5. One could sustain cooperation by matching the players into pairs of two and instructing them to play the same conditional strategy as in the two-player PD. But cooperation can also be sustained collectively. One could seat all six players around a table, and instruct them each to pass the token to the person on their left, with the proviso that, if anyone ever fails to pass the token, then no one will pass any tokens to that person in future rounds. Again, whatever temptation there may be to defect, in order to get the free-rider payoff of $7, is tempered by the expectation of having one's future payoff reduced to a maximum of $2 per round instead of $5.[33] With players who are sufficiently patient, and have a high enough expectation that the game will continue, this makes cooperation the preferred strategy.

The cooperative equilibrium of this new multiplayer version of the game, shown in panel B of Figure 3.1, is based on indirect reciprocity, since the person *to whom* each one gives the token is not the same as the one *from whom* each one expects to receive a token. Thus each player receives a benefit, but no one is in a position to help or harm the person from whom she receives that benefit. None of this makes any difference to the strategic properties of the equilibrium, or to the "cooperativeness" of the overall interaction. There is also no need for the relationships among the players to be long term. One could at the beginning of each round have a random assignment determining which other player each one must pass the token to. One can think of each player who complies, and passes the token, as being in "good standing" with the community, and of each person who fails to pass the token as being in "bad standing." The cooperative strategy could then dictate that, if the person one is paired with remains in good standing, then one must pass the token to that person, but if that person is in bad standing, then one can keep it. This is sufficient to make universal cooperation the equilibrium of the game.

It is a notable feature of this multiplayer version of the game that the system of cooperation is ultimately closed, in the sense that each player both provides a benefit to one other person and receives a benefit from some other person. Perhaps surprisingly, this property is also not an essential feature of the cooperative equilibrium. Suppose that one were to open up the game, so that players are sometimes randomly removed from the game, while others are randomly added. If the probability of being removed is too high, this will of course lead to a collapse of cooperation, because players will not expect to be in the game long enough for the benefits of the $5 stream of payoffs to outweigh the one-time gain of $7. But if the probability of removal is sufficiently low, then cooperation can continue as before. In this case, the flow of new players into the game is important, because it gives those who remain in the game confidence that the $5 stream of payoffs will continue indefinitely (or at least for a long while).

One can already see the relevance of this to the intergenerational case, but it may be helpful to provide a more straightforward example of an overlapping generations model. Suppose that the players are seated around a table. After each round, the person seated at the head of the table is removed from the game (call this "dying"), each player moves over one seat (call this "aging"), and a new player enters the game in the vacant seat (call this "being born"). The cooperative strategy requires that each player pass the token to the player who is one seat "older." Anyone who passes the token remains in good standing, whereas anyone who fails to pass it is in bad standing, and no one is required to pass a token to anyone who is in bad standing. This strategy is sufficient to sustain cooperation, although there is one peculiarity that develops, which is that the oldest player in the game can now defect with impunity. This puts him in bad standing, but since he is also removed from the game (i.e. "dead"), it does not result in any future defections from younger players.[34] It does result in someone not receiving any benefits—in this case it is the youngest player, who receives nothing during the first round. This does not prevent her from cooperating, however, because she wants to be in good standing for the subsequent five rounds, in which she will be able to receive benefits.

Like the previous game, it is an important feature of the stability of cooperation in this interaction that there is an open-ended stream of new players entering the game. Without them, the youngest player would have no reason to pass the token. (Since there would be no one younger to pass a token to her, there would be nothing to be gained from being in good standing, and therefore no reason not to defect.) Thus the cooperative system does not need to be closed, because it extends out indefinitely into the future. This is the (admittedly rather unintuitive) point that Barry failed to observe, which is what permits relations between present and future generations to be genuinely cooperative. In this model, there is a very real sense in which the system of cooperation includes all of the participants, including those who have not yet been "born," because the *expectation* that they will continue to cooperate is an essential part of the equilibrium. If anything is done to defeat this expectation, then the system of cooperation would fall apart.

So here we have a system of cooperation, supported by indirect reciprocity in a multiplayer game with overlapping generations. Older generations receive benefits from younger ones, but are powerless to either help or harm them. Furthermore, this system is one that ties all of the players into a single cooperative system, not just those who are contemporaries. This is because the cooperation of each generation depends, not just upon the anticipated cooperation of the generation one stage younger, but upon the anticipated cooperation of all younger, future generations, including those yet unborn. (With birth and death, the overall system need not be a loop, but can be a line extending out indefinitely into the future. Note that "indefinite" is not the same thing as "infinite." A finitely repeated game in which there is uncertainty about when the game will end is formally identical to an infinitely repeated game.)[35] What makes the system of cooperation hang together is not just the actions chosen, but the expectations players have about which actions *will* be chosen. Thus future generations may show up as virtual participants in the cooperative system, simply because expectations about what they will do, when they finally get a chance to act, may be essential to holding together the equilibrium in the present.

To see this, consider what would happen if a referee were to announce at some point that the game was to continue for three more rounds, but that it would then end. Obviously this would generate universal defection in the final round, since there would be no possibility of punishing those who defected at that point. But if everyone is going to be defecting in the final round, then there is also no way of punishing those who defect in the penultimate round. Since everyone can see this, everyone can be expected to defect in that round as well. But this makes it impossible to punish anyone who defects in the antepenultimate round, and so everyone will defect then

as well. As a result, the announcement that the game will end at some specific point in the future immediately transforms universal defection into the
sole subgame-perfect equilibrium (simply by changing expectations).[36] This
is true regardless of how far off in the future the end of the game is. Thus the
referee could just as easily say that the game would end in 10 or 20 rounds,
and it would have the same effect. This is why cooperation in the present is
sustained, in a very real way, not just by the participation in the cooperative
scheme of those currently alive, but also by the anticipated cooperation of
those not yet born.

3.3. Applications and Objections

The discussion so far has focused on the standard model of intergenerational
cooperation that economists are familiar with. There is however one feature
of this model that is somewhat peculiar, and not unrelated to the fact that
it is represented using a standard game-theoretic conception of rational action. We can think of the benefits of intergenerational cooperation as flowing
in one of two directions: either "downstream," from older generations to
younger, or "upstream," from younger to older.[37] This distinction is a useful
one, although it is important to avoid a potential confusion. A cooperative
scheme is one in which everyone benefits, and so what an "upstream" flow
of benefits really means is just that individuals are expected to act coopera-
tively first, before they can enjoy the benefits of cooperation. Those who receive the benefits of cooperation will be those who have, in some previous
period, already cooperated, which in an intergenerational model typically
means that they will be older. In a "downstream" model, by contrast, players
receive benefits first, and are then expected to provide benefits to the next
generation.

In Western cultures it is standard to regard "downstream" intergenerational obligations as being the most important.[38] Furthermore, when we
consider the two issues that feature most prominently in discussions of intergenerational ethics, viz. "just savings" and environmental preservation, both
appear to involve a downstream flow of benefits. And yet the model presented
in the previous section involves an upstream flow—the cooperative strategy
requires players to pass the token to the person one "older." This is not an
accidental feature of the model. If the direction of benefits were reversed,
and players were instructed to pass the token to the person one younger, it

would be impossible to construct a set of cooperative strategies that was a strategic equilibrium of the game. The problem is that there is nothing to stop the oldest person (i.e. the one about to "die") from defecting, but because of that, with a downstream flow of benefits the second-oldest would have no expectation of receiving any cooperative benefit, and so would have no reason to cooperate. This gives the third-oldest no reason to cooperate, and so on. Iterate this reasoning and the entire system of cooperation unravels.

The fact that there is no cooperative strategic equilibrium in a downstream model does not mean that systems of cooperation with this structure are impossible to organize; it just means that they must be *institutionalized*. Legal regulation is one option, but there are also downstream schemes of cooperation that are held together entirely through informal social norms, or moral incentives (i.e. such as a "duty of fair play" that requires individuals who have received a benefit to do their part in maintaining the cooperative scheme).[39] These are often referred to as "pay it forward" schemes.[40] There is however room for debate over whether such arrangements should be called cooperative, as opposed to just systems of generalized altruism. Consider, for instance, the Kula ring on the Trobriand Islands described by Marcel Mauss in his classic treatise *The Gift*.[41] Although he refers to it as involving "gifts," a term that implies altruism, his goal is actually to show that there is a system of reciprocity underlying it. Each community passed along a set of valuables, some rotating clockwise, others rotating anticlockwise, receiving gifts and then making new ones. None of this was optional though. (Indeed, Mary Douglas's foreword to the English-language edition of Mauss's book is entitled simply "No Free Gifts.") The traditional distinction between an exchange economy and a gift economy, Mauss argued, was misleading, since they were both systems of indirect reciprocity. The central difference was merely that the former has a "pay it back" while the latter has a "pay it forward" structure. Of course, because participants in the gift economy receive the benefit first, and then later pay it forward to the next person, the arrangement is intrinsically more vulnerable to defection than one with an upstream flow of benefits (which is presumably why these arrangements are less common, especially in large-scale societies where interactions are likely to be one-time and anonymous). But it does not change the fundamental nature of the interaction.

It is presumably because of these differences in incentive structure that systems of cooperation with a downstream flow of benefits are not all that common. Systems of cooperation that rely upon an upstream flow, by contrast, are ubiquitous. Again, it is worth emphasizing that just because a particular system of cooperation has an upstream flow does not mean that younger and future generations do not benefit from it. It is, after all, still a system of cooperation, and therefore a source of mutual benefit for all participants. It is just that new entrants must pay their dues before becoming eligible to receive benefits from others. Perhaps the simplest example is that of a pension scheme financed on a pay-as-you-go (PAYGO) basis. In such a system the pension deductions taken from active employees are not saved, or invested; they are simply used to make payments to those who are currently retired. Some money will be set aside in order to handle anticipated increases in expected liabilities, due to factors such as variation in cohort size, but overall the plan will not have savings adequate to cover its liabilities.

This example should be a familiar one, because many universities have a pension scheme of this sort. In my own case—stylizing only somewhat— every month a fairly large portion (over 5 percent) of my salary is deducted from my paycheck and essentially handed over to one of my emeritus colleagues.[42] What I get in return, at least for the moment, is very little. At the end of the year, I am sent a letter, telling me what my own anticipated monthly pension will be, based on my current salary and accumulated years of service. But this is nothing more than a promise, and a weak one at that. It is hostage to all of the vicissitudes of employee-employer bargaining over the next 15 years. So why do I tolerate this? It is because I am confident that when I am older and retired there will be a new generation of young professors who are willing to do the same thing for me that I am currently doing for my emeritus colleagues. But why should I expect the next generation to be willing to hand over 5 percent of their salary to me? Certainly not out of gratitude for the fact that I am doing so now, to the benefit of my older colleagues. It is because I expect them to expect that they will someday have younger colleagues, who will do the same for them, i.e. that the chain of cooperation will continue on into the future, unbroken (or that the pension scheme will remain, as they say, a "going concern").

The value of this example is that a pension system like this obviously has the structure of a system of intergenerational cooperation sustained by indirect reciprocity. It can therefore be appealed to as a way of disarming many of the doubts that have been raised by critics of contractualism about the realism or appropriateness of the relevant game-theoretic models. First of all, it is clearly a system of cooperation. The "defect" strategy, in this case, is for each individual simply to save for his or her own retirement. This is Pareto-inefficient, because uncertainty about one's own time of death creates a risk of either over- or under-saving. If everyone saves, using average life-expectancy at the time of retirement as a guide, this will result in a significant fraction of the population outliving their savings. If individuals agree to pool their retirement savings, however, they can take advantage of the law of large numbers in order to drastically reduce the chances of anyone running out of savings. This is the insurance mechanism at the core of a life annuity (and a defined-benefit pension is essentially a bundle of collectively purchased life annuities). Hence the value to me of the pension scheme. If life expectancy increases by 10 years between now and the time of my retirement, whatever individual savings plans I may have had would be thrown into disarray, yet the promise made to me by the pension plan would still stand.

The second important thing to observe about this pension scheme is that the cooperative arrangement cannot be sustained by any proper subset of the total set of cooperators—such as those who are alive right now. Closing it off at any point would leave the younger generation contributing but not receiving anything. Thus one could not set up a pay-as-you-go pension scheme that was designed to run for 100 years and then stop; it must continue indefinitely into the future. Because of this, the total set of cooperators must be taken to include future generations, including those yet unborn, and this means, in turn, that the circumstances of justice obtain between us and them, and so the system of cooperation in which we participate must be governed by appropriate principles of justice.

Consider now some of the objections that have been directed toward this sort of "overlapping generations" model of intergenerational cooperation:

3.3.1. Doubts about the punishment scheme

Gardiner has expressed reservations about whether, in an arrangement of this sort, any future generation would ever really defect (and thus, whether the threat of defection is really doing any work in sustaining the supposedly "cooperative" system). He writes, "It is not clear that it would actually pay the later group to withhold cooperation for the sake of past bad treatment once it actually achieves causal parity, when this withholding damages its interests still further."[43]

Two responses are in order here. First, whether or not the interaction is cooperative does not depend upon whether there is a credible punishment scheme in place, but upon the underlying structure of the interaction having the properties of a PD. In other words, what makes it cooperative is the fact the individuals have an incentive to free-ride but are able to achieve a Pareto-superior outcome if they all refrain from doing so. This is why experimental subjects brought in to play a one-shot public goods game are described as "cooperating" and "defecting," even though threats are precluded by the very structure of the game. The major reason for the emphasis on punishment schemes in the game-theoretic literature is that this solution to the problem of cooperation is amenable to formal modeling. But this doesn't mean that it is the only way to sustain cooperation. Internal constraints, such as the duty of fair play, or an aversion to free riding, are often able to accomplish the same thing, and if they are, so much the better for everyone involved. However, when dealing with large-scale, anonymous interactions, people do tend to act more instrumentally, and so external incentives are likely to play an important role in sustaining any such system of cooperation. (This is presumably why large-scale systems of cooperation with an upstream benefit structure are more common than those with a downstream structure.)

The second response is to observe that the nature of the punishment mechanism in the game-theoretic model is often misunderstood or mischaracterized. Under the "trigger" strategy, the expectation is not that the younger generation is going to pursue a non-utility-maximizing course of action, or that they will go out of their way, harming their own interests, in order to retaliate against their elders. The punishment, in other words, is not spiteful. The suggestion is simply that they will stop cooperating, so that defection by an older generation would have the effect of shifting expectations toward a non-cooperative equilibrium. In the case of a pension plan, this is not such an unreasonable expectation.[44] If one cohort of faculty members refused to make any payments into the pension plan for a period of a decade, and reduced the benefits being paid to current retirees to cover the shortfall, it is quite unlikely that they would be able to themselves retire with full pensions, merely by imploring younger generations to ignore the breach of trust, or to let "bygones be bygones." The more likely outcome of such flagrant defection would be the dissolution of the pension scheme, which is to say, a reversion to the non-cooperative baseline of individual retirement savings (or a defined-contribution plan). This is just another way of describing the trigger strategy.[45]

3.3.2. Doubts about backward induction

Arrhenius has expressed some doubts about the validity or realism of the backward induction argument that makes present cooperation dependent upon the anticipated cooperation of all future generations in the iterated PD model.[46]

But again, the pension example would seem to belie these concerns. Suppose the government were to announce that my university was to be closed down in 20 years (with no provision for taking over the pension scheme). Would I continue to make pension payments, knowing that by the time I retire there will be no more younger professors around to give me 5 percent of their salary? Certainly not. But suppose the date of closure were pushed back to 40 years. Does this change anything? I can now expect that there will be younger faculty around when I am retired. But will they be willing to give me 5 percent of their salary? It is hard to see how I could have any reasonable expectation of this, since they know that by the time *they* are retired, there will be no more younger faculty around. The same logic applies no matter how far back the closure date is pushed. And even if some faculty members have doubts about the validity of backward induction, their investment advisers and benefit negotiators most certainly do not. Thus it would not be unreasonable to expect to see "preemptive defection," which is the tangible manifestation of backward induction reasoning. Naturally, longer stretches of time introduce greater uncertainty, so one might see cooperation continue under the assumption that "things might change."[47] (It is worth recalling, in this context, that a finitely repeated game with uncertainty about when the game will end is equivalent to an infinitely repeated game.)[48] People may continue to cooperate simply because they think there is a good enough chance that they will benefit from the system of reciprocity. But if there really is no uncertainty, it seems plausible to suppose that cooperation will completely unravel. (This is the scenario envisioned in the P. D. James novel *The Children of Men*, in which the sudden onset of infertility brings to an end the ongoing production of new generations.[49] The youngest generation, or "omegas," wind up becoming dangerous outlaws, essentially outside the social contract, because they recognize that they have nothing to gain from participating in various systems of cooperation. This has a number of ripple effects for the society as a whole, including the widespread adoption of elective suicide as an alternative to retirement.)

These arguments about unraveling, however, are largely just empirical speculation. Even if one finds this backward induction reasoning implausible, it is irrelevant to the question of whether there are systems of intergenerational co-operation in place in our society. After all, cooperation need not be sustained by the threat of defection among the parties; it can be sustained by an honor code, or by the force of law, or some other mechanism entirely. And regardless of the specific mechanism, it is simply a fact about pay-as-you-go pension schemes that they involve cooperation between present and future generations. To see this, consider the criticism that is often made by critics of these pension arrangements that they are nothing but glorified Ponzi schemes (since both use money from new members to pay off old members). What makes the comparison misleading is the fact that in a pyramid scheme, the last generation of "investors" foreseeably loses its money. Pay-as-you-go pensions are not like this, because there is no reason to think that any one generation is going to wind up losing money (or being suckered). But this is true only because the chain of cooperation extends indefinitely into the future.

It is also worth keeping in mind that critics who express doubts about the possibility of intergenerational cooperation have an obligation to provide some alternative account of what motivates individuals, like myself, who are participating in institutions that, at least superficially, seem to have this structure. If my pension scheme is *not* a system of cooperation extending out indefinitely into the future, then what is it exactly? In order for it not to be a system of intergenerational cooperation, it would have to be the case that my contributions are not motivated by the expectation of reciprocity (i.e. the benefits that I will someday receive). But how could that be? Since I am giving away over 5 percent of my salary, it cannot be self-interested in any obvious way. The only alternative hypothesis is that I am acting altruistically, to the benefit of my emeritus colleagues. This implies that, should the pension scheme be abruptly canceled, leaving me to live out my retirement in penury, I would not feel in any way cheated, since it was not the expectation of reciprocity that was driving my contributions. This is (needless to say?) not the self-understanding of myself or anyone I have ever discussed pension issues with.

3.3.3. Time bomb arguments

There is also the claim, made by several critics, that contractualists are unable to exclude the possibility of one generation constructing a so-called time bomb, "where an action beneficial to the present generation has a devastating effect on some distant future generation, but no direct effect on intervening generations."[50] The availability of this option is thought to undermine the possibility of cooperation between generations, because it does an end run around the generational overlap that is supposed to keep each generation on the cooperative path. A sufficient number of critics find this argument persuasive that it deserves a detailed response.

Unfortunately, despite the fact that many critics find it intuitively quite powerful, the argument has never been given a precise formulation. There are three possible interpretations that could be given. First, critics might be claiming that construction of a time bomb is a type of *free-rider strategy*, which defeats the effectiveness of generational overlap at sustaining cooperation in a repeated PD. For example, one might imagine adding a third strategy to the PD (call it "E") that gives the person who plays it a free-rider payoff right away, yet instead of giving the sucker payoff to his immediate opponent, gives it to some other person several rounds later. The intuition, then, is that a player could get away with playing such a strategy because it does not give anyone an incentive to punish him right away (since no one is injured), and by the time someone *is* injured, it is too late to do anything about it. This intuition, however, rests on a misunderstanding of the trigger strategy. It implies that individuals switch to the punishment sequence only out of a desire to retaliate for injuries they have received. This is not the case (since retaliation would be a form of non-utility-maximizing action). One can see this clearly in a system of indirect reciprocity, where the person doing the punishing is typically *not* the person who has suffered from the defection. The trigger strategy does not impose punishment, it merely shifts expectations away from cooperation toward defection in response to free riding. Thus one could easily sustain cooperation in response to the introduction of a time bomb option to the PD by amending the trigger strategy to read: "Play C, and continue to play C as long as the other player plays C; if the other plays D or E, then play D forever." This strategy is obviously able to sustain a cooperative equilibrium, and makes it impossible for anyone to free-ride by constructing a time bomb.

Second, critics may be claiming that, outside of game theory, with real cognitive-warts-and-all human agents, cooperation may not be sustainable, because this free-rider strategy is likely to be adopted by present generations, who anticipate that it will go unpunished. The proper response to this claim is the one made earlier: that real human agents might not adopt such a strategy even if it were available (for the same reason that they cooperate in one-shot public goods games), and that one cannot count upon real agents being more forgiving of free riding (for the same reason that one cannot count on people to accept lowball offers in ultimatum games). Thus the time-bomb argument fails to show that cooperation cannot be sustained intergenerationally; it is simply an intuition that falls apart when any attempt is made to give it precise formulation.

This leaves us with a third possibility: that the time bomb may not be a free-rider strategy at all, but merely a move *within* the system of intergenerational cooperation, toward an arrangement that favors the interests of present generations over future ones.[51] But then this is no longer a criticism of the basic contractualist framework, since the critic is essentially granting that intergenerational cooperation is possible. The suggestion is simply that the contractualist does not have the normative resources required to condemn cooperative arrangements that are self-serving from the standpoint of present generations. In response, the contractualist merely has to show that the time bomb is unjust, according to whatever principles of justice are being used to determine the proper distribution of the benefits and burdens of cooperation. In other words, the contractualist need not show that construction of a time bomb in this sense is contrary to the self-interest of present generations (it isn't); it only needs to be shown that it is unfair to future generations.

This point is sufficiently subtle that it perhaps merits further elaboration. The folk theorem for the iterated PD establishes that there are an infinite number of possible "cooperative" equilibria (i.e. equilibrium outcomes with payoffs better than the non-cooperative baseline). For the payoff matrix in Figure 3.1, with appropriate discounting and normalization, any payoff in the space bounded by ($2,$2), ($2,$7), ($7,$2), and ($5,$5) can be produced through some combination of strategies that will be in equilibrium.[52] For instance, a payoff of ($5½,$3¾) can be obtained by the strategy profile ([C,C,C,D,C,C,C,D . . .], [C,C,C,C,C,C,C,C,C . . .]) backed up by the "trigger" threat of universal defection in response to deviation from the pattern. This equilibrium is obviously nice for player 1, since it allows him to exploit player 2 with impunity every fourth round. We are therefore inclined to describe it as cooperative, yet unfair. Many critics, however, seem to think that it is incumbent upon the contractualist to show that the adoption of an unfair arrangement of this sort will lead to the collapse of cooperation. Thus, for example, Arrhenius criticizes the iterated PD model of intergenerational cooperation on the grounds that it "leaves room for" strategies that "involve negative consequences that only stay in effect for a limited time" and there-fore are still better than universal defection.[53] Gardiner also criticizes the model on the grounds that mere "unfairness to future generations" cannot be shown to provoke universal defection.[54] But this is not a bug; it is a feature of the model (and the existence of strategies such as the one Arrhenius sketches out need not be demonstrated; that they exist is a trivial consequence of the folk theorem).[55] To exclude the possibility of such strategies would be to commit oneself to the self-evidently false claim that cooperation can only be sustained if the terms are perfectly fair.[56] Thus one suspects that the crit-icism involves some misunderstanding of the contractualist model. Mutual advantage is intended only to fix the proper scope of the principles of justice. It does not purport to explain how people choose a cooperative arrangement (on the contrary, as the folk theorem shows, it radically underdetermines that choice, since there will typically be an infinite number of "mutually advanta-geous" arrangements, some of them perfectly fair, most of them quite unfair). It is the willingness of all parties to accept common principles of justice that serves as the basis for choosing a particular one of these arrangements. Thus if one player comes along and proposes some arrangement that is mutually advantageous, yet in violation of the principles of justice, then the problem with the proposal—and the basis for its rejection—is simply that it violates the principles of justice.

3.3.4. Doubts about the significance of upstream cooperation

Critics have questioned whether systems of intergenerational cooperation with an upstream structure matter all that much. Gardiner, for instance, makes the significant admission that if "earlier groups know that they will eventually be at the mercy of their successors," this would undermine his claim that intergenerational cooperation is impossible. "Still," he says, "it is doubtful to what extent it characterizes many contemporary relationships between generations."[57]

This ignores the fact that government pensions, such Social Security in the United States, are all unfunded, or PAYGO. Furthermore, public health insurance has very much the same structure—most of it involves a pure transfer from the young and healthy to the old and infirm. This is quite explicit in the case of Medicare in the United States, where almost the entire population pays into the program throughout their working lives, yet are unable to draw benefits until they reach age 65. Again, the reason that people are willing to pay into it (i.e. largely oppose attempts to dissolve the system) is that they anticipate being able to draw from it when they are retired. But since the plan is unfunded, there is no basis for this anticipation other than the expectation that younger workers will continue in their willingness to contribute. (In the case of Medicare the cooperative dimension is even more apparent, since it is based entirely upon an implicit promise. Unlike state pensions, where you at least get a piece of paper saying how much you can expect to draw, in the case of Medicare there is not even a piece of paper.) Since Social Security and Medicare spending together make up more than 8 percent of GDP in the United States, pay-as-you-go systems of intergenerational cooperation— which leave elderly groups entirely "at the mercy of their successors"— constitute a significant fraction of the economy, and of individual lifetime consumption.

More important for our purposes, however, is the fact that the entire system of savings and investment in our society is a system of cooperation with an upstream flow of benefits. This is a tricky point, because economic growth *looks* like a giant pay-it-forward scheme. Each new generation inherits from the previous generation a capital stock, which is the product of past investment. It could choose to consume the entire social product, slowly drawing down the capital stock. And yet what previous generations have done, and what we are currently doing, is not just renewing the capital stock, but also expanding it, so that future generations can enjoy an even higher standard of living. This arrangement is obviously cooperative, in the sense that it is mutually beneficial and yet vulnerable to free riding. If one generation were to "live off its capital," using inherited infrastructure and machinery while failing to repair or replace it as it broke down, and therefore lowering the standard of living of future generations, we would not hesitate to describe them as having failed to "pull their weight" or "do their part." They would have violated an obligation to future generations. Certain appearances, however, can be deceiving. While it is true that investment and economic growth are the product of a system of intergenerational cooperation, it is not actually a pay-it-forward scheme, because the direction of cooperative benefit is almost entirely upstream, not downstream. This is rather unobvious, and so merits its own discussion.

3.4. Just Savings

While economists have not traditionally been confused about the issue of intergenerational cooperation, there has unfortunately been a great deal of confusion over the question of savings, and thus over the just savings rate. The issue of savings is, of course, closely tied to that of economic growth, because saving is the mechanism through which the pool of funds becomes available for investment (via the banking system), and so, classically, the savings rate is what partitions the economy between the production of investment goods and consumption goods. For example, with a savings rate of 10 percent, one can think of the present working population as consuming 90 percent of what it produces, while setting aside 10 percent to serve future needs. What makes it important to save is that, by setting something aside, we can increase the amount produced—holding all else constant—so that we can consume *more* in the future by consuming less in the present. The question of how much we should set aside then starts to look like a rather tantalizing optimization problem, which is, unsurprisingly, how economists have traditionally construed it. The foundational work in this vein is Frank Ramsey's 1928 paper "A Mathematical Theory of Saving."[58] Ramsey adopts the standpoint of a utilitarian planner and then asks what the optimal savings rate would be if one wanted to maximize consumption over time. From this perspective, the intergenerational dimension of the choice is not important. Since one person's consumption is as good as anyone else's, it does not matter for Ramsey's purposes whether "society" is composed of overlapping generations of finite agents, or a single cohort of agents who live forever. But of course, since real societies are composed of finite agents, Ramsey's utilitarian framing of the problem has the effect of treating savings as essentially *altruistic*. On his model, present generations are sacrificing consumption opportunities so that other people, sometime in the future, can consume more, all in the interest of having "society" maximize consumption (i.e. in the aggregate).

Normatively, this is not a very compelling way of thinking about the issue (particularly because, given that investment produces an open-ended stream of future benefits, it is very easy to come to the conclusion that present generations should be saving almost everything that they produce).[59] Ramsey's way of thinking was partly due to the intellectual climate at the time, in which economists were just beginning to think seriously about the economics of socialism. In a centrally planned economy, the savings rate would be determined by the state, which would have to decide how much economic output should be channeled into investment.[60] In a market economy, by contrast, the savings rate is not determined by the state, and so there is no utilitarian planner in a position to make these choices. Nevertheless, the way that Ramsey *framed* the problem became extremely influential, even among those who rejected his normative perspective. In particular, the idea that savings are essentially altruistic—some people give up consumption now, so that others can enjoy greater consumption later—became an unquestioned presupposition in much of the subsequent debate.[61]

This is unfortunately a mischaracterization, one that has created enormous problems for later theorists. For instance, the reason that Rawls had such difficulty with the just savings problem is that he accepted the idea that saving was essentially altruistic. This is what led him to assume that the way to maximize the consumption of the worst-off generation would be for them to adopt a savings rate of zero (since any positive rate of savings would worsen the situation of those required to save, while benefiting later generations, who are, by hypothesis, better off). The "eternal poverty" reductio follows rather immediately from this, since if the first generation cannot save, this means that the second generation will be just as badly off, which means that they cannot save either, and neither can the third, and so on.[62] Rawls's attempt to avoid this conclusion involved fairly ad hoc modifications, and there is good reason to suspect that he was never satisfied with his own view of the matter.[63] (It is noteworthy that the most extensive revisions made to *A Theory of Justice*, between the original and revised editions, occurs in the section on intergenerational justice and just savings.)[64] His revised solution, which is to conceive of individuals behind the veil of ignorance as contemporaries, but burdened with the responsibility of acting as heads of families that extend over multiple generations, has struck most commentators as particularly ad hoc if not positively bizarre.[65] What Rawls never questioned, however, was the assumption—inherited from Ramsey and the utilitarians—that the act of saving involves some form of self-sacrifice.

It is this last point that represents the most serious problem with this way of thinking. If one looks at the savings decisions that people make, both today and throughout history, they have almost never been motivated by any altruistic concern for future generations.[66] People occasionally try to save something for their immediate (contemporaneously living) children, but the overwhelming reason that they save *is to finance their own future consumption*. In particular, most personal savings take the form of retirement savings, which are entirely self-interested. These savings are used not only to finance ordinary consumption, including personal care and assisted living arrangements, but also healthcare expenditures (two-thirds of which occur in the last five years of life). Because of this, no generation would ever find it in its interest to adopt a zero rate of savings during its working life. Thus the "eternal poverty" reductio that Rawls worried about from the application of the difference principle was entirely illusory.

Not only are private savings decisions basically self-interested, but the state also plays no role in promoting savings for future generations. Despite the fact that some public expenditure can be characterized as investment, the net effect of the welfare state is to depress the rate of savings.[67] Since the wealthy have the highest propensity to save, progressive income taxation will tend to decrease the level of private investment, compared to private consumption, by taking more money away from those who are more likely to have saved it. Furthermore, the programs that this tax revenue are used to finance are mostly pure consumption. Most obviously, redistribution to the poor, the unemployed, and the elderly, all of whom have a propensity to save of zero, will increase consumption. Healthcare spending is also consumption. Furthermore, welfare states usually have structural deficits, which means that they are drawing directly from the pool of private savings (i.e. by selling bonds), and using the funds to finance consumption rather than investment. Thus the net effect of the welfare state will be to shift the ratio of consumption to investment in the direction of present consumption. As a result, if there is some kind of a "just" rate of savings, it does not seem to be playing a role in anyone's decision-making—neither that of private individuals nor the public authority. Indeed, it suggests that the welfare state has radically mistaken priorities when it comes to promoting economic justice.

Thus the Ramsey model of savings, which sees present generations forgoing consumption so that future generations can consume more, is deeply misleading. Structurally, the way that private savings work is much closer to that of a PAYGO pension scheme, where each individual is actually just doing her part in an ongoing system of cooperation, while receiving benefits from her participation. Individual retirement saving—either through funded pensions schemes or individual savings accounts—is sometimes presented as though it were radically different from unfunded PAYGO schemes. In the latter case the elderly are portrayed as being a burden upon younger generations, receiving unproductive transfers, while in the former case they are presented as being self-sufficient, able to provide for their own needs without burdening the young. Yet if one looks at it from the standpoint of economic fundamentals, one can see that the two arrangements are, of necessity, equally burdensome. No matter how you organize it—whether through PAYGO or savings—the fact remains that a certain percentage of the population at a certain age stops producing, and yet continues to consume (in many cases at a very high level) until death. In both cases there is a huge transfer of goods and services from the young to the old.

In the past, people either continued working until death or relied upon the support of their children in old age.[68] Yet in our society the elderly for the most part rely upon strangers to supply their needs. They are able to do so because of a system of intergenerational cooperation, which allows them to produce more than they consume when they are young, save the difference, and then cash in these savings when they are old. The important thing to see is that the money with which they pay for goods and services later in life is not itself a good. It is nothing but a promissory note, not so different in kind from the annual letter that I receive from my pension fund telling me how much I can expect to receive in monthly payments. When people save for their retirement, what they are essentially doing is accumulating promissory notes—not just from the bank, or the mutual fund, which promise to allow them to redeem their savings, but from society more generally, which assures them that these banknotes can later be exchanged for food, clothing, nursing care, medical treatment, etc. It is the credibility of these assurances that, in turn, makes it economically rational to save a fraction of one's income.

The system of intergenerational cooperation involved in the savings system does not really have a special name, other than simply *capitalism*, if one interprets this in terms of its core feature, viz. the ability of individuals to take money, invest it, and then receive the proceeds. To see how it motivates saving, consider a case where, instead of participating in a pension scheme, I decide to save 5 percent of my income in order to support myself during retirement. What this means, in practice, is that during a certain period of my working life I will have to produce 5 percent more than I consume. But what shall I do with the extra 5 percent? One strategy is to become a *hoarder*—I could start stockpiling all the consumer goods that I expect to need when I am older, or some set of goods that I think I have a reasonable chance of trading for consumer goods that I will need. Another alternative is to become an *investor*. Instead of spending the money on consumer goods, I could lend the money to someone who wants to produce capital goods. In a sense, I am taking the extra 5 percent and using it to produce goods that will not be consumed right away, but will instead be used to produce consumer goods in the future. What I accept, in return for the loan, is a promise that I will be able to get a share of the consumer goods that are to be produced. The advantage of this arrangement is that forgoing consumption now, in order to make productive investments, reduces the amount of time and energy that will be required to produce consumer goods later on. This makes it possible for the borrower to promise to repay me, not just the principal that I have lent him, but also accumulated interest. Thus the advantages of investing, rather than hoarding, correspond precisely to the advantages that come from putting money in a bank, rather than hiding it under a mattress. This is what it means for money to serve as capital.

What is the cooperative benefit produced by this system? It is increased labor productivity, and more generally, economic growth.[69] To see this, consider a simplified example. Suppose that I am a farmer. One way of saving for my retirement would be to take 5 percent of my grain and hide it away in the basement. Suppose that this is the prevailing practice in the community. Yet suppose also that my neighbors and I could benefit from an irrigation system, which would increase crop yields by 30 percent. Unfortunately, the main canal would take many years to dig, and as a result no one has any individual incentive to do it. One way of solving this problem would be for all of us to take the time that we had spent growing extra grain, and spend it instead digging the canal, in return for a promise to the effect that anyone who has spent time working on the canal will be entitled to a grain ration once the project is complete. Furthermore, imagine that the ration is more than could be produced through individual hoarding (a promise not difficult to honor, given the way that the irrigation system increases crop yields). Thus an intergenerational arrangement could be proposed of the form "We agree to invest now, if you promise to support us later."[70] This would be mutually advantageous, because shifting production from consumption to investment goods makes it easier to produce consumption goods later, and this "cooperative surplus" can be divided in a way that makes the arrangement positive sum. Yet the younger generation has an obvious free-rider incentive. Once the canal has been dug, what is to stop them from using it, yet refusing to honor the promises made to those who dug it?

One solution is to create a completely generalized system, where people are credited, not for having made specific investments, but for having produced more than they consumed, thereby making resources available for investment. And rather than issuing them ration cards, or promissory notes, one can simply give them money and pay them a return on their savings. The canal can be constructed by hiring people to build it, paying them wages that they may opt to exchange for consumption goods right away, or else save for later. Younger generations have an incentive to continue this system of co-operation as long as there are still productive investments to be made, and to the extent that they themselves stand to gain from the integrity of the monetary system. Defection is, of course, still possible, either through debt repudiation or, more subtly, through inflation. Indeed, the central vulnerability that the elderly have toward the young is due to the effects of inflation, which can easily wipe out the value of all savings. Hyperinflation is therefore the central mechanism of intergenerational defection. It says, in effect, "Even though you produced more than you consumed, in order to accumulate this cash balance, you will no longer be entitled to exchange it for consumer goods equivalent to what you forewent." The elderly are often hardest hit by inflation, but they are not singled out. Because inflation destroys the value of savings generally, and weakens the integrity of the money system, young people wind up being denied access to a system of mutually beneficial cooperation. Thus the consequences of defection, with respect to the value of money, are not so different from the consequences of dissolution of a PAYGO pension scheme.

Thus the entire savings-and-investment system in our society contains important features in common with a PAYGO pension system. This structure is obscured by a money illusion, which leads us to think of money that is saved as having been hoarded, rather than having been invested by someone else in the production of capital goods. Cooperation is required in order to allow savers to "cash out" at a later point in time, converting the money balance into consumption goods (the honoring of such requests is what deserves to be labeled "C" in the game tree diagram). The advantage of having such a system in place is that it enlarges the pool of funds available for investment, making it individually rational to save (or, more accurately, increasing the incentive to save), and thus promoting economic growth. Young people in general have no incentive to deviate from this arrangement, because they themselves stand to benefit from it, when the next generation takes over the major productive tasks in the economy.

This analysis casts doubt upon the way that the question of just savings has traditionally been posed, with saving being treated as essentially an altruistic activity. If one attempts to transpose this directly onto a contractualist schema, by treating savings as "cooperative" rather than "altruistic," problems

develop. Because saving is the behavior that generates the benefits for future generations, there is an inclination to think that people must be saving
for moral reasons, and so any contractualist theory must be able to explain
saving as a form of cooperative behavior. If one looks at the actual motives
that people have to save, however, these are almost entirely self-interested, or
narrowly altruistic. Many people save nothing at all. Among those who do,
the savings typically take the form of owner-occupied housing and retirement savings, both of which are essentially self-interested. Some people save
more in order to provide a legacy for their living descendants, but almost
no one saves in order to benefit "future generations" generally. Most legacy
savers do not even concern themselves with their own future (i.e. currently
unborn) descendants.[71]

Thus we arrive at the somewhat surprising realization that the benefits provided to future generations through economic growth are largely byproducts
of the savings and investment system that are not really "costing" us anything.
As individuals we want both savings and economic growth so that our own
retirement portfolio will appreciate in value; the fact that others will benefit, in the distant future, is entirely incidental to our concerns. Put in non-
monetary terms, our goal is to induce younger generations to provide us with
consumption goods, not just equivalent to what we gave up while working,
but vastly exceeding that amount (equal to the compounded returns on our
savings). The way we do this is by contributing to the production of capital
goods, which allows younger generations to produce consumption goods
more easily, both for themselves and for us. The economic growth that results
is almost entirely a byproduct of this self-interested action.

All of this raises serious doubts about whether we have any obligation to
save for the sake of future generations. The obligation imposed by the system
of cooperation is not to save, but rather to honor the savings of others, when
the time comes that they want to exchange these money balances for goods.
Accordingly, what justice determines is the entitlement that individuals acquire through their saving activities. Cooperation creates an *incentive* for
individuals to save, but the amount that they will save is determined by their
own self-interest, not by a moral obligation to realize potential benefits for
future people. This accords with common sense, to the extent that it is very
difficult to find any trace of a general obligation to save, in either individual
behavior or institutional arrangements.

Of course, one should not rule out a priori the possibility that everyday
morality is quite mistaken in the way that it thinks about obligations toward

future generations, and that utilitarians are correct in thinking that we are obliged to maximize their welfare. Although this is radically revisionary with respect to everyday morality, that should not be treated as disqualifying. Historically, utilitarianism has been a powerful source of moral change in our society. Attitudes toward corporeal punishment, for instance, have changed dramatically over the past two hundred years, largely under pressure from the utilitarian moral sensibility.[72] Furthermore, the entire point of Singer-style "expanding the circle" arguments is to change our everyday moral views. Expanding the circle to include future people is an unusually bold and radical form of expansionism, but it should not be ruled out on those grounds alone. Even though neither the state nor private individuals currently recognize any obligation to save in order to maximize the welfare of future generations, we may nevertheless be under such an obligation.

What should give us pause, however, is the implication that our concern over climate change, being tied to this conception of just savings, depends upon our willingness to accept a radical revision of our everyday moral views. This runs the risk of transforming climate change mitigation into a niche issue, like vegetarianism, that appeals to only a small segment of the population.[73] More generally, it makes it seem as though climate change mitigation is wedded to a specific "private comprehensive view" about the metaphysics of future persons. And yet, in most other domains, environmental regulation is not thought of as resting on exotic and revisionary moral commitments. It is, on the contrary, considered a fairly straightforward consequence of our prevailing ideas about justice. As a result, if we can find a way of addressing the problem of climate change without positing some new moral obligation that most people do not realize they have, then there is a lot to be said for adopting such an approach.

The central advantage of the contractualist formulation, I will argue, is that it is able to motivate a carbon abatement policy without requiring any major revisions in either our moral views or our political conceptions of justice. It is, however, an approach that is initially rather unintuitive, because it takes as its point of departure the suggestion that we do not actually have any obligations of justice to future generations per se (or at least not in the way that utilitarians have conceived of them). The *mere* fact that there are going to be people in the world, sometime in the distant future, does not entail that we have obligations of justice to improve their well-being. Whatever obligations we do have arise from the systems of cooperation in which we participate. Outside such systems, we may have duties of benevolence, or be prohibited from imposing harms, but the strict obligations that we have, which require a commitment to efficiency and equality, arise only within cooperative relations. The market economy, along with the system of savings and investment that sustains economic growth, is a system of intergenerational cooperation, but as we have seen, it is not one that imposes a particular savings rate as a duty of justice. This raises doubts as to whether "just savings" is the correct framework for thinking about climate change policy at all.

3.5. Conclusion

The view that climate change is not a collective action problem, because the "absence of mutuality" between generations precludes cooperation, has become widespread in the philosophical literature. I hope to have shown, in the preceding discussion, that this view is based on a misunderstanding of several key features of both contractualism and the operations of a modern economy. There are, as far as I can tell, three major reasons why so many philosophers have failed to see the system of intergenerational cooperation that underlies many core features of the basic structure of our society:

1. The system is based upon indirect reciprocity, and so those who think of mutual benefit in very narrow terms, on the model of direct reciprocity, have difficulty discerning it.
2. The cooperative benefits flow upstream—from younger to older. The downstream benefits to future generations are all byproducts. This is counterintuitive.
3. The most important system of intergenerational cooperation, the one that underlies the savings and investment system, is hidden behind a money illusion.

Getting clear on these three points allows us to set aside the more radical critique of the contractualist perspective—that it has nothing to say about intergenerational justice, because there can be no such thing as intergenerational cooperation. The economy is not like a river that we are standing beside, which we might choose to add or subtract value from, in ways that will affect those downstream. We are all participants in an ongoing scheme of cooperation. It is more like a parade, in which we are all participants, and where if any group fails to do its part in sustaining it, the entire procession will stop.

The system of savings and investment that produces economic growth over time is the core feature of this structure of intergenerational cooperation.

This insight does not do much to resolve the problem of how we should respond to climate change, it merely sets aside another ill-conceived objection to the standard policy framework. It suggests that the temporal structure of the problem is largely a distraction. Climate change is a collective action problem, not different in any fundamental way from a variety of other environmental problems involving atmospheric externalities.

4

The Case for Carbon Pricing

On the standard economic view of savings, investment is something that is costly *now*, but produces greater benefits *later*. The intertemporal trade-off is the fundamental feature. Since carbon abatement is something that will generate costs in the present, along with significant benefits in the future, it is tempting to think of it as a type of investment. And yet in the case of global warming the intertemporal trade-off is not fundamental. It is in fact an accidental feature of the climate change problem that there is a significant delay between the abatement efforts undertaken and the climate change mitigation that will occur. Many other environmental problems that are otherwise indistinguishable do not have the same delay, and indeed, it is difficult to see how the climate change problem would change in any fundamental way if climatic inertia suddenly disappeared, and the effects of any abatement efforts were experienced immediately. Motivationally, of course, this would make it easier to get people to change their consumptions habits, but *normatively* it is not clear how it would make any difference. Every environmental regulation imposes compliance costs, in order to achieve some benefit—and if the regulation is cost-benefit justified, then those benefits will exceed the costs. None of this leads us to think of these regulations as investments. This suggests that it is misleading to think of climate change as a problem of intergenerational justice, just as it is misleading to think of it as a problem of distributive justice. It is, first and foremost, a collective action problem, the solution to which merely *raises* issues of distributive and intergenerational justice.

The question, therefore, is what sort of normative principle can account for the wrongness of greenhouse gas emissions, if not a general duty to promote the welfare of future generations. The answer, I will argue, is that it is the same principle that is violated by any other negative externality, which is the obligation to act on the basis of norms that could be accepted by all. In order to articulate this, it is helpful to restate some of the fundamental ideas underlying environmental regulation. This will allow us to see more clearly the sense in which climate change is a standard environmental problem,

Philosophical Foundations of Climate Change Policy. Joseph Heath, Oxford University Press. © Oxford University Press 2021. DOI: 10.1093/oso/9780197567982.003.0005

responsive to the standard remedies. I will begin by articulating, at a fairly high level of abstraction, the reciprocity constraint that is institutionalized by the pricing system, with respect to goods and services generally. This makes it possible to state more precisely the way in which negative environmental externalities, such as greenhouse gas emissions, constitute a form of free riding. The attempt to eliminate this provides the central normative rationale for a system of carbon pricing. This can be defended without reference to intergenerational savings, or an obligation to maximize the welfare of future generations. What matters is simply that we participate in an ongoing system of cooperation, the structure of which can be evaluated in terms of certain basic principles of justice. I will pause briefly to rehearse some of the general arguments for a pricing approach to carbon abatement, as opposed to other forms of environmental regulation, before in the final section considering the issues of distributive justice raised by this approach and how they can be addressed.

4.1. Market Reciprocity

The market, as we have seen, is a complex system of cooperation based on indirect reciprocity. In the simplest case, person a may provide a service to b, and receive a benefit from c, who in turn is compensated by b. With the introduction of money, however, the task of following these chains of exchanges becomes infernally complex, because each one branches at almost every point (e.g. person a might be paid \$100 by b, which she in turn spends procuring goods from c, d, and e). Thus it makes more sense to adopt a systemic perspective on the market as a whole. Abstractly, we can think of "the economy" as an enormous pool of goods and services (which we can refer to, for simplicity, as "goods"), which individuals can make both contributions to and withdrawals from. Reciprocity is satisfied when the value of each individual's contribution is roughly equal to that of the person's withdrawals. With the division of labor that exists in modern societies, however, the set of goods become extraordinarily, almost mindbogglingly, heterogeneous. This naturally creates a problem, of determining how the value of any particular contribution or withdrawal from the common pool should be assessed, in order to determine whether the general constraint of reciprocity is being satisfied. Some type of common metric is required that will allow us to express the value of any one thing in terms of another, and vice versa, in order to ensure that what individuals take out is matched by what they put in.

Such a metric is known generically as a price, although it is important to recognize that a price is just a number, and different principles of comparison can generate different numbers, or different price systems. For instance, there is the well-known Marxist proposal to have a system of "labor prices," where the price of everything will reflect the average amount of social labor required for its production.[1] The price system that prevails in our own society is known as *scarcity pricing*, or sometimes "social cost" accounting. The idea, roughly speaking, is that the price of a good should reflect a balance of two things, viz. the amount of satisfaction that it is able to produce, and the amount of inconvenience involved in its production. The first half is just the idea that the amount of credit individuals get for producing particular goods for others should be a function of how much other people actually want those goods. The second is the idea that individuals should get more credit, the more difficult these goods are to produce. Consumption is just the reverse. If one consumes something that others also really want, or that was more inconvenient to produce, then one should have to give up more credit.

These are, of course, the relationships that are traditionally represented using supply and demand curves. I have chosen not to use those terms here, though, because they immediately summon up the image of markets, whereas the underlying normative ideas are more general than their particular institutional realization in markets. Most notably, a number of socialist economists have endorsed the principle of scarcity pricing, and some socialist schemes, such as the *parecon* (or "participatory economy") proposal, preserve scarcity pricing despite having abolished markets.[2] In part this speaks to the attractiveness of the principle, which has both a "micro" and a "macro" rationale. On the micro, or transactional level, as Ronald Dworkin has observed, scarcity pricing basically equates the cost of one's consumption to the level of welfare forgone by others.[3] This strikes many people as intuitively fair. If the good is relatively easy to produce (does not require much labor or skill, uses few raw materials, can be assembled quickly, etc.), then one is simply not imposing as much of a burden upon society by consuming it, and as a result, it should not cost as much. Similarly, if few other people want it, or they do not want much of it, then one is imposing less deprivation on others through one's consumption, and so again, it should not cost as much.

On the macro level, scarcity pricing has the attractive property of generating—in principle—a Pareto-optimal allocation of both inputs and outputs in the economy. The proof of this is the greatest technical achievement of 20th-century welfare economics, and I will not be delving into the details of it here.[4] I will simply state the intuitive basis of the conclusion. On the production side, scarcity pricing gives everyone an incentive to invest more effort in the production of goods that people want more, *modulo* the level of inconvenience involved in their production. On the consumption side, it gives everyone an incentive to consume goods that are less onerous to produce, or that others do not want. The attempt by everyone to satisfy all of these constraints simultaneously has the effect of "squeezing" both production and allocation, in such a way as to eliminate waste (i.e. the production of goods people do not want). Waste, however, is just another way of describing Pareto inefficiency. (If an allocation is Pareto-inefficient, this means that it should be possible, by rearranging the assignment of goods to people, to make one person better off without making anyone else worse off. This means that, in the prior allocation, some good was being wasted, because its consumption was producing less welfare that it otherwise might have.)

The primary way that we institutionalize scarcity pricing in our society is through markets, or more specifically, through the combination of free labor, private property, voluntary contracting, and competition. This generates scarcity pricing because it grants individuals control over their own labor and goods, which they must be induced to give up voluntarily. The more reticent they are to do so, the more they must be offered in exchange. If there is significant demand for a particular good, and yet others are unwilling to supply it, competition among those attempting to acquire it will increase the price, which will both increase the willingness of others to supply it as well as winnow down the numbers of those seeking to purchase it. If there is less demand for a good, the price will decline, with the opposite effects.

This institutionalization of the scarcity-pricing principle will be more or less imperfect, depending upon how much control the individual is able to exercise over the elements of the environment that affect her welfare. Perfect institutionalization would require that it be impossible to impose welfare losses upon a person without contracting, and thus, without providing compensation. Only then would the price precisely match the level of forgone welfare. In reality it is difficult to provide individuals with that level of control. Private property rights work well, as we say, over land and medium-sized dry goods. With other aspects of the environment, such as water and the atmosphere, they work poorly or not at all.

This is what gives rise to so-called externality problems. On the negative side, these arise when individuals are able to impose welfare losses upon others, but where the absence of property rights makes it such that they are not obliged to offer any compensation. (Another way of putting it is to say that the transaction is involuntary, which is what accounts for the economist's traditional, but overly narrow, definition of an externality as the effect of a transaction upon an uninvolved third party.) The absence of compensation means that the price is going to be wrong, from the standpoint of the scarcity-pricing ideal and, thus, the reciprocity principle. In particular, consumption goods that are associated with the production of negative externalities (i.e., that involve the production of negative externalities anywhere along the supply chain, up to and including the point of final consumption) are going to be priced too low, which is to say, at a level that does not reflect the true social cost of their consumption. This produces two problems. On the macro level, the existence of negative externalities will tend to result in inefficiency, or the misallocation of resources. This is the so-called tragedy of the commons, or market failure, where every member of a group may engage in the activity that produces the externality, because they have no incentive to stop, even though it leaves everyone worse off in the end. On the micro level, the problem with the externality is that it allows people to take more out of the economy than they are putting in. In other words, it upsets the balance of reciprocity in the system. In the presence of negative externalities, the price of a good will be lower than it should be, and so individuals will be able to consume too much of this good, relative to the value of the contribution they themselves have made to the cooperative system. Negative externalities, therefore, allow for a type of free riding to occur in the economy, and are objectionable for the reason that all free riding is objectionable.

One way of thinking about this is in terms of what I have referred to elsewhere as the "apology model" of market pricing.[5] Every act of consumption generates a cost to society, in the sense that it results in there being less of some good in existence than there previously had been. When I purchase a cup of coffee and drink it, this results in there being one less cup of coffee in the world for others to enjoy. Under a system of autarkic production, this cost would always be "internalized," in the sense that the only way for me to get a cup of coffee would be to grow my own coffee beans, roast and grind them, then collect water, boil it, brew the coffee, and so on. The fact that I have to do all this work ensures that I will only drink coffee if it is genuinely worth my while, which is to say, if the benefit that I derive from the cup of coffee exceeds the cost, which in this case involves both the time and energy that I put into producing it, as well as the forgone value of all the other things that I could have done with those resources.

The central advantage of the complex division of labor that exists in modern societies is that I am able to secure goods like coffee without having to do all of this work, because I am able to get other people to do it for me. Unfortunately, this also creates an obvious opportunity to free-ride. If others are doing all the work for me, of tending the coffee plants, collecting the beans, roasting and grinding, etc., it eliminates the primary constraint that stops me from overconsuming. "Overconsumption," in this context, means that the benefit I derive from drinking the coffee does *not* exceed the cost. The only reason I am drinking it, in this case, would be that the benefit accrues to me, while the cost is borne by others. The purpose of the pricing system, apart from just coordination, is to ensure that this sort of overconsumption does not occur. The major reason I do not overconsume coffee is that I am obliged to *pay* for it, which requires that I be willing to give something up in return for it. This internalizes the social cost of my consumption, so that again I will only drink coffee if it is genuinely worth my while. Thus a well-structured market retains the basic principle governing consumption in an autarkic system of production while nevertheless allowing individuals to reap the benefits of an extensive division of labor.

The obligation to pay for something, such as a cup of coffee, can be thought of as a way of apologizing to all of the people who have been inconvenienced by my consumption. For example, I am able to spend a lot less money on coffee on days when I am willing to boil water for myself and prepare it at home. This is, however, somewhat inconvenient, and so often I would rather have someone else do it for me (as well as acquire the beans, provide the brewing equipment, etc.). In other words, I would rather inconvenience them than do the work myself. As a result, I am obliged to compensate them for their troubles—which is why a cup of coffee served to me at a coffee shop costs more than one prepared at home. The extra amount that I pay constitutes, in effect, my way of apologizing to those who spent time, energy and resources making coffee for me, rather than doing something else that they might have preferred doing. Through the supply chain of the coffee shop, the amount that I pay gets divided among all those who have been inconvenienced in this way (some fraction to the landlord, some to the employees, some to the coffee wholesaler, some to the shipping company, some to the farmer, etc.).

This is the ideal. When it is realized, it allows us to get all of the benefits of a complex division of labor (including the massive economies of scale in activities like coffee production) while simultaneously ensuring that no one is taking advantage of anyone else by overconsuming. It is easy to see, however, that the ideal will only ever be imperfectly realized. In particular, there are many circumstances in which an act of consumption can inconvenience others, but where the person doing the consuming need not compensate all those who are inconvenienced. Typically, this will be because that person lacks rights against the particular intrusion, and so need not be contracted with as part of the supply chain involved in provision of the good. For example, while I am obliged to compensate most of those inconvenienced by my consumption of a cup of coffee, when I buy a liter of gasoline and burn it, I am not obliged to compensate many of those affected. I must apologize to all of those who are involved in the extraction and refinement of the petroleum, but the burning of the fuel produces atmospheric gases that affect others who are in no position to demand compensation from me. Because of this, the system of property rights is not able to prevent overconsumption of gasoline as effectively as it prevents overconsumption of coffee.

This way of framing the issue makes it easier to see why externalities generate collective action problems. The presence of a negative externality means that, as a consumer, I can get others to do things for me (i.e. be inconvenienced by my consumption), but I am not obliged to compensate them for their trouble. As a result, I lack the incentive to restrain my consumption. And yet when everyone does this, it produces an outcome in which the quality-of-life degradations we suffer, including the work that we must do to the benefit of others, are unjustified, given the consumption benefits that they support. If we had to pay the full cost of our consumption, we would rather scale back our consumption than persist with the current allocation of time, energy, resources, and effort.

It is noteworthy that almost every environmental problem—e.g. ozone depletion, deforestation, overfishing, biodiversity loss, nitrogen imbalance, water contamination, plastic and chemical pollution—involves the production of negative externalities. In one sense of the term, the word "environment" simply refers to aspects of the world that are not under human control, and in particular, under the control of property rights, and so there is a conceptual connection between the idea of an externality and that of the environment. It is certainly not an accident that environmental problems are highly concentrated in oceans, lakes, and the atmosphere, where there is no possibility of enforcing property rights, and that major environmental issues involving land involve spillover effects, such as fertilizer runoff, or common pool resources, like aquifers. This is why Garrett Hardin's tragedy of the commons article is at the heart of liberal environmental thinking.[6]

At the same time, it is important to avoid a widespread misunderstanding of Hardin's argument. The fact that environmental problems are less common with resources subject to private property rights does not mean that everything should be privatized. On Hardin's view of things, most environmental problems are instances of market failure. All this means is that they are collective action problems, which would not occur if individuals were forced to pay the full social cost of their consumption (and thus, would not occur in an ideal market economy). The fact that it is a market failure, however, says nothing at all about the correct way of fixing it. In the case of a common pasture, which is the example that Hardin chose, the most widespread solution was through "enclosure," or the dividing up of the shared resource into private property. There are many other instances, however, in which the creation of a system of private property would either be impossible or undesirable. In such cases, there are a variety of other devices in the institutional toolkit, any one of which may be used, severally or in combination. Calling something a market failure in no way implies that one must seek a market solution.

This misunderstanding of Hardin is what underlies the argument, which one encounters on occasion, that Elinor Ostrom's discussion of informal commons management represents a counterpoint or refutation of his analysis.[7] What Ostrom observed is that the mere presence of a commons (and thus the absence of private property) does not necessarily generate a tragedy. If it is possible for the group to prevent free riding through other institutional arrangements, such as informal norms or moral constraints, then that constitutes a perfectly reasonable way of resolving the (potential) problem.[8] The "tragedy" is what occurs if individuals are left free to act in an entirely self-interested fashion, and this is not the situation that Ostrom encountered in her famous study of village irrigation systems in Nepal, where there were a set of informally enforced norms governing water use. The problem with these informal commons-management arrangements is not that they do not work, but that they generally lack scalability. In other words, they work well in small-scale sedentary communities where people know and trust one another, but they become more difficult to sustain as the number of people involved becomes larger, and become practically impossible to sustain among strangers.

Because of this, the standard remedy for commons problems, in cases where private property systems are impractical, is legal regulation (i.e. coercive imposition of rules by the state). This type of regulation can take many forms. The simplest approach, in the case of negative externalities that are extremely harmful, is simply to prohibit their production. In the case of lead as a fuel additive, for instance, the negative health consequences associated with its introduction into the atmosphere are sufficiently severe, relative to the benefits of improved engine performance, that the simplest course of action is simply to ban its production and sale. The case with PCBs (polychlorinated biphenyls) is the same. In other cases, the appropriate regulatory approach, in managing a common-pool resource, is to impose limits on the quantity that individuals can withdraw. In the case of fisheries management, for instance, the standard approach is to require that anyone fishing have a license to do so, and that each license be associated with a quota, specifying the maximum quantity of fish that can be caught. And finally, there is the possibility of imposing a price correction, basically forcing individuals to pay for production of the negative externality. Such an arrangement is known as a Pigovian tax, after the economist Arthur Cecil Pigou, who was the first to suggest the strategy.[9] Greenhouse gas emissions are an ideal candidate for the imposition of such a tax, which is why carbon pricing has emerged as the recommended policy solution for the problem of climate change.

This Pigovian remedy is in fact the origin of the "polluter pays" principle. As I mentioned earlier (perhaps somewhat cryptically), this is not a principle of distributive justice, but rather an efficiency principle. This can be seen most easily by considering the natural follow-up question to the principle, which is "How *much* should the polluter pay?" The answer normally given is that the polluter should pay a price equal to the social cost of the pollution. In cases where markets fail to price it correctly (and typically, when we talk about pollution, we are talking about negative externalities, and hence unpriced costs), government should step in and impose an additional cost in the form of a tax, equal to the difference between the market price and the social cost, bringing the total price into alignment with the true social cost. One way of thinking about this, recommended by Ronald Coase, is to imagine that the parties did actually have a complete set of property rights, such that it would be impossible for one person to produce the externality without securing the consent of all those who are affected by it. This would force the parties to find a contractual solution, which would in turn determine a price level at which those who suffer the externality would consider themselves sufficiently well compensated that they would allow it to be produced. When the state imposes a Pigovian tax, it is essentially trying to estimate what this price level would be and then impose it. Such an arrangement satisfies the two normative properties that serve to recommend the scarcity-pricing system as a whole: consumption will satisfy the reciprocity principle, and the system will tend toward a Pareto-efficient allocation of goods. Thus the purpose of the "polluter pays" principle is not to punish polluters, it is rather to ensure that the optimal level of pollution is produced, which is to say, the pollution level that the parties themselves would have accepted if they had the capacity to contract over it.

So far this is all very standard environmental economics. Some have felt that it does not apply to the case of climate change, however, because of the temporal dimension of the problem. After all, many of the people who will be negatively affected by our actions have not been born yet. How can we apologize to those who do not yet exist? And since they are not around to sign contracts or exercise their property rights, how can we expect the market to take their interests into consideration? Indeed, one can imagine a system of cooperation, of the sort that could be characterized as a market, that had a fixed beginning and endpoint. In an agricultural society, for instance, one can imagine time being divided into a series of discrete seasons, with all exchange transactions involving the set of seasonally produced goods (and no futures contracts). Under such an arrangement, a person who sold goods during one season would be entering into a system of cooperation that included only those who participated in the market that season. As a result, it might be thought unreasonable to impose obligations of justice on individuals who participated in the market one season, toward a different set of individuals who participated in the market in some later season, because cooperative relations did not extend over the two time periods. Thus, there would be nothing in this institutional arrangement that precluded individuals from externalizing costs onto individuals in a later season.

But as the analysis in the previous chapter showed, our economy is not a series of discreet exchanges, but rather a system of intergenerational cooperation that extends out indefinitely into the future. An important feature of the economy is that it encourages savings, which is to say, it permits individuals to overcontribute during some periods, with the assurance that they will be able to undercontribute, i.e. make net withdrawals, in later periods. (It also allows people to borrow, and thus to undercontribute in the earlier period, in return for a commitment to overcontribute in a later one.) The possibility of using this willingness to overcontribute, on the part of some, as a source of investment, and thus as a way of producing the capital goods that drive economic growth, has become an essential feature of our economic system. This savings and investment system transforms what might have been a discrete set of cooperative exchanges into an open-ended cooperative system, which includes future generations. As a result, the pool of goods that we contribute to, and make withdrawals from, when we engage in production and consumption is one that also extends out and is renewed over time.

The fact that we are currently emitting greenhouse gases with little or no restraint means that we are taking out more from this pool than we are entitled to, given the value of what we are putting in. We are therefore violating a reciprocity condition in the system of intergenerational cooperation that ties us to future generations. This is why we should be imposing a Pigovian tax on greenhouse gas emissions. This issue is not that we are saving too little, or that we are failing to maximize the welfare of our descendants. The problem is that we are free riding.

All of this is a somewhat complicated way of establishing an essentially negative conclusion: there is nothing special about the problem of greenhouse gas emissions; it is a perfectly ordinary negative externality problem, no different in principle from the problem of sulfur dioxide emissions. The temporal structure of the problem—the fact that the negative effects associated with burning fossil fuels are felt much later than the benefits are obtained—makes no essential difference, because those who feel those effects are a part of the same system of cooperation as us, and so our relations with them are governed by the same principles of justice as our relations with contemporaries. Thus climate change is not a problem of "intergenerational justice," as the latter has typically been conceived by consequentialists; it is simply a question of ordinary justice. Like every such question, it does have an intertemporal *aspect*, because we need to find some way of rendering commensurable costs and benefits that occur at different points in time. This is what the debate over the social discount rate concerns (which will be discussed in greater detail in Chapter 6). But this is a subsidiary question. The temporal structure of the climate change problem means that the choice of discount rate winds up having a more important impact on policy, but this does not change the fundamental nature of the problem—every regulatory initiative subjected to cost-benefit analysis is assessed using a discount rate. And so again, climate change shows up as just an ordinary externality problem, which can be analyzed using the same conceptual tools as any other, and addressed using the same regulatory instruments.

4.2. Carbon Pricing

These reflections suggest that the problem of climate change is caused, fundamentally, by the fact that individuals who engage in greenhouse gas-emitting activities—first and foremost, consumption of fossil fuels—are not obliged to pay the full cost of these activities, as a result of which there is systematic overconsumption. Again, the problem is not that we are burning fossil fuel, it is that we are burning too much fossil fuel relative to the benefits that we are deriving from it. The most widely endorsed policy response to this is to increase the price of fossil fuel, or of greenhouse gas-emitting activities more generally, so that these costs are fully internalized. Any regulatory approach that attempts to achieve this is referred to as *carbon pricing*. The objective of a carbon-pricing regime in the near term is not to stop global warming from occurring. It is intended to reduce warming to the point where that which does occur is *justified*, by the magnitude of the benefits associated with the activities that produce it.

There are two basic approaches to carbon pricing: cap-and-trade systems and carbon taxation. While often considered equivalent policies by economists, the basis for this is not always obvious to the layperson. That is because economists are accustomed to looking at supply and demand curves, which are drawn on graphs that represent the relationship between the price of a good and the quantity that is expected to be exchanged (shown in Figure 4.1). Cap-and-trade systems involve regulatory intervention on the quantity axis, while carbon taxes intervene on the price axis, but both are intended to achieve movement away from the status quo (s).[10]

A standard demand curve (D) slopes downward, based on the expectation that as the price of a good declines, the quantity demanded will increase, and vice versa. This suggests that if one wants to reduce the quantity of a good transacted (from q to q'), one could do it either directly, by legally constraining the quantity that can be transacted, or indirectly, by intervening to change the price. So in the case of cap and trade, one can change the quantity (by imposing a permit scheme that caps emissions) and then let the price adjust (as trading determines the permit price); or with a carbon tax, one can

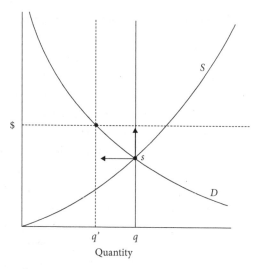

Figure 4.1 Two carbon-pricing strategies

change the price (by imposing a surcharge on carbon-emitting activities), then allow the quantity to adjust. Thus cap-and-trade systems price carbon indirectly, while carbon taxes do so directly. In both cases, however, it is the carbon price that is important, because this is what provides the incentive to reduce emissions. The idea that cap-and-trade systems allocate emissions rights, or that they impose hard caps on emissions, is a bit misleading, since a firm can always emit more by purchasing more permits, and there is always *some* price at which other emitters would be willing to sell. Thus it is the price that does the work; the quantity restriction is just a way of generating the price. This is also why, in a cap-and-trade system, it does not matter so much whether the emissions permits are auctioned off or are given away. Once the trading mechanism is in place, using up one's permit has an opportunity cost, because even if one has not paid for it, one always has the option to sell it, which one cannot do if one chooses instead to produce the emissions. Thus firms can be expected to reduce their emissions at the margin, even when they have not paid for their permits (just as firms that receive a fixed quantity of carbon tax rebates can still be expected to reduce emissions to avoid paying further taxes).[11]

Unfortunately, the fact that the price in a cap-and-trade system is achieved indirectly, through the trading of permits, has generated the impression among some that it represents an objectionable form of commodification of nature (e.g. that it represents the "neoliberalization of nature," or that it involves "performing market governance in capitalist world ecology," etc.).[12] This may or may not be a legitimate concern, but it should not be taken as an objection to the principle of carbon pricing, because the latter could just as easily be achieved through carbon taxation, which is typically immune to the accusation that it is commodifying, or even particularly neoliberal. The best way to assess any normative critique of cap and trade is to imagine an equivalent system implemented through carbon taxation, and then see whether the same objection holds. If the answer is no, then the objection is really just an argument for carbon taxation, not an objection to the principle of carbon pricing. (My own view is that it does not matter whether it is private actors who make polluters pay, through a market for emissions permits, or the government, through a carbon tax, so as long as *someone* makes polluters pay.) For example, the most common objection to cap-and-trade systems in the philosophical literature seems to be simply that the price generated by the emissions-trading system is too low (typically because governments are tempted to issue too many permits).[13] This is not even a principled objection to cap and trade, much less the general principle of carbon pricing. If the problem with the pricing system is that the price is too low, the solution is to raise the price.

It is important to acknowledge that the creation of a pricing system is not *always* the correct approach to controlling environmental externalities. In some cases, it is better for the state to engage in direct prohibition or emissions control. This is the approach that Naomi Klein refers to, in a memorable phrase, as "planning and banning" (or as others have put it, more prosaically, through "mandated technology," or "specification standards").[14] The major feature of this approach is that it is inflexible—the rules take a long time to formulate, they are highly specific, and are inevitably difficult to revise. With some environmental problems, this inflexibility is a desirable feature. As John Braithwaite observes, "The rigidity of specification standards is needed in many areas critical to human and environmental health. It would be intolerable to regulate a nuclear power plant according to the performance standard of how much, if any, radiation escapes the plant. . . . Governments justifiably specify that nuclear plants incorporate the most modern and effective technologies, standard operating procedures, in-process controls, and checks and balances to avert malfunctions."[15]

In other cases, however, inflexibility is a problem. If technology is changing quickly, for instance, mandating specific technology may inhibit development, and in some cases may lock in a higher pollution level than would otherwise be achievable. Sometimes the problem is simply that a set of specific rules are difficult to formulate. In the case of carbon emissions, for instance, fossil fuels are used in so many different ways, in so many different areas of the economy, that it would be impossible to draw up a set of regulations directly controlling their use. Perhaps the only plausible regulation would be a ban on burning coal. Beyond that, however, things become complicated rather quickly. The development of the "sport utility vehicle" (SUV) as America's preferred automobile, for instance, is a perfect example of how rigid environmental regulation can generate perverse effects.[16] The US federal government regulates fuel economy directly, using a typical planning-and-banning approach. Instead of imposing heavy taxes on fuel, the way it is done in most other jurisdictions, the US government instead attempts to prescribe to automakers the fuel economy that their vehicles must achieve. This already creates problems, because there are many different vehicles, of different sizes and with different uses, which makes it difficult to say what a reasonable fuel consumption standard would be. As a result, the United States regulates only the average fuel economy of the entire "fleet" of vehicles sold, allowing automakers to make their own decision about specific models. Even with this flexibility, however, it was necessary to exempt some vehicles entirely. Delivery trucks, for example, and other commercial vehicles, could not be held to the same emissions standard as passenger automobiles. This created a gray region, which included the "pickup" trucks traditionally driven by farmers, which are used for transportation, but also for commercial purposes. As a result, these trucks were classified as commercial vehicles and exempted from the fuel economy standard for passenger vehicles. This turned into an enormous loophole in the fuel economy law, as consumers began to purchase these trucks as passenger vehicles. Automakers were in turn happy to sell these SUVs and to discontinue the older station wagons, because removing the large vehicles from their fleet of passenger vehicles allowed them to sell less fuel efficient vehicles in *every* category. Thus US environmental regulation had the perverse effect of *lowering* the average fuel economy of American vehicles on the road, despite the fact that changes in technology were permitting dramatic improvements in fuel efficiency.

This is just one of many, many examples of traditional environmental regulation having perverse or unintended effects. None of this would have happened if the US government had simply increased fuel taxes, rather than trying to mandate specific fuel economy standards. So while planning and banning may offer a certain moral satisfaction, it is important to recognize its very significant practical limitations. Furthermore, when debating different approaches to environmental regulation, it is important not to think of the state as the institutional embodiment of the public interest, or as a fantasy government that always does the right thing. The track record of actual governments at achieving various policy objectives using different instruments is an important data point when it comes to assessing the merits of these instruments. Vehicle fuel economy is a relatively simple problem. In other domains, it is even more difficult to say precisely how much emission reduction should occur and where. Thus in jurisdictions where the planning-and-banning approach to carbon abatement has been attempted, emissions regulations focus only on the largest industrial emitters, because this is the only area in which the development and enforcement of rules is practicable. To date, every proposed policy alternative to carbon pricing is less comprehensive (in that it ignores a large range of emitting activities), more difficult to enforce, and less effective at targeting actual emissions (as opposed to a set of proxies). The crudeness of these efforts, along with the inefficiencies caused by the price distortions that they introduce, is what underlies the strong consensus in policy circles that a comprehensive carbon-pricing system represents the best way of controlling greenhouse gas emissions.[17]

In order to see why pricing enjoys such strong support in the case of carbon abatement, it is perhaps easiest to consider the circumstances in which the price mechanism is *not* considered an appropriate policy instrument for resolving environmental problems. Braithwaite has offered an excellent summary of these cases:[18]

1. *The pollutant is threshold-sensitive.* This is typically the case when the concentration level of a pollutant matters more than the total quantity emitted. Pollutants such as mercury, arsenic, and lead are not appropriate candidates for pollution pricing, because the damage that they cause are due to local concentration levels. Thus one would not want a large manufacturer to purchase rights to emit large quantities into a particular sink. At the same time, there may be no reason to impose constraints on low-level emissions. Carbon has neither property—it is rapidly dispersed, and so there are no concerns about local concentration levels. At the same time, it contributes to global warming even in small quantities, and so there is reason to want to limit *all* emissions.

2. *Ease of enforcement favors regulation.* In many cases, "mandated technology" regulations are the least difficult to enforce, and thus produce the highest levels of compliance. Actual pollution is, in general, difficult to measure, whereas the installation of appropriate pollution-abatement technology can easily be verified and enforced. In the case of carbon, however, we are not obliged to measure the emissions directly, because we can easily calculate the carbon content of fossil fuels prior to burning them. Furthermore, the liquid fuel supply chain has an important bottleneck, which is that all production passes through a relatively small number of refineries before being shipped to retailers.[19] So even though consumption is highly dispersed, it is possible to price it indirectly by imposing the pricing scheme at the level of refineries. Fossil fuel-based electricity production has a similar bottleneck at the generating facilities.

3. *The loss of deontic force of the regulation reduces compliance.* The punitive nature of a fine should not be confused with the deterrent effect of a tax. The fine associated with violation of an environmental regulation ought not be treated as just a cost of doing business, and indeed, firms expose themselves to criminal liability in some jurisdictions for doing so. Replacing that fine with a tax changes this, and so should only be done in cases where one *wants* firms and consumers to adopt an essentially instrumental orientation toward the regulation. Much of the criticism of tradable pollution permits, which compares them to the sale of indulgences by the Church in the Middle Ages, evokes this concern about loss of deontic force.[20] Digging up carbon, however, is not a sin. It becomes problematic only when we dig up *too much* of it. If the goal is not to prohibit, but just to control quantity, then a pricing solution is perfectly appropriate.[21]

4. *Desire is to control risk, not emissions.* With some pollutants, the desired level of emissions is zero, or the damage is so great that after-the-fact punishment is inappropriate, or the goal of regulation is to prevent risky or unsafe behavior. In all of these cases, an argument can be made for mandated technology, because of its preemptive character. With carbon, however, the desired emission level is relatively high, there is nothing wrong with after-the-fact penalties, and there is no single act so catastrophic that risk management becomes the primary concern (unlike, say, the situation with nuclear power and radiation leaks). All of these considerations speak in favor of pricing.

Considerations such as these have, for most policy analysts, settled the question in favor of carbon pricing, the only remaining question is which way of implementing the pricing scheme is best. Again it is worth emphasizing that there is little that can be achieved with cap and trade that cannot also be achieved with a carbon tax.[22] (For instance, the practice of giving away permits in a cap-and-trade system—which leads many politicians to favor those schemes—can be mimicked by offering polluters a fixed quantity rebate on their carbon taxes that is below their emission level.[23] As long as this does not affect incentives at the margin, it will not reduce the incentive to curtail emissions.) There are, however, certain factors that favor carbon taxation. Each policy instrument basically creates greater certainty in the dimension that it controls directly, along with greater uncertainty in the other. Thus the choice of instrument should be dictated—absent other considerations— by which dimension one cares about most, or is more concerned to get right. Cap and trade therefore provides considerable certainty about what the overall emission level will be, but uncertainty about the price, whereas carbon taxes provide certainty about the price, coupled with uncertainty about how much carbon abatement a given price level will actually produce.

Many people, it should be noted, prefer cap-and-trade systems because of their willingness to adopt the technocratic goal of limiting climate change to 2°C warming, from which they infer the carbon budget and then let the market sort out the price.[24] This focus on the quantity of emissions, however, has several perverse consequences. First of all, there is the simple fact that the climate does not exhibit particularly high sensitivity to emissions, and if there are tipping points, we do not know with any precision where they are. As a result, there is good reason to think that an extra gigaton or two of carbon emissions is not going to make all that much difference to the overall picture. This alone should serve as an early indication that achieving certainty about the quantity of our emissions should not be the dominant priority in climate change policy. The second problem is that cap and trade has perverse consequences during periods of economic recession. In 2009, for instance, in the wake of the financial crisis, the price of carbon permits on the European Trading System declined precipitously. This was due to the overall slackening of economic activity caused by the recession. Under such circumstances, however, a cap-and-trade system has the perverse effect of stimulating increased emissions (by lowering the permit price), in order to keep the overall system on track to hit its quantity targets. Our inclination, however, is to think that a recession, or the slackening of economic activity due to the COVID-19 pandemic, should be used as an opportunity to make even *deeper* emissions cuts. This is because the harms caused by carbon emissions are unaffected by cyclical features of the economy, and thus, it would seem, we should be discouraging them at all times, rather than only during periods of economic expansion.

The standard response to these concerns has been to put a floor on the permit-trading price. However, businesses are also rather averse to the uncertainty generated by the movement of permit prices, which makes it difficult for them to engage in long-term planning in energy-intensive sectors. As a result, the best practice in cap-and-trade systems is now to put both a floor *and a ceiling* on permit prices, so that they cannot fall below, or rise above, certain price points. Imposing these constraints on the trading price, however, makes the cap-and-trade system much more similar to a carbon tax, and suggests that a certain amount of time and energy, not to mention bookkeeping, could be saved just by moving directly to direct specification of the carbon price.

All of this goes to suggest that if it is the harm caused by climate change that concerns us, first and foremost, then we should target that harm directly, by imposing a carbon tax that forces a price adjustment. It is far more important that the full costs of our actions be internalized than that we hit a particular emissions target. In this context, it is worth keeping in mind that not only will people in the distant future be harmed by our carbon emissions, but they will also derive a benefit from our economic activities. The objective of policy is to balance the two, so that the economic activities we undertake are justified, given the costs that they impose upon society. The way that we achieve this objective is by ensuring that the price is right. Controlling the quantity of emissions is just an indirect way of adjusting the price, and so if there is a more direct means of doing so available, it seems only logical that we should take it. Thus carbon taxation is not only a very practical solution to the problem of greenhouse gas emissions, but one that directly guarantees satisfaction of the reciprocity principle that, I have claimed, is at the heart of our system of production and consumption of goods.

It is worth observing that, whatever one makes of the argument I have been advancing here, almost no one in the philosophical literature *opposes* carbon taxes. Most consider it a matter of course that we should institute some form of carbon pricing, but think that this is just a prelude to more serious action, of the traditional planning-and-banning variety. In many cases, this is a consequence of what Joseph Stiglitz has referred to as the "control fallacy," of overestimating the power of centralized, command-and-control institutions, while underestimating the power of decentralized systems.[25] Contrary to popular perception, the ability to influence energy prices is by far the most powerful lever that the state possesses, because it allows it to affect *almost every transaction that occurs in the entire economy*. Not only that, but a carbon tax is easily enforceable, and so it confers effective control of the price. Traditional environmental regulation, by contrast, is extremely weak as an approach to climate change policy, because it is able to influence only a relatively small number of decisions, and in many cases is very difficult to enforce. So while planning and banning sounds much tougher than mere pricing, in practice it is usually a weaker policy tool.

These are, it should be noted, all rather uncontroversial policy prescriptions. Dale Jamieson, for instance, despite believing that climate change reveals nothing less than the failure of the Enlightenment project, nevertheless puts forward a set of policy proposals that fall squarely within the mainstream welfare-liberal tradition:

> The first policy priority is to integrate adaptation with development. The second is to protect, encourage, and increase terrestrial carbon sinks while honoring a broad range of human and environmental values. The third is to adopt full-cost energy accounting that takes into account the entire life cycle of producing and consuming a unit of energy. The fourth is to raise the price of emitting GHGs to a level that roughly reflects their costs. The fifth priority is to force technology adoption and diffusion. The sixth priority is to substantially increase research, especially in renewable energy and carbon sequestration, particularly air capture of carbon. The seventh priority is to plan for the Anthropocene.[26]

There is nothing objectionable on this list. The problem is that most philosophers have normative positions that fail to motivate policy prescriptions in this vein. Jamieson, for instance, after suggesting that the price of GHG emissions should be raised "to a level that roughly reflects their costs," goes on to develop a sustained critique of cost-benefit analysis, concluding that "no number seems right because the costs of climate change damages go beyond economic damages."[27] Thus his sensible policy prescription winds up being undone the moment that he returns to his philosophical reflections. One cannot sensibly demand the implementation of a carbon-pricing scheme, then declare that there is no way to determine what the price should be.

Similarly, Henry Shue is certainly not wrong to draw a distinction between what he calls "subsistence" and "luxury" emissions, and to note that, while the latter are expendable, people are entitled to produce the former.[28] Underlying this is the correct observation that climate change would be a lot more defensible if all of the things that we were using fossil fuels to accomplish were in some way essential to human survival. The problem is that we are also engaged in a wide range of frivolous or wasteful uses, such as driving oversized vehicles whose added bulk serves no useful purpose, or installing heaters on outdoor patios, or disturbing the placid calm of a lake by firing up a jet ski. In other words, a huge amount of our consumption is *wasteful*. The major purpose of climate policy is not to keep the most disadvantaged people in the world locked into a perpetual state of energy poverty, but rather to eliminate the unnecessary consumption that is going on all around us.

These are sensible observations, and yet, from a policy perspective, they all push in the direction of *pricing carbon*. First of all, there is the fact that most emissions fall somewhere on the spectrum between subsistence and luxury, and so drawing the distinction between what is necessary and what is unnecessary involves making a number of very fine judgments. Second, there is the fact that what is necessary to one person may seem unnecessary to someone else. (When talking about luxury consumption, philosophers should perhaps keep in mind that a solid majority of the population regards academic philosophy degrees as a frivolous indulgence.) As a result, it is much better to impose the obligation to cut back upon individuals themselves, who are in the best position to examine their own consumption and decide where to cut back. This is precisely what carbon pricing accomplishes—forcing individuals themselves to decide whether they should drive less often, turn down the thermostat, turn up the air conditioner, eat less beef, attend fewer conferences, or any of the other myriad changes that can have a significant impact on their carbon footprint.

Shue, however, refrains from drawing this conclusion. Putting on his philosopher's hat, he declares carbon pricing to be unacceptable, on the grounds that it does not also guarantee distributive justice in the outcome. Because endowments are unequal, a simple pricing policy might allow wealthy westerners to continue driving their luxury cars while peasants are forced to plant less rice. "Does it make no difference that some people need those rice paddies in order to feed their children, but no one needs a luxury car!?"[29] But what is the alternative? If decisions about where to cut back are not going to be made by individuals, in response to price changes, then they must be made by governments. How are governments supposed to decide what to cut? Shue endorses, with some qualification, the suggestion that economic activity be partitioned into two sectors, agricultural and industrial, and that emission controls target only the industrial. "Better still, if it is practical, would be a finer partitioning that left the necessary industrial activities of the developing countries uncontrolled . . . and brought the unnecessary agricultural services of the developed world, as well as their superfluous industrial activities, under the system of control."[30]

Anyone familiar with the challenges involved in environmental regulation will recognize the absurdity of this proposal. (To point out only the most obvious problem, industry and agriculture combined make up only 20 percent of US GDP, while close to 80 percent is the service sector. Thus Shue's proposal is one that would leave most of the US economy untouched.) Even the regulation of highly dangerous substances, such as asbestos, mercury, or arsenic, which are used only in restricted applications, has required the development of a fiendishly complex system of rules, in order to reflect the varying circumstances of their use. Regulation of carbon emissions, which are largely anodyne, and are produced in literally every sector of the economy, would require something closer to central planning. The idea that state officials—much less negotiators of global climate treaties—are going to start compiling lists of agricultural goods that are deemed "unnecessary," and then prohibit farmers from producing them, is otherworldly in the most pejorative sense of the term. (Even the history of administration of luxury taxes provides many cautionary examples of how difficult it can be to produce an operational definition of simple distinctions, such as that between a luxury car and one that is expensive for some other reason.)

Thus Shue is basically willing to scuttle an efficient carbon-pricing regime in order to push for a disconcertingly crude and politically unworkable regulatory alternative. The problem stems ultimately from the fact that he has a philosophical view that rules out any workable scheme. Prioritizing one's egalitarian commitments over the concern for efficiency, although a common stance in the philosophical literature, in a policy context winds up producing unworkable recommendations. If the goal is to punish the rich for driving expensive cars, this can be accomplished more directly by the tax system (e.g. through luxury taxes); it is not necessary to hold the climate hostage in the process.[31] Furthermore, the inefficiency is not even necessary, because whatever distributive justice concerns one has can be addressed at the "back end" of a carbon-pricing system, in the way that revenue is spent. Thus there is an air of unreality around the entire discussion. It is difficult to see how anyone can seriously maintain, on one page, that climate change is a looming environmental catastrophe and yet, on the next page, reject the most efficient carbon abatement regime, merely because it fails to satisfy an ideal conception of equality that no other institution in the world satisfies. Carbon pricing fails to satisfy distributive justice concerns for the same reason that every other price in the economy fails to satisfy those same distributive justice concerns—because the market is designed to achieve efficiency, not equality. We typically rely upon tax-and-transfer schemes to make market outcomes more palatable from the standpoint of distributive justice. This system can easily be extended to address any concerns about the distributive impact of carbon pricing.

My central objective in the first two chapters of this work was to show that Jamieson and Shue are not alone in having philosophical views that translate poorly into policy advice. This is in fact an endemic problem in the literature. To date, only welfare consequentialists have managed to articulate a coherent normative position that can motivate a policy position that falls somewhere within the realm of the feasible. My objective in this chapter has been to show how contractualism can also motivate such a position. This can be achieved by stepping back from large-scale claims about what we owe to future generations, and focusing on the simple mechanics of externality production, along with the principles that we use to argue for internalization of these externalities in a wide range of environmental cases. Central to this argument is the insistence that climate change be diagnosed as an ordinary collective action problem. The only thing extraordinary about it is the scale on which it is occurring. Because of its scale, climate change stands in some sort of a relationship to almost everything else that is going on in the world today.[32] Both its causes and its effects are, from an empirical perspective, incredibly complex. But this does not mean that the problem is *conceptually* complex. Indeed, focusing on the complexity risks impeding our ability to work toward a solution, by bringing up tangential and peripheral issues that serve as little more than a distraction from the main problem.

4.3. Example: Food

At the risk of belaboring the point, I would like to examine a particular case in greater detail, in order to show how carbon pricing provides an elegant solution to a problem that has been the object of considerable consternation.

A great deal of social activism in recent years has been focused on issues related to food, especially the environmental impact of food production and
distribution systems. This has given rise to a number of food movements,
favoring local production (locavorism), rejection of agricultural technology (organic, anti-GMO), as well as a concern for animal welfare (free
range, vegetarianism, veganism). Although many of these movements are
organized around concerns that are essentially orthogonal to the issue of climate change, many have also claimed the mantle of environmental virtue.
Examining their effects more closely shows, however, how difficult it is to
formulate simple dietary rules that will effectively target environmentally
harmful practices.

Consider, for example, the locavore movement.[33] To the extent that it
has any surface plausibility, it is usually justified by appeal to the environmental consequences of transport, particularly the carbon footprint. "It is
madness to be eating grapes imported from Chile," people say, "instead of
the ones grown in vineyards just down the road." This is the sort of argument that sounds plausible on first pass but cannot survive closer scrutiny.
First, it focuses entirely on the distance that the food has traveled, while
paying no attention to how it was transported, which is what matters the
most from an environmental perspective. As far as wholesale transportation is concerned, the most serious environmental offender is trucking.
Trucks produce 10 times more greenhouses gases, per ton-kilometer, than
trains (180 tons of CO_2/t-km compared to only 18).[34] Trains, in turn, produce about twice as much greenhouse gas as ships (11 for container, 7 for
tankers). Ships, in fact, produce close to nothing by comparison to all other
modalities. As a result, the international dimension of the global food trade
is the least important, from an environmental perspective, simply because
most of it occurs by ship. A person who lives anywhere near a container
port can eat food from anywhere in the world with a clean conscience.
Finally, the locavore movement also fails to focus attention on the most
important source of emissions in food transportation, which is the "last
mile," between the retailer and the consumer's own home.[35] This is the least
efficient segment, because the food is no longer being transported in bulk.
As a result, taking a car or a bicycle to the grocery store winds up being a
far more consequential decision than buying fruit that is grown locally or
shipped in from across the world.

The second major difficulty lies in the assumption that transportation
represents an important component of the environmental impact of food

production. This is largely untrue. How far one's food has traveled is far less important that what kind of food it is, how it was produced, or even how it is cooked.[36] According to one calculation, transportation in North America accounts for only about 11 percent of emissions associated with food supply; the other 89 percent arises from the production process. Whether vegetables are grown in heated greenhouses, for instance, has a much more significant environmental impact than where they came from. Furthermore, different kinds of food produce dramatically different levels of carbon emissions. Red meat and dairy are the worst; they produce about 2.5 kg of CO_2 per dollar spent, compared to less than 1 kg for most other products (fruit, vegetables, poultry, fish, etc.). It is calculated that if the average American reduced red meat consumption by around 20 percent (for example, substituting chicken or fish for red meat once or twice per week) this would achieve a reduction in carbon footprint equivalent to adopting a zero-mile diet.[37]

Thus the idea of restricting diet to locally produced food does not map onto any environmentally significant distinction. The proposal made by some locavore groups that there should be mandatory labeling on food, specifying the distance that each item has traveled, is clearly inadequate from an environmental standpoint. If the objective were to adopt a diet that minimizes carbon footprint, then consumers would need much more de-tailed information than simply a measure of how far food has traveled. For any given item, one would need to know the exact social cost associated with its consumption, which would have to include not just information about how it was transported, but also what went into its production.

But of course, we already have a label on food products that provides us with this information: the price tag. Every time a farmer runs a tractor, or ships grain, or heats a greenhouse, it costs money. The price is supposed to take all of the inconveniences that everyone suffers, in order to get food on the table, and put it into a single measure. If something is too inconvenient, it will cost more, and fewer people will buy it. Thus the fact that one can import grapes from Chile, and they are not outrageously expensive, is the first clue that it is not madness to be importing them from halfway around the world. Fuel is a major cost in transportation. Transport by ship is incredibly cheap, in part because it takes so much less fuel than by truck. Of course, the price that is charged for food in most jurisdictions will be too low, because all of the spending on fuel generates an environmental externality. But the solu-tion to that lies in imposing a carbon tax that prices the externality. Once that is done, then the price immediately gives the consumer all the knowledge that he or she needs to assess the social cost of a good, including its carbon footprint.

Growing awareness of the arbitrariness, from an environmental point of view, of major dietary trends has led to interest in the development of a more targeted low-carbon diet. Considerable effort has been invested in the task of calculating the average carbon-intensity of various popular dietary restrictions, such as pescatarianism and vegetarianism.[38] A more precise way of accomplishing this would simply be to impose a carbon tax and then try to eat inexpensively. Figure 4.2 shows the average expected price increase on foods in different dietary categories (in the United States) from the imposition of a $30/tC carbon tax.[39] As can be seen, the major offenders do not fall neatly into any of the traditional moral categories. Red meat is an obvious target for reduction, but so is milk production and grain farming. By contrast, fish and poultry are quite low, and hunting is the lowest of all (for somewhat obvious reasons). It should be noted that these are just estimates, based on the carbon-intensity of production prior to imposition of a tax. Once the tax is imposed, producers have an incentive to alter their methods in such a way as to reduce emissions, and thus taxes paid. Since opportunities for emissions reduction will be unequally distributed across the production methods of different food groups, these relative price increases will change. Note also that with a carbon tax, producers who succeed in reducing

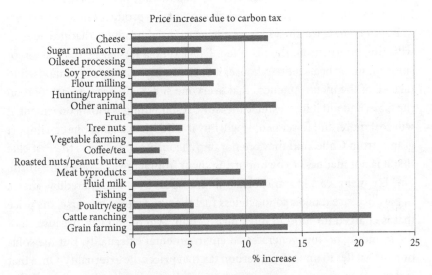

Price increase due to carbon tax

Figure 4.2 Effects of $30/tC carbon tax on food prices

emissions will automatically be rewarded by consumers (as lower prices attract greater sales), in a way that is not the case with regulatory approaches or self-imposed dietary restrictions.

When contemplating the data presented in Figure 4.2, it is worth considering how difficult this fine-grained adjustment of incentives would be to achieve using only regulatory means, much less how difficult it would be to update as circumstances changed (not to mention how vociferously ranchers and dairy farmers would lobby against policies that targeted their products). Climate change is already a motivationally demanding problem to solve, in the sense that doing so will require average citizens to make significant changes in their lifestyle in order to avoid abstract harms that they will never live to experience. The least we can do, under these circumstances, is assure them that none of the sacrifices they are being asked to make are gratuitous or unnecessary. For example, while limiting climate change clearly requires that we eat a great deal less beef, it does *not* require that anyone give up eating chicken or venison. Climate policy should be able to make this distinction. And yet carbon pricing is the only policy that can make an even halfway credible claim to satisfy this constraint. This is one of the major reasons that it enjoys such strong support among policy analysts.

4.4. Complementary Policies

Nothing is ever perfect, and there are a number of issues that remain outstanding even with a carbon tax regime. The most obvious is that pricing deals only with the negative externality of carbon emissions; it does nothing to promote the positive externality of carbon sequestration.[40] Ideally the latter should be subsidized. In some cases this can be integrated into the carbon-pricing system. For instance, fossil fuel-based power plants that invest in negative emissions technologies (NET), such as carbon capture and storage (CCS), can be rebated a fraction of their carbon taxes. In other cases, such as protection of forests from land use changes, there is little alternative but to implement direct subsidies. In the future, the development of more efficient carbon-scrubbing technologies might create an arrangement under which NET firms can sell offsets that can be used to rebate carbon taxes. This is, however, analytically equivalent to an arrangement under which the state just collects carbon taxes and uses some fraction of the revenue to finance NET activities. At the moment, however, carbon extraction is nowhere close to being relevant. The most efficient NET technology available today is called *not burning coal*, which in some parts of the world removes carbon from the atmosphere at a bargain-basement price of less than $10 per tC. It will be a long time before NET is in a position to compete with carbon abatement. Rather than pay several hundred dollars per ton to scrub CO_2, one can achieve vastly greater reductions by paying people not to burn fossil fuel.

There remains also a great deal of room for more specific state initiatives, aimed at overcoming path-dependencies and network effects that lock people into fossil fuel use. For example, despite the clear cost advantages associated with using LED lights over incandescent bulbs, many consumers are put off by the higher upfront costs of LEDs, and so persist in purchasing inefficient bulbs. Instances such as this, where the marginal cost of abatement of carbon is negative, are truly the low-hanging fruit of climate change policy! In this case, a regulatory ban on incandescent bulbs is justifiable on moderately paternalistic grounds. Similarly, fossil fuels enjoy significant network externalities in transportation, because of the refueling infrastructure already in place. An obstacle to the transition to electrically powered vehicles is the unavailability of charge stations. So again, direct subsidization may be warranted in order to create comparable network effects in support of zero-emission technologies.

It is also possible that the imposition of a carbon-pricing regime, while lowering the probability of a climate catastrophe, still would not reduce the chances to zero. Thus it may be prudent to invest some fraction of the tax revenue in a combination of research and disaster preparedness, in order to manage this risk. As I suggested earlier, it does not make sense to engage in broad-based carbon abatement as an approach to catastrophe-avoidance. It makes more sense to investigate the specific scenarios, along with the feedback effects that would generate rapid changes, in order to find ways of detecting them in the earliest stages and if possible disrupting them. Conceptually, however, this should not be confused with the core set of policies designed to mitigate the predictable damages caused by the negative externality of greenhouse gas emissions. While there is a great deal to be said for bundling the two policies together, our concerns over fundamentally uncertain events should not be allowed to cloud the debate over ordinary policies, based on our best estimate of the relevant risks.

Once a carbon tax has been introduced, it does raise the question of what should be done with the revenue (assuming that it is not all spent on NET or research). This is the distributive justice issue that, as I suggested, is secondary to the central question of how the collective action problem should be resolved. (It is also the point that Simon Caney hit upon, when he observed that the only major distributive justice question that arises, in a permit allocation system, is how the proceeds of the permit auction are to be distributed.)[41] One of the less-than-ideal features of carbon taxes is that they are, for now and the foreseeable future, collected by national governments, which puts limits to how much global redistribution one can expect.[42] In a private market, when one person purchases something, the money goes to the person who is giving up the good, ensuring that no one is harmed (and that the outcome is a literal Pareto improvement). When the market price is equal to the social cost, it means not just that the correct amount of the good is consumed, but that all those affected are fully compensated for all of the costs that this consumption imposes upon them. The major deficiency of Pigovian taxes is that, while they bring us closer to the correct consumption level, they are collected by the government and not the individuals involved in the transaction, and so they do not necessarily compensate the people who bear the costs of the consumption. Ideally, the revenue raised by carbon taxes should go to those who are negatively affected by climate change, and in particular, those who must undertake costly adaptation measures. Thanks to the absence of world government, or of any supranational authority capable of imposing taxes or mandating international redistributive transfers, full compensation is unlikely to occur.

Ideally one would have a system of international redistribution, so that tax revenue raised in predominantly northern, developed countries would be used to finance adaptation projects in the predominantly southern, less developed countries, which will experience the most significant ill effects of climate change. There are some international funds (the Green Climate Fund, the Adaptation Fund, the Climate Investment Funds, and the Global Environmental Facility), financed primarily by wealthy countries, that achieve some international redistribution.[43] In 2009 in Copenhagen, countries agreed to a target of $100 billion per year in contributions to such initiatives. These are important objectives, even though they fall well short of what an ideal conception of distributive justice would entail.

Another possibility for transnational movement of carbon tax revenue could occur through subsidization or support of carbon sinks. This could include protection of dense tropical forests against land use changes, and more controversially, implementation of population control policies.[44] In the future it could also include active carbon scrubbing (e.g. in countries with very low-cost solar power). In order for these activities to occur at appropriate levels, those who undertake them must be paid at a level equal to the social cost of carbon. Carbon tax revenue represents an obvious source of funds to finance these subsidies.

It is also worth noting that while increased investment in research and development of new forms of renewable energy will occur naturally as a consequence of carbon pricing, such investment also stands poised to generate significant positive externalities, and so the use of carbon tax revenue to provide additional subsidy to such research is another way in which benefits could be generated that will extend beyond national borders (and into the future). Underdeveloped countries are almost always late adapters of new technology, and so they benefit the most from diffusion of the relevant knowledge. Thus the financing in developed countries of energy research and development, along with absorption of the costs associated with scaling up production of new technologies, followed by low-cost transfer to developing countries, represents an implicitly redistributive arrangement that is likely to generate much less resistance than an explicitly redistributive one.

Finally, it is important not to forget that most of the harm that will be caused by climate change will be in the future. This means that, in principle, the revenue being collected from carbon taxes now should be used to defray the costs of those who will suffer damages in the future. As we have seen, however, there is no forward transfer of funds or goods across time, and so we cannot do intergenerational transfers in the way that we do intragenerational ones. The mechanism we have for compensating people in the future is economic growth. We could, of course, earmark the funds raised by carbon taxes for a special investment fund, used to increase the rate of growth in developing countries, or spend it on durable infrastructure. None of this is necessary, however, because future generations are already expected to derive such massive benefits from general economic growth—especially if we adopt an optimal carbon abatement regime—that we do not owe them very much further. This is a point on which differences in philosophical perspective matter. The fact that future generations are not *entitled* to the baseline rate of economic growth, but instead receive its benefits as a byproduct of our self-interested savings decisions, is important, because it allows us to say that they are already being compensated for the effects of climate change that will occur, without our having to make transfers specifically aimed at dealing with the damages caused by climate change. Thus it is legitimate for states to use carbon taxes simply as a way of raising general revenue, and as an opportunity to reduce more inefficient taxes, thereby promoting economic growth. In particular, there is no reason that a significant fraction of carbon tax revenue could not be used to lower income taxes. This is particularly important in developing countries, where it is sometimes argued that people "cannot afford" to pay carbon taxes at the same level as in developed countries. If these taxes are fully rebated to the population, in the form of reductions in other taxes, it *makes* them affordable for anyone willing to reduce emissions below the average level.

In addition to these distributive justice issues, there are a whole host of compliance problems that arise at the international level, due to the fact that countries that do not adopt a carbon-pricing policy are in a position to free-ride off of those that do. The Kyoto Accord model was to create an international treaty under which states would not only accept emissions targets, but also agree to penalties in case they failed to live up to their commitments. Whereas the United States earned a certain measure of opprobrium for its refusal to sign the treaty, it is actually Canada that undermined the model more comprehensively, by signing the treaty, making no effort to meet its emissions reduction targets, but then withdrawing from the treaty in order to avoid the penalties that it had previously committed itself to. The powerlessness of the international community in the face of such intransigence, along with the fact that Canada suffered little reputational damage from its behavior, had a great deal to do with the subsequent failure of this model in Copenhagen 2009. And yet the Paris framework adopted in 2015, in which it is left up to individual states to determine what emissions reductions they feel obliged to make, is clearly inadequate. It is like trying to pay the restaurant bill for a large party by asking individual diners to contribute what they think they owe (a method of collection that, in my experience, seldom generates more than half the amount required). Thus the international free-rider problem remains an important unresolved issue.

One attractive proposal for addressing it is the idea of a "climate club," defended by William Nordhaus and others.[45] The plan is intended as an alternative to the suggestion that countries with a domestic carbon price should impose a set of tariffs reflecting the carbon content of goods imported from countries that do not price carbon. This is to avoid a situation like that of the European Union, where a significant fraction of the emissions reductions that have been achieved are the result of offshoring production of emissions-intensive goods (such as steel and aluminum) to China. The major problem with this proposal is that the attempt to assess such tariffs would be a logistical nightmare, and the rates could only ever be approximately correct. What Nordhaus recommends, instead, is that countries that are committed to climate change mitigation get together and agree upon an appropriate price for carbon, then form a club. Any country willing to impose that carbon price domestically is welcome to join the club. Members of the club then impose a flat tariff on *all* imported goods from all nonmembers of the club. This creates a highly flexible, decentralized enforcement mechanism for carbon abatement policy, which does not require international agreement in order to be effective.

Without getting into too much detail about the feasibility of these arrangements, the example shows how a single-minded focus on the carbon price, as the essential requirement of justice in climate change policy, can serve as the organizing principle of the international response. The question of who is entitled to produce what quantity of emissions has proven to be incredibly divisive at the international level (as it no doubt would be domestically as well, if any country decided to start parceling out caps on emissions to different constituencies within its jurisdiction). The insistence on a uniform carbon price narrows the basis of necessary agreement considerably, by requiring only a consensus value on the social cost of carbon. The climate club structure even allows countries that disagree to go their own way, and form their own clubs, although the costs involved, in the form of forgone opportunities for trade, are significant enough that one would not expect a high degree of factionalization.

4.5. Conclusion

The arguments that I have made in support of carbon pricing are not particularly controversial. The basic principle of carbon pricing is widely accepted, even among environmental philosophers. What is controversial is the claim that I have been making for the *sufficiency* of carbon pricing. Carbon pricing, in my view, is not a mere prelude to the adoption of more serious environmental policies; it is the policy. Supplementary policies may be required in order to promote carbon scrubbing and sequestration, or to deal with cases in which individuals do not respond appropriately to the incentives provided by the carbon-pricing regime, but there is no need for supplementary carbon abatement policies. There is, of course, no reason to discourage supererogatory actions aimed at reducing emissions beyond what the carbon price would require. But our basic obligation toward others is satisfied once the negative atmospheric externality is fully internalized by the pricing system.

The strongest argument for this conclusion lies in the insistence that climate change is a straightforward example of a tragedy of the commons, which should be addressed using the same policy framework that has been used successfully in related domains of environmental regulation. This is the signal that is unfortunately at risk of being lost amid all of the noise that has been generated on the question. Climate change is an unusual problem only in the number of people that it affects and in the delay that occurs between production of the externality and the realization of the harmful effects. Neither of these affects the basic structure of the problem or its solution. They do, however, raise a narrow problem, about whether we should worry less about damages the further removed from us they are in the future. This issue will be addressed when we turn to the question of the social discount rate in Chapter 6. Apart from this, it is important to recognize that there are no significant differences between greenhouse gas emissions and any other diffuse atmospheric externality.

5

The Social Cost of Carbon

Given the choice of carbon-pricing regimes, between cap and trade and carbon taxes, I have argued that a carbon tax is the better policy. That is because there is no fixed quantity of greenhouse gases that can serve as our target for emissions. The amount that we can permissibly emit depends in part on the value of what we are producing, keeping in mind that future generations will inherit some fraction of both the benefits and the costs of our current economic activities. The reciprocity principle, described and defended in the previous chapter, suggests that it is the social cost of our greenhouse gas emissions that should be the focus of policy. The problem with our current rate of emissions is not that we are emitting too much greenhouse gas, in some absolute sense, but that we are emitting too much relative to the value of what we are producing. We are, as it were, taking more out of than we are putting into the cooperative system. Since it is the task of the price mechanism to create the appropriate balance between these two, the best response to the problem of GHG emissions is to impose a price upon these emissions that reflects their true social cost. The social cost of carbon (SCC) calculation is what establishes the latter.

Critics of the policy stance I am recommending have sometimes observed that the SCC approach is, in fact, a way of soft-pedaling what amounts to the use of cost-benefit analysis (CBA) as a way of determining climate policy. This is true, although the relationship between the two is actually somewhat more complicated than it might first appear, since the SCC value is not actually a statement of the cost, but is rather an indirect way of stating the benefit of any carbon abatement policy. To the extent that the policy reduces carbon emissions by a certain amount, then the SCC value of that reduction is a statement of the damages that will *not* occur, thanks to the policy, which is just another way of stating the benefit. The "cost" of the policy, on the other hand, is slightly more subtle. With a carbon tax, the cost is not the tax paid, but rather all of the economic activity that does not occur because of the tax. Once a carbon tax is imposed, it essentially forces everyone who is considering engaging in a carbon-emitting activity to do a private cost-benefit analysis. If

Philosophical Foundations of Climate Change Policy. Joseph Heath, Oxford University Press. © Oxford University Press 2021. DOI: 10.1093/oso/9780197567982.003.0006

the activity that is associated with the emissions is valued at more than the amount of the tax, then the activity will continue, the tax will be paid, and the benefits expressed by the SCC will not be realized. But if the value of the activity is lower than the tax, then the activity will be canceled. This constitutes a loss, and is thus the "cost" of the policy. But now the benefit expressed by the SCC is realized, since the carbon emissions do not occur. Thus a carbon tax, set to the SCC, is essentially a way of ensuring that each carbon emission decision made, throughout the economy, is cost-benefit justified. This decentralization of the cost-benefit calculation is one of the major advantages of pricing schemes over other regulatory approaches.

Needless to say, CBA is not held in particularly high regard by environmental ethicists, or even philosophers more generally.[1] Nevertheless, the past few decades have seen a steady expansion in the use of cost-benefit analysis as a tool for policy evaluation in the public sector. In 2003, for instance, the *Green Book* was introduced in the United Kingdom, with its requirement that "all new policies, programmes and projects, whether revenue, capital or regulatory, should be subject to comprehensive but proportionate assessment, wherever it is practicable, so as best to promote the public interest. . . . This is achieved through: identifying other possible approaches which may achieve similar results; wherever feasible, attributing monetary values to all impacts of any proposed policy, project and programme; and performing an assessment of the costs and benefits for relevant options."[2] And in the United States, where many progressive groups have historically refused to participate in consultations involving CBA (based on the perception that is part of a broader anti-regulatory agenda), it is increasingly being argued that they need to learn to accept the approach and work toward using it to their advantage.[3]

This slow, steady creep has been a source of consternation to many philosophers and political theorists, who are inclined to view cost-benefit analysis as simply a variant of utilitarianism, and consider utilitarianism to be completely unacceptable as a public philosophy.[4] This interpretation of cost-benefit analysis as a type of applied utilitarianism is, of course, exacerbated by the fact that many of its most prominent defenders are utilitarians (or moral consequentialists), and defend it on the grounds that it approximates what they take to be the correct comprehensive moral view.[5] Furthermore, there is the obvious fact that cost-benefit analysis has at its core a calculation that is consequentialist, aggregative, and appears to presuppose a welfarist theory of value—all positions that are associated with utilitarianism. Finally, and perhaps most problematically, cost-benefit analysis seems to exhibit what many people take to be the central flaw of utilitarianism, viz. that it does not respect the distinctness of persons. This is reflected in a variety of ways, including an apparent commitment to distributive neutrality (the view that, as long as the total quantity of welfare is constant, it is a matter of indifference from the standpoint of society who gets that welfare) as well as its willingness to place a valuation on human life (suggesting that, if one person's death would produce a sufficient compensating benefit to others, then society should be happy to let that person die).

These observations do conspire to present what is, admittedly, a rather damning circumstantial case. The situation, however, is more complex than it may at first appear. The first and most important thing to appreciate is that cost-benefit analysis is not literally a decision procedure, which is to say, it is never applied "baldly" to any particular policy question. It is always embedded in a set of more complex institutional decision procedures, which impose a set of constraints that reflect essentially non-utilitarian concerns.[6] Once this institutional context is taken into consideration, one can see that CBA is best understood, not as a species of applied utilitarianism, but rather as reflecting a commitment to a set of concerns that are shared widely by political liberals. The primary use of CBA is to determine what outcome private individuals would have contracted to—which is to say, agreed upon—absent some market failure.[7] The underlying rationale is therefore Paretian and contractualist, not utilitarian. The basic commitment is to resolving collective action problems, not to promoting social welfare in the aggregate. My central objective in this chapter will therefore be to defend the use of CBA in climate change policy, by showing that its underlying philosophical presuppositions are far less controversial than they have traditionally been taken to be.

5.1. Embedded CBA

It is important to recognize, from the very beginning, that CBA *as institutionalized* is very different from the abstract decision procedure described in introductory public finance textbooks.[8] (I will use the term "bald" CBA to refer to the textbook version, and "embedded" CBA to refer to the procedure within its broader institutional context.)[9] The overall impact of CBA, for instance, is strongly affected by the choice of problems to which it is applied. To take one, particularly obvious example, when "comprehensive" CBA was first introduced in the United States by the Reagan administration, it was applied to all new major regulatory initiatives, but was *not* applied to deregulatory initiatives.[10] So, for example, a proposal to "save drivers money by adding lead to gasoline" would have been exempt from scrutiny, because it could be achieved simply by eliminating various regulations on fuel additives. Yet if later one wanted to reintroduce the ban on leaded gasoline, a full-scale CBA would be required. Thus CBA was introduced with the not-so-subtle goal of making it harder to regulate than to deregulate, which in turn produced the impression that CBA was part of a neoliberal conspiracy to impose "market norms" on government action. Sorting out what is true and what is false in these allegations requires setting aside the partisan and ideological passions that CBA has aroused, which admittedly can be rather difficult to do.

What this example shows is that a *prior* decision about which problems one should apply CBA to can be very important. And while the decision to apply it asymmetrically, to regulation but not to deregulation, reflects little more than an ideological hostility to government, there are many other forms of selective application that have a more principled basis.[11] For example, governments do not apply cost-benefit analysis to programs that are purely, or even primarily, redistributive. It is not difficult to see why. Suppose that one were to propose a new program called "taking money away from the rich and giving it to the poor." Using a standard willingness-to-pay (WTP) willingness-to-accept (WTA) framework for measuring the benefits and costs (respectively), this program would be guaranteed to be at least CBA neutral. (The amount that a person should be willing to pay, to receive a transfer of $1,000, should be exactly the same as the amount that a person would need to be paid, in order to accept a loss of $1,000, viz. $1,000.) Also, because money is subject to diminishing utility, the UK *Green Book* recommends imposing a set of distributional weights on the WTP/WTA amounts, stipulating that benefits to the lowest income quintile ("the poor") be subject to a multiplier in the range of 1.9–2.0, while benefits to the upper quintile ("the rich") be discounted, with a multiplier in the range of 0.4–0.5. Using the higher end of these values suggests that taking $1,000 away from a rich person and giving it to a poor person produces a net benefit to society of $1,500.

Of course, the fact that bald CBA has this consequence should come as no surprise to anyone, since this sort of derived egalitarianism is a well-known implication of utilitarianism (when applied to a world in which money, or consumption generally, is subject to diminishing returns). So the fact that CBA is never applied to this type of problem should serve as the first hint that, underlying CBA, there is not actually a commitment to utilitarianism.

When one looks at the problems that CBA is applied to, what one finds is that they are almost exclusively instances of market failure. In the United States, Executive Order 12866, issued by President Clinton, makes this reasonably explicit with respect to regulations: "Federal agencies should promulgate only such regulations as are required by law, are necessary to interpret the law, or are made necessary by compelling public need, such as material failures of private markets to protect or improve the health and safety of the public, the environment, or the well-being of the American people. In deciding whether and how to regulate, agencies should assess all costs and benefits of available regulatory alternatives, including the alternative of not regulating."[12] Despite a somewhat complex history of subsequent amendment, the impact of this order has been to put the burden squarely upon agencies proposing new regulations to specify the market failure they are responding to.[13] Thus the goal of a regulation cannot simply be to increase social welfare; it must be to solve a collective action problem that private parties are unable to resolve through voluntary contracting. This suggests that the underlying rationale for CBA is actually Paretian (i.e. aimed at producing Pareto improvements), rather than utilitarian.

Given the importance of this distinction, it is worth dwelling on it for a moment, in order to ensure that it is not subject to any misunderstanding. Consider Figure 5.1, which shows the difference between Paretianism and utilitarianism in their application to a two-person distribution problem. The line U shows the set of other social states that contain the same aggregate utility level as the status quo.[14] Utilitarianism therefore ranks any point to the northeast of this *line* as superior to the status quo, while the Pareto principle ranks every social state to the northeast of the *point* representing the status quo—including those strictly north and strictly east—as superior. Informally, this is the set of win-win (or at least no-lose) outcomes. Changes that make one person better off but leave the other worse off are Pareto-noncomparable, and are not ranked by the Pareto principle. Thus it has nothing to say about the more controversial set of proposals, to the northwest and the southeast of the status quo. Utilitarianism makes the controversial claim that it is acceptable to harm one person to the benefit of another, so long as the magnitude of the benefit is greater than or equal to that of the harm. The Pareto principle, by contrast, is about as attractive and uncontroversial as a normative

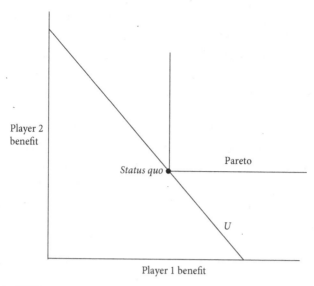

Figure 5.1. Utilitarianism vs. Paretianism

principle can get (although it achieves this through the somewhat dubious means of simply saying nothing about the set of cases most likely to arouse controversy).

In textbook treatments, the way that CBA is usually presented to students is not by starting with the moral case for utilitarianism. Typically the argument starts with the Pareto principle, which is often treated as either self-evident or self-recommending. The argument then moves along quickly to the observation that, under real-world conditions, we are unlikely to find many pure Pareto improvements. For example, no matter how magnificent and useful a particular infrastructure project would be, there always seems to be one landowner who refuses to sell at any price. We do not want it to be the case that a harm to a single person is able to hold up a vast project, which could benefit many others. So we need some way of articulating the idea that the benefits to the many should, on some occasions, outweigh the harm to the few. The Kaldor-Hicks principle is then proposed as an alternative criterion—that if the gains to the winner are great enough that they *could* be used to compensate the loser, then the proposal is ranked better than the status quo. This is sometimes known as the "potential Pareto" principle, because whenever it is satisfied, it means that one could, in principle, carry out the compensation, which would then make the outcome an actual Pareto improvement.[15] CBA is then presented as a method for determining whether a proposed policy satisfies the Kaldor-Hicks principle—if the benefits, measured in monetary terms, are greater than the costs, then the beneficiaries of the policy could, in principle, compensate the losers.

There is often some mild sleight of hand involved in the presentation of this argument. The way that the Kaldor-Hicks principle is presented often makes it sound like only a slight modification of the Pareto principle. Once it is clear, however, that compensation need not actually be paid, it is obvious that the principle is practically equivalent to full-blown utilitarianism (they diverge only in cases where, even though a particular change would increase aggregate utility, compensation is not possible because of differing attitudes toward the medium in which compensation would be paid). This observation has led some critics, such as Amartya Sen, to argue that the Kaldor-Hicks principle is either redundant or unacceptable.[16] If the intention is actually to pay the compensation, then there is no need for a potential Pareto principle, all one needs is the Pareto principle, because after the compensation has been paid an actual Pareto improvement will have been achieved. So in order for the principle to have any purpose, there must not be any intention to pay compensation. But if that is the case, then the principle is subject to all of the traditional objections to utilitarianism. In particular, it seems difficult to justify imposing a loss on someone just by saying, "Well, someone else is gaining more than you are losing."

The major claim I want to make is that, even though bald CBA uses the Kaldor-Hicks criterion, embedded CBA does not (in practice), but is instead used in the pursuit of actual Pareto improvements. Technically, the purpose of the procedure is to rule out proposals that are demonstrably not on the Pareto frontier. Restriction to the space of Pareto improvements is achieved primarily through the choice of problems to which CBA is applied. This can be represented by subjecting the basic calculus to two "screening" procedures, one on the input and the other on the output side.[17] On the input side, policy proposals that are not essentially a response to collective action problems (such as questions about the progressivity of the tax code, or changes to the criminal law), are set aside as unsuitable for CBA. Once the CBA is conducted, then the results are weighed against other considerations (such as rights that might be violated). In cases where there would be significant costs imposed upon some people, one of the most important questions to ask would be whether they will *actually* be compensated. If it is impossible to compensate them (e.g. because they cannot be identified, or because one cannot exclude impostors), then one might decide not to pursue the policy, and so the potential welfare gain would be screened out on the output side. In other cases, it might be decided that compensation ought not be paid, because the original gains were ill-gotten.

Thus the major commitment at the heart of CBA can be expressed in the form of a principle of efficiency:

1. *Pareto improvement.* The state should seek to resolve collective action problems that private actors are either unable to resolve, or unable to resolve at reasonable cost, through voluntary contracting.

This is combined with a second set of commitments, which are central to liberalism, and closely connected to the privileging of the Pareto standard. The first is a commitment to liberal neutrality, broadly construed. The idea is that when it comes to controversial questions of value, where there seems to be some sort of reasonable disagreement, the state should refrain from using its powers of coercion to the benefit of just one party. A great deal of the resistance to CBA among environmentalists is just a downstream consequence of a more general rejection of liberal neutrality, and in particular, the difficulty many have accepting the constraints imposed by the principle of liberal neutrality when it comes to their own, most deeply held values.[18] It is, after all, much easier to say that the state should not take a particular set of values as a basis for policy when they are values that one considers woefully mistaken than it is when they are values that one considers sacred. For example, in the area of environmental policymaking, many of the objections to CBA are, at bottom, objections to the fact that the procedure treats a very broad range of values—including those that instrumentalize nature—on par. Thus there is underlying CBA a commitment to the following:

2. *Liberal neutrality.* The state should avoid adopting policies that can only be justified through appeal to the correctness of particular, controversial values.

There is a close connection, of course, between this principle of neutrality and the Pareto principle. The central attraction of a Pareto improvement is that it brings about a state of affairs that is deemed better for all, each from his or her own perspective—which is to say, according to each individual's own conception of the good, or system of values. Pareto efficiency is therefore the paradigm instance of a normative principle that the state can act upon without presupposing the correctness of any first-order conception of the good, and thus without violating neutrality.

And yet if the state avoids taking a position on the correctness of certain questions, and still it needs to regulate in that area, what is it supposed to do? The answer is that, as much as possible, it should defer to citizens' own judgments about these questions, and to the extent that it does so, it should assign their views and interests *equal* weight. This is very closely connected to the idea of neutrality. In the same way that the state should not just pick a winner among a contested set of values, neither should it privilege some particular subset of these values by assigning them greater weight in its deliberations. Thus the third principle, which also expresses a very standard liberal commitment, is to a certain form of equality:

3. *Citizen equality.* In areas of reasonable disagreement, the state should assign equal weight to the values and interests of each citizen in deciding which policies to pursue.

It is this commitment to equality that explains one of the most controversial features of CBA, which is that it uses money as a metric to compare the magnitude of the gains and losses that would be produced by a given policy. It is a well-known feature of utilitarian moral systems that, in order to aggregate happiness, they require some basis for making interpersonal comparisons of utility. The difficulty of finding such a basis is often thought to provide a powerful argument against utilitarianism. Less seldom noted, however, is the fact that egalitarianism also requires some way of making interpersonal comparisons, since in order to say that two people are equal with respect to some endowment, one must be able to say how much each person possesses. If the endowment in question consists of utility, or welfare, then the egalitarian confronts the exact same problem as the utilitarian. If the endowment involves a complex bundle of goods, then the problem is only slightly different, because a person may have more of one good and less of some other, so in order to compare this to the endowment of another person, one must be able to come up with some sort of score that reflects how much the entire bundle is worth. This is not a metaphysical exercise, but rather a pragmatic one—the purpose is not to say anything deep about the true value of each person's endowment, it is simply a matter of finding some way to compare these endowments against one another, in order to say whether, or when, equality has been satisfied.[19]

One of the most important claims I will be making in this chapter is that the use of money as a metric of value in CBA is not a consequence of any underlying commitment to utilitarianism, neoliberalism, or economism, but rather a consequence of an underlying commitment to equality. The need for CBA typically arises when the state must choose between employing some bundle of resources to achieve one outcome rather than some other, in which many people want x while many others want y. This may reflect just a difference in interests, but it may also reflect a deeper disagreement about questions of value. (Consider, for instance, the conflict between those who want wilderness preservation and those who favor resource extraction.) Neutrality prevents the state from simply declaring one set of values to be correct, while equality commits the state to assigning equal respect to both concerns. But how is this to be done? Treating citizens equally does not mean assigning exactly the same weight to their interests, because citizens themselves regard some interests as more important than others. Indeed, one of the problems with voting as a decision procedure is that it gives equal weight to everyone's preferences, even though some may have a great deal at stake in a particular question, whereas others have only a trivial interest. A better procedure would be one that recognizes that interests have a certain weight, within each individual's system of values, and so respects individual equality by assigning interests that have comparable weight the *same* weight in the decision procedure regardless of who holds them.

In order to do this, it will be necessary first to figure out how strongly people feel about a particular question, or how much they care about the values that are generating the conflict. One strategy for getting this information is to ask them what they would be willing to sacrifice, in order to get their way—or more specifically, how much they would be willing to give up of something else, in order to get their favored outcome. If one can find a "something else" that one has good reason to think will be valued roughly the same by everyone, then it can be used as a metric of value, to ascertain the relative intensity of the value commitments in conflict. There are various candidates for such a metric, but one that has many attractive qualities is *money*. This is the primary reason that the costs and benefits in a CBA are expressed in monetary terms.

Each of the three principles listed above has been formulated in a way that is intentionally loose (for example, through inclusion of a "reasonability" constraint). That is because my goal is not to provide an airtight formulation of the central tenets of liberalism. On the contrary, my goal is to show that the commitment to a vague form of welfarist liberalism, of the sort represented by these three principles, is sufficient to motivate the use of cost-benefit analysis in most if not all of the policy domains in which it is currently being applied. This is, I will argue, what explains its appeal—departures from CBA seem to violate one of these three principles, and are, to that extent, widely perceived within policy debates to be unacceptable (or else motivated by political ideologies that are, at some level, illiberal).

It is worth noting, perhaps just in passing, that none of these three principles involves any commitment to consequentialism. While most consequentialists—particularly utilitarians of one stripe or another—would be inclined to endorse them, all three principles can more naturally be justified in a contractualist fashion. Of course, the principles create a normative structure within which consequences will be *relevant* to the evaluation of a policy. A collective action problem, after all, is defined in terms of the consequences of individuals' maximizing choices. This does not mean, however, that the Pareto principle must be justified through appeal to these consequences (and nothing *but* these consequences, as the consequentialist would have it). As we have seen, it could be justified by its capacity to bring about agreement.[20] Thus the framework that is presented here is perfectly compatible with all but the most implausibly strict deontological moral views. This is just to accentuate my basic point, which is that there is a lot more daylight between CBA and utilitarianism than is commonly realized.[21] Indeed, to use the Rawlsian term, one might say that CBA is a good candidate for being a "freestanding" component of a more general conception of justice for a liberal state. (There is some irony in this, since one of Rawls's stated objectives, in the early passages of A *Theory of Justice*, was to displace utilitarianism from its dominant position in public decision-making. This was normally taken to imply opposition to CBA. My claim here is that the basic conceptual framework of Rawls's work, far from displacing CBA, actually provides support for it.) Rawls's difference principle is of course in tension with CBA. But if we take Rawls's more abstract conception of social cooperation, and the approach to justice that it implies, it leads fairly naturally to cost-benefit analysis as an approach to state regulation.

5.2. Basic Principles of CBA

The aversion to economism is so great, in some quarters, that CBA is often dismissed summarily on the grounds that it imposes a market logic on government decision-making. Setting aside whatever normative implications critics might think follow from this claim, the descriptive characterization is not entirely wrong, although somewhat backward. CBA is, in a sense, a "market-simulating" exercise, in that, confronted by a market failure, an attempt is made to determine what outcome would have been selected had that market failure not occurred. If the market underproduces positive externalities, and so the state decides to provide them, it uses a CBA in order to determine the level of provision that would obtain if the externality were internalized. When the market overproduces negative externalities, and so the state decides to limit them, it also uses a CBA in order to determine the level of production that would prevail if the externality were internalized. In neither case is this because the market, as such, is taken to have any particular moral authority. On the contrary, the normative authority comes from what I have been referring to as the reciprocity principle, which the market represents an attempt to institutionalize. When that institutionalization fails—or fails egregiously—an attempt may be made to institutionalize it in some other way, such as a regulatory intervention, or state provision of a good. In such cases, the same type of calculation must be performed, explicitly, that markets perform implicitly, in order to determine the level of provision.

To take a simplified example, suppose that a municipality comes into possession of some land (perhaps it seizes a derelict property for nonpayment of taxes) and must decide what to do with it. Two proposals are made for its use: first, the land could be cleared and converted to a neighborhood park, or second, it could be sold to a developer. It is the responsibility of the planner to decide which option constitutes the best use. If the only two options were between selling it to a developer who wanted to build a shopping mall, and selling it to a developer who wanted to build a condominium complex, then there would not really be any difficulty. One could simply sell it to the highest bidder, knowing that in so doing, one would actually be ensuring that the land was being put to its best use. The underlying question is what "people" want most. The market is the mechanism through which their desires are transmitted and aggregated. If there are a lot of people who really want to live in that neighborhood, and there is not enough housing, relatively speaking, then they will be willing to pay a lot for the condominiums, making that venture the more profitable one. If there is instead a shortage of retail stores in the area, then the shopping mall will be the more profitable venture, and so the commercial real estate developer will be willing to bid more for the land. Competitive bidding for the resource is a method for discovering what constitutes its best employment. The procedure is subject to numerous, well-known difficulties, but it also has some significant advantages, which can best be appreciated by considering the alternatives to market allocation.

A noticeable feature of the property market is that there are never any developers lined up to buy land and convert it to public parks. The reasons for this are obvious—they could never get their money back, because they could not charge admission, or if they did, very few people would come. Thus much of the benefit associated with the construction of a public park takes the form of positive externalities. A public park, like the public beach or a public festival, is a distinct type of good, where part of the enjoyment that people get from its consumption arises precisely from the fact that entry and exit is free and open to all.[22] Because of this, markets will systematically underproduce these goods (this is the "market failure"). We refer to them as public goods in the informal sense, of being *relatively* non-rival and non-excludable. Voluntary contribution will also not generate the optimal level of provision, because of free-rider problems (and the fact that it is impossible to identify some potential users—such as tourists, or people who live far away but occasionally pass through the neighborhood). The desire that people have for a park will not be transmitted faithfully into market demand for a park. Even *actual* willingness to pay for a park through a contributory arrangement may fail to generate market demand. Thus competitive bidding for the resource will produce an undersupply of public parks.

The fact that the good generates this problem is in one sense quite accidental. The park itself has no special moral property, such that it must be provided by the government; it is just an ordinary good that people enjoy, which happens to be very difficult to organize the provision of through any institution other than government. Thus there is no reason to think that there is a morally correct level of provision, other than what people themselves want to consume. After all, there is no point providing parks if people are not going to use them (just as there is no point lowering people's taxes if what they would really like to buy is more parks). And if some strange technological intervention were to make the underlying market failure go away (or if some burst of public-spiritedness made voluntary financing by "passing the hat" feasible), there would be no special objection to allowing market forces (i.e. supply and demand) to determine the level of provision.

Thus what the state needs to engage in, when deciding whether to build the park, is essentially a market-simulating exercise. It needs to produce the outcome that the market would have produced had the market failure not occurred. (Again, this is not because the market is special, but because the market is a general mechanism for channeling resources to their best employment, and in this particular case, the state shares the same objective.) Of course, we know how much people want the condominium project, because this takes the form of actual market demand, which is what generates the price that the developer is willing to pay. What we do not know is how much people want the park, because it does not show up as market demand. The only way to get at it is to try to ascertain a hypothetical willingness to pay. This can be done using a variety of methods, the main ones being the stated preference method (e.g. do a telephone survey of the neighborhood, asking people what they would be willing to pay to have a new public park) and the indirect valuation (or "revealed preference") method (e.g. do a comparative study, to look at what impact public parks have on property values in surrounding neighborhoods—this allows one to infer how much people are *actually* willing to pay for access to parks).

Once this is done, we must compare one value to the other, to see which project is more desirable. But this is, of course, just what it means to do a cost-benefit analysis, since the "cost" of any one project is just its opportunity cost, which is to say, the benefit that is forgone when a decision is made to *not* do any of the alternative projects. The cost of building a park is the fact that the land is then not used for private development, which is reflected in the price of the land, which constitutes the revenue that the state will *not* be receiving if it decides to build the park. Thus the focus on "cost" in CBA is potentially misleading: when performing a cost-benefit analysis, what one is really doing is comparing the benefits of a particular project to the benefits of the presumptively second-best use of the resources, in order to ensure that the former is greater. Thus it could be called "benefit-benefit comparison" or "relative benefit assessment," which would in some ways be more confusing, but would have the advantage of emphasizing the fact that the comparison of costs and benefits in a CBA is not the same thing as the balancing of "gains to the winners" and "losses to the losers" that one would see in a utilitarian calculus.

This characterization of the decision procedure—which is admittedly not the standard way of framing it—can help us to see the shortcomings of many of the standard criticisms that are made of CBA. For instance, perhaps the most global and persistent critique of CBA is the argument that it "commodifies" public goods and therefore represents a fundamentally wrong-headed way of thinking about goods that are, as Elizabeth Anderson puts, "not properly regarded as mere commodities."[23] (Similarly, Frank Ackerman and Susan Heinzerling criticize CBA on the grounds that, "by monetizing the things we hold most dear, economic analysis ends up cheapening and belittling them.")[24] There is no doubt some basis for suspicion here, since when calculating a WTP value, then using it as the basis for a decision, we are in effect asking, "What would the market do?" where "the market" in question is a hypothetical market in which all goods are available, and all costs and benefits are fully reflected in their prices. Yet if all that it meant to "commodify" or to "monetize" something was to subject it to such acts of the imagination, then it is not clear where the harm would lie in "commodification." After all, no one is talking about turning public parks into actual commodities. The plan is still for it to be provided as a public park, free and open to all. Asking people "How much would you pay for a new park?" as a way of deciding whether to build it, or how large it should be, does not really change anything about the eventual status of the park as a public good.

This suggests that the commodification charge needs to be made a bit more carefully. When it comes to debates over, say, the buying and selling of transplant organs, the question is whether a particular good should become an *actual commodity* that can be bought and sold on private markets. But when the government performs a CBA in order to determine whether to build a park (or, for example, to decide whether the public health service should provide a vaccination without charge to the public), there is no question of actually commodifying the good in question.[25] Just thinking about the good as if it were a commodity is not itself a harm, and does not violate any taboo against the buying and selling of it. In order to show that there is harm, those who make the criticism need to show that this way of thinking will distort decision-making, leading to incorrect levels of provision of the good (including perhaps, in some instances, failure to provide it at all), compared to some other decision procedure.

This is what Anderson tries to show, although there are significant problems with her argument. She starts by providing a helpful definition of what she means by commodification: "A good is treated as a commodity if it is valued as an exclusively appropriated object of use and if market norms and relations govern its production, exchange and distribution."[26] From this definition, it seems clear that the park is not being commodified. But she then goes on to claim that CBA "measures people's valuations of these goods in market transactions and hence, only as they are valued as privately appropriated, exclusively enjoyed goods. This assumes that the public nature of some instances of these goods is merely a technical fact about them and not itself a valued quality. The possibility that national parks . . . might be valued as shared goods does not enter into its evaluations."[27]

This argument provides a good example of a problem that makes the philosophical literature on CBA so difficult to navigate, which is that critical discussions are so replete with false or misleading claims.[28] It is simply not true that calling someone up on the telephone and asking, "How much would you pay to have a public park in your neighborhood?" is the same as asking, "How much would you pay to have a private park that only you can use in your neighborhood?" (this would be equivalent to asking, "How much would you pay for a larger backyard?"). Apart from the fact that one would not need to ask this (one could just look at how much people actually pay for larger backyards), what one is asking about specifically is the private value (i.e. the value to the individual) of a *public* amenity. It is not the case that CBA tries to estimate the value of public goods as if they were "privately appropriated" or "exclusively enjoyed."

The same is true of revealed preference methods, such as looking at property values in neighborhoods with and without parks. Even though one is looking at the value of private commodities, one is doing so as a way of estimating the value of public amenities. It is a well-known feature of the real estate market that houses in "nice" neighborhoods sell for a lot more money than houses in not-so-nice neighborhoods. Many cities have a housing stock that is architecturally fairly homogeneous, so it is possible to find what amounts to the same house in a number of different neighborhoods. Because the "private good" (i.e. the actual house and yard) in each case is the same, the differences in price must be due to externalities, both positive and negative. With a good data set and sufficiently refined statistical techniques, this allows one to determine how much these externalities contribute to the value of the home, and thus, how much people value them. (This will include public goods, such as proximity to good schools and clean parks, the availability of transit, etc. as well as public bads, such as crime, the noise from a nearby freeway, traffic congestion, etc.) This allows one to determine how much people really care about these various things (so that if it turns out that access to transit makes a huge difference to property values, while parks only modestly so, it might make sense to invest more money in transit, rather than parks). There is simply nothing commodifying about this. More generally, there is nothing paradoxical about the idea of eliciting people's private valuation of a public amenity, or in thinking that the *public* valuation of public amenities should be a function of these private valuations.

It should be mentioned that, when the numbers come back, the results can sometimes put pressure on one's commitment to liberal neutrality. That is because the planners who do these studies all tend to belong to roughly the same social class, and therefore tend to have views on these questions that are both similar to one another and strongly held. For example, there is a marked tendency to think that people ought to value neighborhood green space over proximity to a freeway.[29] So when it turns out that people in certain neighborhoods have the "wrong" preferences, according to one's own perfectionist values, there is a temptation to want to impugn the methodology of the study, or the validity of a decision procedure that assigns those "incorrect" preferences the same weight as one's own. It is important to observe, however, that the problem here lies not with the CBA; the source of the tension is actually the more fundamental liberal commitments that are embodied in CBA. No matter how passionately one may be committed to the virtues of urban green space, the fact that not everyone shares this passion means that one must be willing to compromise, and should not seek to impose one's values on others.

5.3. CBA and Regulation

The case of a public park is in certain respects atypical. Since it involves the provision of a public good, it is not difficult to imagine circumstances in which it would be a Pareto improvement. Most applications of CBA, by contrast, involve regulation, which appears to have much more of a win-lose structure. Consider the conflict between a factory that is polluting the river with mercury and the people living downstream who would like to be able to go fishing and eat their catch. Any restrictions on pollution that are introduced are clearly going to impose a cost on the factory, while generating a benefit for the residents living downstream. The question that CBA can be used to answer is whether the benefit to residents is large enough to justify the cost imposed on the business. This seems very much like a utilitarian calculation—and indeed, it is often described that way (e.g. "regulations seek to counteract externalities by restricting behavior in a way that imposes harm on an individual basis but yields net societal benefits").[30] These appearances, however, are misleading. The purpose of CBA, when examining a regulation, is actually to determine where the Pareto efficient outcome lies (and thus, whether the status quo arose only through market failure). The same is true with respect to climate change and carbon taxes, although the point is a very subtle one, and so needs to be addressed with care.

A good place to begin is with Coase's argument, in his 1960 paper "The Problem of Social Cost," which laid much of the groundwork for contemporary thinking about regulation.[31] One of Coase's ambitions in this paper was to criticize the assumption, made by Arthur Cecil Pigou, that *merely* because there is an externality being produced in a particular market, that the resulting outcome must be inefficient.[32] Coase showed that market outcomes may still be efficient, even in the presence of externalities, depending on how important those externalities are. He illustrated this with a classic example of a railroad track that passes through several farmers' fields. The faster the trains travel, the more sparks they throw off, and the more likely they are to set the farmers' crops ablaze. Under the status quo, the sparks are a negative externality, the cost of which is manifest in the form of crops lost to fire. There are two solutions to the problem: either the railroad could run the trains slower, or the farmers could increase the setback between the railbed and the land where their crops are grown. Naturally the farmers prefer that the trains run slower, while the railroad prefers that the setback be increased.

Coase used this example to make one obvious point, which no reader has failed to grasp, and several far less obvious ones, which are sometimes overlooked. The obvious point is that one need not necessarily impose a *regulation* in order to resolve this problem; all it would take is a clear assignment of rights. Less obvious is his observation that, even if the rights are assigned to the farmers, it does not necessarily mean that production of the externality will stop. Suppose, for example, that the farmers would suffer a loss of $80 from increasing the setback, while the railroad would suffer a loss of $100 by slowing down its trains. If one grants the railroad the right to run its trains as fast as it likes, then obviously it will continue to do so, and farmers will have to absorb the $80 loss. But if instead one gives farmers the right to have their fields protected from sparks, it actually won't change the outcome—the railroads can obtain permission to run the trains fast by negotiating with the farmers and offering to pay them $80 to increase the setback. Either way the parties can be expected to contract to the efficient outcome. The difference between the two outcomes is purely one of *distribution*—which party has to bear the $80 loss. If the right is given to the railroad, the farmers bear the loss; if the right is given to the farmers, then the railroad has to bear it (by paying the farmers), but in both cases the externality will continue to be produced.

This separation of the distribution issue from the efficiency issue is the first rather counterintuitive point that Coase makes.[33] The second point is that the mere presence of a negative externality (in this case) is not necessarily inefficient; it all depends on how much the externality costs the person who suffers it, and how much it benefits the person producing it. In the case of the railroad and the farmers, the status quo ante, with the production of the sparks, is the efficient outcome—as witnessed by the fact that, even if the farmers are given the right to block production of the externality, they would choose not to do so, by selling the right to the railroads. By contrast, imposing a speed limit on the trains, even though it would eliminate the externality, would create an inefficient outcome. So while the presence of the externality creates what many people are inclined to regard as a *distributive injustice* (i.e. the farmers are given no compensation for their lost farmland or ruined crops), it was not actually producing an inefficiency.

Now there are many circumstances in which externalities are being produced, but the parties will not be able to negotiate a solution (typically because of transaction costs, understood broadly). If the railroad is hundreds of kilometers long, for instance, it would be very costly and difficult for all the farmers to get together to bargain collectively with the railroad. Thus the state cannot just assign rights and let the parties decide; it must actually impose an outcome. In doing so, it should be guided by some conception of what the parties *would* have decided, had they been able to negotiate freely and costlessly. First and foremost, it must decide whether the parties would actually have negotiated a limit on the externality (and thus, whether the state should even regulate at all), and then, if it determines that they would have agreed to reduce it, what level of output they would have settled on. This is precisely what a CBA does. The important point is that this calculation is independent of the distributive justice decision. For instance, if the state decided to look at the situation between the farmers and the railroad, with an eye to regulating the externality, a CBA would recommend against it, on the grounds that the cost of slowing down the trains ($100) exceeded the benefit ($80). It might, however, also decide that the farmers were owed some compensation by the railroad, but this would be on the basis of separate, distributive justice considerations, not the CBA.

The important point, from Coase's perspective, is that we should not use our judgment about the distributive question as the basis for the decision whether or not to regulate. The naive tendency, when one sees the trains producing sparks, is to conclude that, since the railroad is harming the farmers, the state should prevent production of the externality. In reality, what we refer to as an externality is always the joint product of an interaction between the parties. In one sense, the railroad produces an externality for the farmers by running the trains too fast, but in another sense, the farmers produce an externality for the railroad by growing their crops too close to the track. Either party can eliminate the externality by changing its behavior. The question of who should be obliged to change behavior, Coase claims, should not be answered by any conception of prior entitlement, but rather by asking which party can do so at least cost. The latter is the outcome that the parties would have contracted to.

In the case of the polluting factory and the residents downstream, the logic is much the same. Suppose that in this case, the residents would be willing to pay $100 to be able to eat the fish, while it would only cost the factory $80 to eliminate the mercury from its effluent. The problem could be eliminated in either of two ways: the residents could stop eating the fish, or the factory could stop polluting the river. In this case, however, the prevailing state of affairs, in which the factory poisons the river, is inefficient—there is a mutually beneficial (i.e. Pareto-improving) transaction between the residents and the manufacturers that could be taking place, but is not happening, because of incompleteness in the system of property rights (i.e. the residents could pay the factory $80 to eliminate the mercury, leaving them with a welfare gain worth $20). Thus the inefficient state of affairs, in which the factory pollutes with impunity, persists only because of a market failure. This is precisely what the CBA shows when it comes back positive. So if the state prohibits the pollution through regulation, it is realizing a Pareto improvement over the true status quo ante, in which it is not yet determined whether there will be pollution or fishing.

The regulation, of course, creates only a *potential* Pareto improvement over the existing state of affairs, because it imposes the cost of pollution abatement on the factory without requiring any transfer from the residents. The question of whether or not to turn it into an *actual* Pareto improvement is a separate issue, which must be decided by considerations that are relevant to the distributive question, such as whether the factory was entitled to produce the pollution in the first place. This might come down to something as simple as who was there first. In general, whenever the state regulates, it has the option of compensating certain parties, based on who should be entitled to do what. In the case of these classic pollution externalities it usually does not compensate, on the grounds that the rights would typically not have gone to the emitter, but rather to those affected, and so the cost impact of the regulation is more like a seizure of ill-gotten gains.

Stated more technically, the decision whether or not to regulate, and if so how much, is motivated by (Pareto) efficiency concerns, and efficiency concerns alone. The subsequent decision, whether to compensate those who are adversely affected by it, is a separate, distributive justice decision. Put the two together, and it looks like a utilitarian decision, but it is actually two separate decisions, one about efficiency, the other about distribution. If there is a loss imposed on one party, this is due to the distributive justice decision. So if the party that suffers the loss were to complain to the government, demanding compensation, the response would not be, "You are not owed compensation, because your losses are less than the gains to the winner" (i.e. state policy is guided by utilitarianism). It would be, "You are not owed compensation, because you should never have been allowed to do what you were doing in the first place."

Figure 5.2 sketches out the situation. The two diagonal lines show the benefits that are obtainable under the two scenarios (regulation or no regulation), and the possible distributions of those benefits between the factory and the residents. One can see that any arrangement that features an absence of regulation is Pareto-inferior to a set of arrangements in which

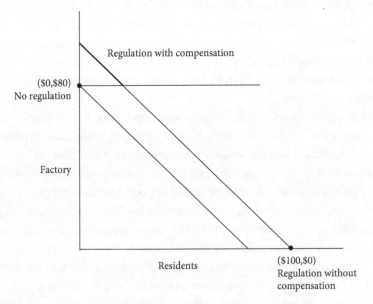

Figure 5.2 The decision to regulate

there is pollution regulation, and thus the parties, had they been in a position to negotiate an arrangement, would never have settled on one in which the pollution was produced. Nevertheless, if one takes the "no regulation" outcome as the point of departure, then the move to regulation without compensation is clearly not a Pareto improvement; it is only a Kaldor-Hicks (or utilitarian) improvement. Starting with this outcome, however, begs the question, because the factory should never have been entitled to produce the pollution—and absent the market failure that prevented negotiations with the downstream residents, it might never have done so. Thus the status quo point from which all regulatory decisions should be assessed is the outcome in which neither party is exercising any of the rights that are in question (or the "status quo prior to the interaction"). From that point of departure, one can see that if one is motivated by the Pareto principle, one would never choose an outcome in which the factory was allowed to pollute. Thus even if the state decides to impose the emissions regulation without requiring compensation, it can do so without making any essential appeal to the Kaldor-Hicks principle. Instead, there are two decisions going on. The first decision, to impose the regulation, is justified by the Pareto principle. The second decision, not to compensate, is made on the basis of a distributive justice consideration.

So despite appearances, the state is really not doing anything very different when it imposes a regulation than what it does when it decides to build a park. In both cases it is trying to bring about an outcome that would have been brought about if the parties had no opportunities to free-ride. Provision of public goods is a response to the real market's underproduction of positive externalities (or goods that produce positive externalities); regulation is typically a response to the market's overproduction of negative externalities (or goods that produce negative externalities). Both interventions are aimed at correcting market failures, and thus are justified by the Pareto principle. They involve asking exactly the same "What would the market do?" question. The fact that someone has been taking advantage of a market failure, illegitimately imposing costs upon others, does not defeat that logic. The case is like that of two players confronting a prisoner's dilemma, in which one person is abiding by a cooperative agreement while the other is free riding. If the state comes along and imposes the cooperative solution, then naturally the transition from the free-rider outcome to the cooperative outcome will not be a Pareto improvement. The state's rationale, however, is still one of Pareto efficiency. The goal is to impose the outcome that the parties contracted to. The fact that one party had been successfully free riding should not be allowed to cloud the issue, much less block the transition.

5.4. Objections and Replies

There are many objections to the use of CBA, some of which are too compli-cated to get into here. I would, however, like to discuss several that come up often in the discussion of environmental policy, and of climate change in par-ticular, and offer what I can in the way of response.

5.4.1. The "Garbage In, Garbage Out" Objection

Proponents of CBA sometimes defend the procedure on the grounds that it is more "rational" than the alternatives, or at least less likely to be subject to cognitive bias.[34] There is a lot to be said for this argument. If the primary al-ternative to CBA is some sort of political decision-making process, and one looks at the quality of political deliberation in most Western democracies, it is not difficult to show that CBA will be much less susceptible to bias.[35] And yet these arguments tend to cut both ways. The value of CBA as a proce-dure depends entirely upon the quality of the valuations that it uses as inputs, which it must take as given. Thus it is susceptible to a version of the "garbage in, garbage out" principle, which states that no matter how good a procedure may be, if the input is worthless, then the output will also be worthless. And if cognitive bias is such a serious difficulty that it impairs ordinary deliberative decision-making, as proponents of CBA such as Cass Sunstein claim, then what reason is there to hope that the WTP values that are taken as input will not also be hopelessly compromised by many of the same cognitive biases?

This is the central contention made by Peter Diamond and Jerry Hausman, in an influential article entitled "Contingent Valuation: Is Some Number Better Than No Number?"[36] They make a number of extremely important observations. The first is that calling people up on the phone and asking, "How much would you pay for x?" is often worthless as a way of ascertaining the value of an outcome. In particular, it is subject to all of the familiar biases, such as framing effects, anchoring effects, etc.[37] Perhaps the most damning example of this comes from a well-known study that showed that the WTP values for rescuing birds was roughly the same, regardless of whether the number of birds to be saved was 2,000, 20,000, or 200,000.[38] Similar problems afflict "revealed preference" methods. For example, the idea that workers' valuations of life and injury will be revealed through the wage premium associated with dangerous employments is solidly contradicted by observation.[39] There are no doubt many factors that contribute to this, but one of them is cognitive bias. Workers suffer from optimism bias and control illusions, which result in widespread underestimation of the chances that they will be victim of a workplace accident.

It is disingenuous for proponents of CBA to respond to these concerns, as they sometimes do, by pleading modesty, and suggesting that their analysis is only one consideration in a complex decision-making process, which no one should assign greater weight to than is warranted by the quality of the underlying data. There are two things wrong with this response. The first lies in a failure to acknowledge that numbers themselves can have a biasing impact (as studies on the anchoring effect have shown), leading to quantitative measures having an outsized role in deliberation, even when this is unwarranted by the reliability of the underlying data. The second important consideration is the fact that, in certain jurisdictions, and in certain policy areas, all major regulations will be litigated, and so must be defensible in court. In an adversarial context such as this, quantitative measures are much more defensible than qualitative ones, and so if both are available, courts will gravitate toward assigning the most authority to a CBA, even if the numbers that it is based on are little more than a guess.

Because of this, the number provided by CBA is not just one data point among many. What Diamond and Hausman refer to as the "some number is better than no number" fallacy arises in part from the fact that information of equal quality will often be given greater credence, or deliberative weight, when presented in a quantitative form rather than a qualitative.[40] Thus if a particular decision requires weighing two considerations, people are more likely to achieve an appropriate balancing if both considerations are expressed quantitatively or both are expressed qualitatively. But if one is presented quantitatively, and the other qualitatively, the former is likely to be overweighted. In this case, it is better to have no number than to have some number.

Thus the mere insistence upon methodological transparency is not an adequate response to the "garbage in, garbage out" objection. A better response lies in the observation that in almost every case in which CBA is used, there is already at least one number available, namely, the number that reflects the *cost.* This will be true, for example, whenever a regulation is imposed on economic activity, or whenever resources are directed to the provision of a public good. With a public park, for instance, the cost is easy to calculate, since it is equal to the market value of the land. Most often, it is the *benefit* that is intangible, because it lacks a market valuation (which is why the state must act to secure its provision). Thus the question is not, as Diamond and Hausman suggest, whether we want to have one number or no number, but whether we want one number—reflecting only the cost—or two numbers—reflecting both the cost and the benefit. The logic of Diamond and Hausman's own position suggests that if there is always going to be one number available, and furthermore, if this number is always going to be on the same side of the scale, then deliberation will be enhanced by generating a second number in order to balance out the effects of the first.

From this perspective, the one thing that can be said for the WTP value is that, whatever its shortcomings—and I certainly have no desire to understate these—it represents at least an *attempt* to measure the right thing. When market prices are used to calculate the cost of a policy, what is essentially being measured is the aggregate willingness to pay for the associated resources in some other employment (keeping in mind that the cost of a particular policy is its opportunity cost, which is simply the forgone benefit of *not* implementing some other policy.) The danger is that, if one does not construct a similar measure, of willingness to pay for the resources needed to enact that policy (i.e. the benefits), then one will not be comparing like with like. Furthermore, misstatement of the benefits will typically err in the direction of underestimation.

One can observe numerous instances of this. Consider, for example, an environmental regulation that we now regard as uncontroversial, such as the ban on leaded gasoline. At the time that it was implemented, the regulation was in fact quite controversial, largely because it was not subjected to proper CBA, which created difficulties for those who supported it. The problem was that the cost of the regulation was easily calculable, because it primarily took the form of "bads" that had market prices. Thus petroleum companies were able to calculate how much their refining costs would go up, in order to achieve the same octane level in their fuels without lead additives (and thus, how much the price of gas would have to go up). By contrast, the benefits of the regulation—primarily health benefits associated with the reduction of airborne lead—had no market value. What CBA recommends in such cases is that we take the primary welfare gain and estimate its value, by asking people, "How much would you pay to avoid exposure to an atmospheric contaminant that gives you a chance of brain damage and/or kidney failure, according to the following risk/exposure schedule?" But instead of doing this—trying, however imperfectly, to measure the direct welfare gain—people at the time chose downstream beneficial consequences that *did* have a market value, such as savings in the healthcare system from not having to treat those conditions, or increased worker productivity. This is completely arbitrary, because the question of which benefits happen to have a market value is determined by arbitrary features of the world, such as the cost of implementing a system of property rights being lower in one domain than another. Furthermore, the effect of doing ad hoc CBA is that it almost always overstates the relative cost of regulation.

Thus the choice is usually not between doing a CBA and not doing a CBA, but rather between doing a proper, complete CBA and doing an improper, incomplete CBA.[41] In most cases of environmental regulation, industry lobby groups can be counted on to do their half of the CBA—the one that adds up the total costs. The question is whether the state will do the other half of the CBA—the one that adds up the diffuse, difficult-to-quantify benefits that flow to the population generally. When it comes to articulating the latter, some number really is better than no number, and a good estimate is better than a bad one. With respect to climate change, for instance, when critics point to the (non-negligible) cost of carbon abatement, supporters often respond by pointing to the number of "green jobs" that will be produced by the move away from fossil fuel. This is problematic on multiple levels. Most annoyingly, it involves taking a cost and misclassifying it as a benefit.[42] (One could just as easily argue that toxic spills are great because they create "green jobs" for those who have to clean them up.) Second, it misstates the purpose of climate policy, which is not to create jobs, but rather to reduce environmental destruction. The correct way to respond to complaints about the cost of regulation is not to point to incidental benefits (or in this case, "costs that are mistaken for benefits") that happen to have a market value, but to express in monetary form the *actual benefits of the policy*, then show that they outweigh the costs.

Finally, it should be observed that most critics of CBA believe there is some *other* decision procedure that is in some way superior. The most commonly cited candidate is participatory democratic deliberation.[43] It is worth observing, however, that instead of replacing CBA, such a procedure could *also* be used to increase the quality of the inputs to CBA. There have been some small-scale experiments that involve using public deliberation as a way of improving WTP values, and there is some measure of enthusiasm for the technique among proponents of CBA.[44] Thus in order to show that deliberation is superior to CBA, it is necessary to show that it is superior to CBA-enhanced-through-deliberation as well. Beyond this, there is increased interest in the development and use of standard valuations, constructed by comparing and cleaning up the data sets from many different studies, in order to develop a model of the impact on the quality of life of the average person from various interventions.[45] Thus there are a variety of ways of improving the inputs to the CBA procedure, and it is difficult to think of a decision procedure that could function as an alternative to CBA that could not also be drawn upon to improve the quality of CBA itself.

5.4.2. Putting a Price on Life

Perhaps the most notorious feature of CBA is the way that it handles loss of life. The standard approach is to assign a monetary value to loss of life, treating it like any other cost. This is a deal-breaker for many people, who argue that human life has an inherent dignity that is incompatible with the assignment of a price. Critics often quote Immanuel Kant, who wrote that "in the kingdom of ends everything has either a *price* or a *dignity*. What has a price can be replaced by something else as its *equivalent*; what on the other hand is above all price and therefore admits of no equivalent has a dignity."[46] The idea that human life might be traded off against something of lesser value, such as a basket of ordinary economic goods, clearly violates this stricture. CBA is therefore thought to diminish the inherent dignity and value of life, and should be avoided for that reason.[47] (In Chapter 1, we saw how this concern informed certain arguments made by Simon Caney and Henry Shue in the case of climate change policy.)

One way of developing this intuition is to interpret it as a prohibition on *aggregationism*. In his influential critique of utilitarianism, Bernard Williams presents a scenario in which an individual (Jim) wanders into a small South American town where the military is about to execute 20 randomly chosen civilians (the "Indians") as an act of collective punishment against their village for supporting antigovernment rebels. The commander, however, makes Jim an offer: if Jim is willing to choose one of the villagers at random and personally execute him, the other 19 will be spared. Williams's intuition (not

universally shared) is that it would be permissible, or perhaps even obligatory, for Jim to refuse. This appears to suggest that there is a deontic prohibition on killing the innocent, even when the foreseeable consequence
of not violating the prohibition is that others will. Thus the aggregationist
procedure, of adding up the number of lives saved or lost under different
conditions, or the number of rights-violations that occur, ignores morally
significant features of the choices that we face. To the extent that CBA is committed to aggregationism, especially with regard to human life, it runs afoul
of these constraints.

The standard response to this argument on the part of moral consequentialists is to say, "Surely there must be *some* point at which you would
be willing to violate the constraint. What if the commander was threatening
to kill 100 Indians, or 1,000?" One can then chip away at the position from
that end, with the goal of showing that there is something irrational or indefensible about any strict deontic prohibition. (Michael Walzer's "ticking
bomb" argument against a strict prohibition on torture has this structure.)[48]

There is no reason, however, for proponents of CBA to get drawn into philosophical controversies of this nature. After all, in claiming that it is permissible for Jim to refuse to execute the innocent, Williams is granting that it is
permissible for Jim to take actions that will foreseeably result in the death
of 20 innocents, including the one that Jim himself was unwilling to kill. So
what is the difference between these two ways of bringing about someone's
death? One way that deontologists have tried to articulate the difference is by
distinguishing *intended* outcomes from merely *foreseen* consequences. If Jim
were to shoot one innocent, that individual's death would be the intended
outcome of an action chosen by Jim. If Jim refuses, the same individual will
die, but merely as a byproduct of Jim's action—a foreseen consequence, but
not an intended outcome. (Another way of articulating the difference between the two cases is to distinguish between doing and allowing, or between
acts and omissions.)

As the subsequent literature on trolley problems has illustrated, everyday
morality imposes deontic constraints in the realm of intended outcomes and
actions, but is much less restrictive of aggregative calculation in the realm of
the foreseen (hence the judgment that it is permissible to divert a runaway
trolley onto a sidetrack where it will kill an innocent maintenance worker, on
the grounds that doing so will save a larger number of innocent passengers
waiting on a platform).[49] Thus the standard defense of CBA lies in the observation that, because of the policy areas in which embedded CBA is applied,

the deaths that it deals with are entirely in the realm of the foreseen, not the intended. Furthermore, to the extent that these deaths do occur, it is typically the result of an omission on the part of the state (i.e. a failure to spend the extra money through which that death could have been averted). Satisfaction of either of these conditions may be sufficient to establish a context in which strict deontic constraints do not apply.

For example, when the state limits the amount that it spends on shelters for the homeless, it has the foreseeable result that some people will freeze to death on the street; when it limits the amount that it spends on chemo-therapy drugs, it has the foreseeable result that some people will be denied access to life-extending treatment; when it limits the amount that it spends on road safety, it has the foreseeable result that more motorists will die in accidents. But this is not the same thing as killing the homeless, or cancer patients, or drivers in order to achieve some other policy objective. These types of decisions simply do not fall under the scope of any deontic pro-hibition. As a result, it is perfectly permissible to ask whether one is really maximizing the number of lives saved by upgrading a highway to provide an additional passing lane, rather than providing expanded outreach services to the homeless.

In many ways, the valuation of life in CBA is simply making explicit something that is implicit in any such decision-making process. Whenever spending decisions are made that affect, in any way, human mortality, or even just longevity, subject to a budget constraint, there is always an im-plicit valuation of life. States that do not perform cost-benefit analysis still have budgets, which they use to improve road safety, as well as to accomplish myriad other objectives, many of which involve merely enhancing quality of life (such as maintaining parks). To the extent that they do so, they are clearly letting some people die so that others can enjoy what may be, taken singly, relatively minor benefits. Cost-benefit analysis merely articulates, or renders explicit, what is implicit in these decisions; it does not provide any independent justification for them.

Furthermore, it is important to observe that the only way to tell whether citizens are being treated equally by the state is to compare the valuation of life across policy domains. Where CBA calculations are not performed, the valuation of life is likely to be quite arbitrary. Thus it would not be surprising to find workplace safety rules that impose over $30 million in costs to save a single life, while the transportation safety administration balks at making in-frastructure improvements that would cost only $1 million per life saved. To the extent that citizens are exposed equally to all of these risks, the arbitrariness of these valuations does not translate into any significant inequality between citizens. But if exposure to certain risks is correlated with other forms of disadvantage, as it often is, then the arbitrariness does have the capacity to exacerbate inequality.[50] In particular, if budgets are determined through the political process, then sympathetic victims are more likely to attract funding for the issues that affect them. To the extent that low-socioeconomic status individuals are less sympathetic, they are more likely to enjoy equal concern and respect from the state if all agencies are forced to use a standard valuation of life that is equal across all projects and policy domains.

Thus CBA does not justify anything that the state is not already doing whenever it makes decisions affecting human life, subject to a budget constraint. There are, however, two further points that are worth making in response to the dignity objection. First, proponents of CBA have tried to explain that when they talk about the value of "life," it is, in most cases, a shorthand way of talking about something quite different, viz. the disvalue that people assign to very small risks of death. So when one says that the valuation of a life is $5 million, it is not actually any known individual's life that is being valued; rather it is the aggregate disvalue of very small risks of death distributed over a very large number of persons. This is flagged by the insistence that the dollar figures in question represent the "value of a statistical life" (VSL), not the value of an *actual* life (VAL). The thought is that if 100,000 people are each willing to pay $100 to eliminate exposure to a pollutant that generates a 1/50,000 risk of death, then the total "budget" of $10 million represents a WTP of $5 million *per life saved* (since the pollutant can be expected to kill two people out of the 100,000 people affected). When deliberating about whether to ban the pollutant, the question that is being asked is not whether it is worth spending the money in order to save these two people's lives, but whether it is worth spending the money to deliver the benefit of a small ex ante reduction in risk for each member of the group.

So another way of responding to the dignity objection is simply to state that, while human life may be priceless, *risks* to human life are not. Even if one could never be justified in killing a person in order to achieve some other good consequence, it is not obvious that one could not be justified in imposing some risk of death on a person in order to achieve some such consequence, as long as the risk is not excessive. After all, every time we get behind the wheel of a car, we are imposing some risk of death on other users of the road. Thus "embedded" CBA can avoid the general concern about human dignity simply by avoiding any application in areas that would require the use of a VAL rather than a VSL. (It is perhaps also worth noting that in climate change policy, any deaths that figure in the SCC calculations are going to be VSLs, not VALs.)

A second line of argument that defenders of CBA appeal to is the fact that individuals themselves make choices, quite often, in which they trade off risks of their own death against other goods. Nobody wants to spend infinite amounts of money on safety, and very few people take living as long as humanly possible to be central to their vision of the good life. On the contrary, safety is just one good among many that make up a balanced and healthy life. (As Jonathan Wolff puts it, "Purchasing—or declining to purchase—devices or services that make small differences to one's safety, or that of one's family, is an ordinary part of life, taking its place alongside other consumer decisions.")[51] The appropriate balance to strike between these various goods is an expression of individual autonomy. Yet if individuals, taken singly, are able to decide how much they are willing to pay to avoid certain risks, then there seems to be no reason that the state, in its "public economic" functions (that are the focus of CBA), should not be able to make choices that reflect these individual preferences.

There are, of course, problems that arise in the transition from individual to collective decision-making, particularly given that the state is often obliged to make "one size fits all" policies that affect the levels of risk to which individuals will be exposed. Setting aside these issues, however, it is worth pausing to consider how dramatically handicapped the public sector would be if it were somehow prevented from performing the ordinary cost-benefit calculations that inform individuals' private choices about safety and risk in their everyday lives. It would mean that a very large number of collective action problems simply could not be resolved, because private individuals are unable to contract their way to a solution (i.e. there is a market failure), and yet the public sector is unable to step in and resolve the issue because doing so would implicitly treat human life as less than infinitely precious. To take just one example, if the dignity of human life really does prevent the state from trading off risks to individuals against other benefits, then the state would be unable to operate public transit systems. Rapid transit is inherently risky, and so one can spend arbitrarily large sums of money improving safety.[52] When choosing private transportation options (e.g. buying a car), individuals trade off the benefits of safety against other considerations. But if the state is unable to make the same calculations, then the cost of public provision of transit would skyrocket. As a result, collective forms of transit like subways or buses would fall through the cracks—too costly to be provided privately, but too unsafe to be provided publicly.

This is, of course, not what proponents of the dignity argument have in mind. Their stated intention is usually to strengthen the welfare state, not destroy it. Indeed, much of the concern over valuation of life seems to boil down to a concern, not over valuation per se, but rather that the use of standard methods of calculation will result in a VSL that is simply too *low* (and therefore that the benefits of certain kinds of regulations will be understated).[53] In particular, critics focus a great deal on the use of wage premiums in risky occupations as a revealed preference mechanism for inferring the valuation that workers put on their own lives. This is a perfectly legitimate concern, but the appropriate response is simply to increase the assigned VSL to a more acceptable level, not to treat it as infinitely large. People's attitudes toward risk are sufficiently irrational, and inconsistent across domains, that the VSL can often be gerrymandered to suit any policy preference. This is why, if one looks at the practice of most states that make extensive use of CBA, one can see that the VSL figure is set through what amounts to an exercise in reflective equilibrium. Most importantly, a standard valuation is selected that is quite stable over time, and is applied in *all* policy domains.[54] This value is typically the result of lengthy reflection and negotiation. In Canada, for instance, the development of a standard VSL began when Environment Canada and Health Canada jointly sponsored a literature review of CBAs, in order to determine the range of VSL values currently in use by the federal government. They found a "mean VSL of $5.2 million with a range from a low of $3.1 million to a high of $10.4 million in 1996 dollars."[55] This mean value ($5.2 million), adjusted for inflation, was picked up by the central agencies and became the standard valuation subsequently used in all departments. When formulated in this way, it does not reflect the average WTA of individuals for mortality risks, so much as the state's commitment to the equality of citizens.

5.4.3. Existence Values

For many theorists who approach these issues from a background in environmental ethics, the use of CBA in environmental valuation is an absolute nonstarter because CBA is self-evidently committed to an anthropocentric theory of value. But as I observed in the first chapter, the way that the term "anthropocentric" is used harbors significant ambiguity. If the claim is that CBA takes into consideration only *human* valuations of the world, then the claim is clearly correct. When a CBA considers the cost involved in the destruction of a natural ecosystem, it does not consider the ecosystem's own perspective on the matter; it considers only the extent to which humans, in one way or another, care about that ecosystem. But if the claim is that CBA necessarily treats nature as only instrumentally valuable, so that the ecosystem is valued only as a source of natural resources, or in terms of the services that it provides, then the claim is false. It is possible to conduct a CBA that considers only instrumental value, but it is also possible to include so-called existence values, which incorporate what environmentalists have traditionally thought of as the intrinsic values found in nature. Since the only *access* we have to those values is through humans, however, the procedure that we use for incorporating those values into our deliberations looks a lot like the one used to incorporate instrumental values.

As we have seen, when it comes to various aspects of nature, there is significant disagreement about what sort of value we are dealing with. Some people value nature instrumentally, others intrinsically; some value it quite highly, while others assign it only negligible value, etc. The liberal state is committed to remaining neutral on these questions, and so is not willing simply to declare one group the winner on the basic question of how nature should be valued. Instead, it looks to how individuals in the society estimate this value, to try to gauge the magnitude of the values they perceive, as well as the intensity of their commitments. For instance, some people may feel passionately about the preservation of an old-growth forest, while others would like to see it logged for lumber. But since there are a variety of alternative sources of wood that would be just as good for lumber (e.g. farmed spruce), it seems incorrect to assign the latter set of interests equal priority to the former. There is no need to take sides on the deeper question of whose use of the land is ethically superior in order to see that the conservationists feel much more strongly about the question than the logging interests do. One way of gauging the strength of these commitments is to determine how much each individual would be willing to give up for the sake of those values (which is to say, what the individual's WTP/WTA is). So in recognition of the fact that certain environmental goods are valued both instrumentally and intrinsically, CBA may take into consideration not just the use value of nature, but also its existence value to conservationists, both expressed in monetary terms.

Thus the inclusion of existence values in a CBA can be considered an appropriate response to the criticism that CBA focuses only on instrumental value to the neglect of intrinsic value. Unfortunately, since money has only instrumental value, the use of money as the metric of comparison is misleading, suggesting that everything is being "reduced" to instrumental value. In fact, monetary valuation is only being used to make instrumental values commensurable with intrinsic values—something that must necessarily be done, no matter what the decision procedure, since no one thinks that intrinsic value categorically trumps instrumental value. This has, of course, been the subject of considerable misunderstanding, and it is entirely possible that the optics of using money and WTP values in environmental CBA are so irremediably bad that some other metric should be chosen. When one scratches the surface, however, what becomes apparent is that much of the opposition among environmentalists to the use of CBA as a guide to decision is not actually opposition to CBA specifically, but really an opposition to liberal neutrality. In other words, it is really a demand that certain first-order environmental values be used as a direct basis for legislation. Formulated in this way, the demand has troubling implications. These concerns, however, take us beyond the scope of the present reflections on CBA and relate back to the more fundamental arguments that have been developed, over the years, for the wisdom of liberal political arrangements.

The use of existence values in CBA is not uncontroversial. For example, the 2009 US Supreme Court case *Entergy Corp v. Riverkeeper, Inc.* dealt with a conflict over an Environmental Protection Agency (EPA) decision concerning a power plant whose cooling intake would have resulted in significant destruction of marine life.[56] The project had passed a CBA, in part because the EPA, in assessing the environmental cost, assigned value only to the lost fish and shellfish that would have otherwise been commercially or recreationally harvested, which amounted to only 1.8 percent of the total affected, while ignoring the existence value of the fish that would have escaped destruction at human hands. This reduced the damages from $735 million to $83 million. The Supreme Court declined to impose the obligation to consider existence values on the agency, although three justices dissented from the decision on this point.

Perhaps the most powerful objection to the use of existence values in CBA is that they are "intrusive" or "external" preferences—preferences that involve controlling the preferences or actions of others—and therefore should not be counted when it comes to determining the level of social welfare produced by a particular change. On this view, the inclusion of existence values represents a form of gerrymandering of the CBA by environmentalists, based on the introduction of a preference that no one would be willing to accept in other areas of decision-making. For example, a person might have very strong views about what color his neighbor's house should be painted, and might become quite distraught when the neighbor decides to change it. Or a person might have very strong views about what books other people should read, or what sort of food they should eat. But unless it is possible to show some tangible harm—as John Stuart Mill argued long ago—then whatever loss of welfare this person suffers should not count as an argument in favor of social regulation.[57]

According to this view, when people claim that the destruction of a natural ecosystem harms them, even though they have never in any way interacted with that ecosystem, nor do they have any concrete intention of doing so, it involves an abusive concept of welfare. Allowing the concerns of urban conservationists to count as a consideration that speaks against a proposal, on par with the concerns of, say, remote communities in need of economic development, opens up a Pandora's box, making the entire CBA framework entirely unworkable. For example, when people eat pork, even in the privacy of their own homes, should the disgust that is evoked among observant Jews and Muslims at the very thought that someone is doing so count as an externality, whose cost must be taken into consideration by the state (justifying, perhaps, a special tax on pork)? Presumably not. However, trying to specify why this disgust reaction, along with the attendant loss of welfare, should *not* count is surprisingly difficult.[58] There are several candidate theories, but what they all have in common is some reference to the lack of close connection or involvement between the event that occurs (the consumption of pork) and the loss of welfare suffered by the individual who is upset by it. And yet this would seem to apply to the environmental case as well. The mere fact that some people do not want a piece of land to be developed and will get upset if it is developed does not seem like the right type of preference to count as an argument against development. In fact, it seems more like a moral judgment, to the effect that the piece of land *ought not* be developed. But this is what the CBA is supposed to be determining. In order to make that determination, the CBA needs to start with some impartial characterization of the *interests* that are at stake in the decision. To allow people's moral attitudes toward the interests of others to count as interests creates a variety of problems, including opening the door to an objectionable form of double counting. So even though the introduction of existence values is a way of defending CBA against its critics, many proponents of CBA are hostile to the tactic.

Supporters of the existence value practice respond, however, by emphasizing that the desire to set aside land for conservation is an extremely common preference, which markets already cater to quite extensively at both an individual and a club level. Many wealthy individuals buy ecologically sensitive land in order to protect it from destruction. It is well known, for instance, that CNN founder Ted Turner purchased 128,000 acres of Patagonian wilderness with this in mind. More significantly, wealthy conservationist Douglas Tompkins owns more than two million acres in Chile and Argentina, with the goal of protecting biodiversity.[59] Many celebrities and wealthy environmentalists have been buying large tracts of Amazonian rainforest. There are also a large number of popular nonprofit organizations, such as the World Land Trust, the Nature Conservancy, the Rainforest Conservation Fund, or the Nature Trust in Canada, which allow less-wealthy individuals to pool their funds to purchase land, which is in turn set aside in perpetuity as a private conservation trust. Most of these people, it should be noted, will never have occasion to visit these properties or to interact with them in any significant way.

Thus unlike the case of the books that other people read, the food that they eat, or the color that they paint their houses, the interest that people have in preserving land for conservation (which includes preservation of habitat to prevent species extinction) is one that private markets already cater to. If people cared strongly enough about the color of other people's houses and were willing to pay them to avoid certain shades, then there might be contracting over this as well. And if there were such a market, it is not clear what the objection to it would be.

Furthermore, while some of the conservationist impulse seems directed toward existence value, a certain element of it can also be construed as the preservation of an "option value," particularly when it comes to protection of biodiversity (an often expressed aim). Ecosystems tend to be rather easy to destroy, but extremely expensive, difficult, and time-consuming to restore. Thus even a very small possibility that one might want to derive ecosystem services from it in the future can make it rational to preserve. (This issue arises quite often with rainforest biodiversity, where many of the plant species have not even been cataloged, and there is some suspicion that they may possess useful medicinal properties). Furthermore, in the same way that having a hospital or a fire station in the neighborhood can be of value to residents, even if they never make direct use of it, the presence of a nature preserve can also generate value for those who never directly enjoy its charms or benefits. Merely the knowledge that it is there can have positive value.

Thus the burden of proof, in the case of ecosystem preservation, is not to show that the relevant existence values are legitimate, it is merely to show that there is some sort of a market failure, which explains why the *state* should be undertaking the conservation effort, rather than a nonprofit organization or an individual conservationist. This does not seem like an impossible burden of proof to discharge, given the enormous transaction costs that would be involved in organizing such efforts among many small contributors through private contracting, along with the free-rider problems that such an initiative would encounter.

5.5. Climate Change

This discussion of CBA is clearly not exhaustive; many other objections have been raised that I have not had the opportunity to respond to. There are, I believe, persuasive answers to them as well, but such a discussion would take us too far afield. What I have chosen to focus on, instead, are the issues most often raised by environmentalists, which account for much of the hostility toward CBA in those circles. Abstractly, one might have thought that the effort to calculate precisely the benefits of environmental regulation would be well received. Since industry groups are constantly complaining about the cost of regulation, the ability to respond to this with a statement of the benefits would appear to offer a welcome counterpoint. Unfortunately, there are powerfully ideological barriers that stand in the way of this acceptance. Because most environmental problems are caused by negative externalities, there is a tendency to think of the market as intrinsically biased against environmental values. Thus the fact that CBA is a market-simulating procedure leads many to think that it must also be biased against environmental values.

For example, in their book *Priceless*, Ackerman and Heinzerling begin by criticizing free-market ideology, deregulation, and privatization, then segue to a critique of the Pareto-efficiency principle, which they describe as "a subtler, more sophisticated form of faith in the free market."[60] From that, they go on to reject CBA, on the grounds that it "tries to set an economic standard for measuring the success of government projects and programs."[61] This reflects a set of "built-in biases" that "tilt strongly toward endorsement of business as usual, and rejection of health and environmental protection."[62] One can see laid out here a set of psychological associations, used to trace out a pattern of moral contagion that they believe starts with the evil of the market. In reality, Pareto efficiency is a self-standing principle of justice, which the market more or less successfully institutionalizes, not the other way around. And CBA does not use an "economic standard" or a market norm to evaluate projects; it uses the Pareto principle. Thus there are no grounds for claiming that it is biased in the direction of market outcomes. Indeed, a socialist state would face exactly the same challenges with respect to regulation and would have good reason to use exactly the same principles to guide its decisions.

Once one moves beyond these negative psychological associations in order to assess the genuine merits of the CBA procedure, it is not difficult to see that many serious difficulties remain. Most are centered on the treatment of existence values, and concern environmental issues that might loosely be classified under the heading of "conservation" (including species preservation). It is important to recognize, however, that as important as these issues are, none are centrally involved in the problem of climate change. Nicholas Stern exaggerated only somewhat when he declared that "the problem of emissions and anthropogenic climate change starts with people and ends with people."[63] I have mentioned this already, but the point bears emphasizing. When it comes to calculating the SCC, the overwhelming majority of costs are uncontroversially human, extremely tangible, and not that difficult to quantify. The intrinsic values that are threatened in nature pale in comparison to the straightforward loss of human welfare that is projected. Consider the following summary, which Dominic Roser and Christian Seidel provide as a way of describing the negative impacts of climate change:

> When temperatures rise, glaciers, which serve as water reservoirs for the summer, begin to melt and the melt water ends up in rivers that supply human beings with water. Without glaciers, there is less water in summer for agriculture, energy production, and daily use. At higher temperatures, the polar ice caps melt and the water spreads into the oceans; owing to the resulting rise in the sea level, land masses contract and the groundwater becomes salinated near the coast where a large proportion of mankind lives. Ocean currents and precipitation patterns change; the resulting increase in the frequency of extreme weather events such as hurricanes, floods, and droughts will make people homeless and destitute, and will aggravate famines resulting from declines in crop yields. Lower crop yields

are synonymous with migration, less (and lower quality) water is synonymous with more conflicts. More frequent heat waves will lead to an increase in suffering and mortality among the old and weak. More people will be affected by tropical diseases because, in a warmer climate, the insects that serve as vectors for these diseases will gain a foothold in new regions.[64]

What is noteworthy about the list of damages described here is that they are all human impacts. One does not get the sense, reading this list, that calculating the SCC is some quixotic endeavor, or an attempt to quantify the unquantifiable. Furthermore, because so many of these impacts already have market values, it is not necessary to use WTP/WTA values, and so many of the concerns that are associated with the use of this estimation method do not arise. Of course, the scale of the damage is enormous, and there are well-known uncertainties surrounding many aspects of climate modeling and economic forecasting. Nevertheless, it is difficult to see how there could be any principled objection to an approach that tries to develop the best estimates, in all of these different dimensions, and then combines them into a best estimate of the expected damage from greenhouse gas emissions.

When integrated into policy deliberations, the primary function of the SCC number is that it allows policymakers to make an educated guess about what the parties would contract to if those who want to burn fossil fuel and those who will suffer the damage from climate change were able to negotiate an agreement. In the case of climate change, mitigation and adaptation correspond to the two sides involved in Coase's analysis of externalities, like the railroad company that wants to run the trains as quickly as possible and the farmers who want to grow their crops right up to the tracks. And the basic Coasian point, that the mere presence of an externality does not determine which party should change its behavior, explains why climate change policy involves balancing mitigation and adaptation. The ideal policy will not be one in which production of the externality (i.e. GHG emissions) ceases entirely, but rather one in which some effort is made in both directions. For example, while some effort should be made to limit sea-level rise, a certain number of people who live on the coast, only a meter or two above sea level, are also going to have to move.

The specific examples that Coase gives are sometimes confusing, because he treats the choices involved in production of an externality as binary—the trains can either run fast or slow, the crops can either be set back or not. In reality, the trains could run at various speeds, and the production of sparks would increase as a function of speed, just as the farmers could vary the set-back of their crops, with the chances of fire decreasing as a function of the distance of the crops from the tracks. Thus the relationship between an action and its cost would be a continuous function, as shown in Figure 5.3. (As the speed of the trains increases, the railroad benefits, but farmers must increase their setback, and so their costs increase. There are limits to the benefits, and so marginal benefit is shown as declining.)

If the rights are initially assigned to the farmers, then they will choose to minimize train speed, and so the outcome will be point *a*. If the rights are assigned to the railroad, it will increase train speed until the marginal benefit of increased speed declines to zero, and so the outcome will be at point *c*. It is easy to see why, if the parties then negotiate an agreement, they will wind up at point *b*. Anywhere to the left of *b*, the losses to the farmers are less than the gains to the railroad, and so the railroad will be willing to pay to increase the train speed—and so if the negotiations start with *a* as the status quo, they will move to *b*. Anywhere to the right of *b*, the losses to the farmers are greater than the gains to the railroads, and so the farmers will be willing to pay the railroad to slow the trains—and so if negotiations start with *c* as the status quo, they will move to *b*.

One can see here that there is some hyperbole in the Coasian claim that "just because there's an externality doesn't mean it's inefficient." When the production level of an externality can be varied, and transaction costs prevent negotiation between the parties, then the existence of the externality

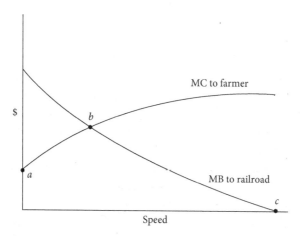

Figure 5.3 The Coasian interaction

will almost always entail some inefficiency. As a result, forcing the parties to contract over production of the externality will almost always result in some reduction in the amount produced. The correct version of the Coasian point lies in the observation that this residual level of externality, which persists even when the cost is fully internalized, indicates that some production of the externality is efficient. This is the case with respect to greenhouse gas emissions, and explains why having a carbon price set to the SCC cannot be expected to eliminate production of greenhouse gases altogether. Indeed, Dieter Helm has suggested that the only large-scale GHG-producing activity that would be ruled out entirely by an appropriate carbon-pricing scheme would be the burning of coal. Most uses of fossil fuels in transportation, by contrast, would continue into the near future, simply because the value to us of the economic product is sufficiently great that we would be willing to pay the full cost in order to see it continue. Again, this is consistent with the basic normative position being defended here, which is that it is only the gratuitous production of GHGs that should offend our moral sensibility and that constitutes unfairness to those who must shoulder the costs. To the extent that we derive sufficient benefit from our actions, our consumption is justified—as reflected in the fact that it would be agreed to if all of the parties were able to come together to negotiate terms of interaction.

Finally, it is again worth emphasizing that if we opt not to use CBA as an evaluation method, some other mechanism must be proposed. Because there are inevitable trade-offs in climate change policy, between different mitigation strategies, as well as between mitigation and adaptation, some method of comparing costs and benefits must be employed. The alternative to CBA most often proposed by liberal philosophers is some form of democratic deliberation.[65] No serious proposal for institutionalizing such a procedure has yet been made for a problem as extensive as climate change, although some attempts to characterize the emergence of a "global public sphere" on this question have been made.[66] For now it may suffice to observe that the temporal dimension of the climate change problem creates unique difficulties for such a proposal. CBA makes no distinction between welfare impacts on those who are currently alive and those who are not yet born (even though it may incorporate a discount rate, which takes into account the passage of time). Democratic deliberation, by contrast, is only open to those who are currently alive. Indeed, one of the reasons that democracies have been laggards in their response to climate change is that elected politicians respond to organized constituencies, whereas those who will feel the most significant impacts of climate change are not yet alive to press their concerns. The only way that a deliberative procedure can incorporate concern over future generations is by having other people, presently alive, advocate for their interests. This violates an essential equality condition, which is that the force of each individual's claim should not depend upon how much *others* care about that claim. Either way, in order to advocate for their interests, we would have to conduct something like a CBA in order to determine what the impacts will be on them. The best way to ensure that they are being treated equally is therefore to integrate this into a comprehensive CBA that treats these effects as commensurable with those on people who are currently alive and able to advocate for their own interests.

5.6. Compensating the Losers?

Both Nicholas Stern and John Broome have been troubled by the fact that, under any responsible climate change mitigation policy, present generations seem to come out as "losers" overall. We are being asked to make emissions cuts and thereby to sacrifice some fraction of our standard of living, but most of us will not live long enough to experience the benefits.[67] The concern is essentially the same as the one discussed in section 5.3, which motivated the suggestion that the Kaldor-Hicks standard is an essential feature of CBA, because regulation generates both winners and losers. Both Broome and Stern note that if climate change is a market failure, or a collective action problem, then any solution to the problem should constitute a Pareto improvement. But if it is a Pareto improvement, then it must be possible to make *everyone* better off. And yet, as Broome observes, because the costs of mitigation are upfront, "It is very commonly assumed that, in order to deal with the problem of climate change, a sacrifice is required from the present generation, for the sake of improving the lives of people in the future." If the market failures analysis is correct, however, then it should be "possible to eliminate the inefficiency without a sacrifice on anyone's part."[68] The solution, therefore, must lie in the creation of a compensation scheme that reimburses people for the carbon taxes that they pay.[69]

Now if something like the utilitarian "just savings" framework were correct, the creation of such a compensation scheme would not pose much of a problem. According to this view, saving and investment are essentially an eleemosynary activity, aimed at producing benefits for future generations.

When present generations engage in climate change mitigation, they are making an investment whose benefits will only begin to be felt after several decades. One way of compensating them for the upfront costs would be to relieve them of the responsibility for some *other* investment they are making, to the benefit of future generations, and to shift these funds into immediate consumption. This is what Broome proposes. On his view, we are passing along a wide range of different "gifts" to future generations. Climate change mitigation will be costly, "But we could compensate ourselves for this cost by leaving fewer gifts to our descendants: we could reduce the amount of other natural and artificial resources that we bequeath them."[70] This can be accomplished, Broome thinks, by increasing government borrowing.[71] Similarly, Aaron Maltais has suggested that "current generations can be motivated to start financing mitigation at much higher levels *today* by shifting these costs to the future through national debt."[72]

Both of these proposals get into somewhat perilous territory, since there is a common economic fallacy lurking in these formulations. A society cannot literally "borrow from the future" in the same way that individuals can borrow from a bank. Money is just paper, used to keep track of things. What matters are the real goods, and, needless to say, one cannot borrow goods from people in the future. So what is the government doing when it runs up a debt with the intention of repaying it later? The state borrows by selling bonds, which are typically purchased by private individuals and institutions, as well as foreign states. These bonds are ultimately repaid with revenue raised through taxation, which is transferred to the bondholder upon maturity. And so it is not the case that when government debt is issued, "present generations" acquire an asset, while "future generations" are left with the liability. When the state borrows, it creates both an asset (the bond, held privately) and a corresponding liability (the debt, which must ultimately be paid by taxpayers). Both the asset and the liability are transferred forward across generations (the former through inheritance, the latter by staying on the government's books), and so the transaction is a wash as far as intergenerational justice is concerned. What the state is really doing when it borrows money is creating a redistributive obligation that binds people (ceteris paribus) in the future. When the state sells a bond with a 10-year maturity date, it means that 10 years from now, someone (i.e. the taxpayer) is going to have to transfer a specified sum of money to someone else (i.e. the bondholder). There is no intergenerational redistribution going on here; it is all intragenerational.

Economists often try to explain this point by saying that the national debt is not really burdensome because it is one that we "owe to ourselves." Thus we cannot compensate ourselves, at the expense of future generations, by increasing government debt. The exception, of course, is if the debt is owed largely to foreigners, but this also tends to be canceled out by the fact that domestic investors hold a great deal of foreign debt. In any case, these qualifications are not relevant in the case of climate change since we are talking about intergenerational relations on a global scale, where there are no foreigners and so debts are necessarily ones that we owe to ourselves.

If government debt does not actually cost future generations anything, one might wonder what stops governments from running up arbitrarily large debts. The answer has to do with the impact of government borrowing on investment, and thus on the rate of economic growth. Government spending, as has been noted, almost all takes the form of consumption. Indeed, taxation is in some ways best thought of as "mandatory consumption." And yet bonds are traditionally used as a savings vehicle. Thus what the state is doing, when it sells bonds, is soaking up some of the private savings that might otherwise have been used to finance investment, and using them instead to finance consumption. The state also competes with private investment for funds, thereby raising the cost of borrowing. All of this has the effect of depressing the level of investment and thus lowering the rate of economic growth. In the limit case, state borrowing could crowd out private investment completely. This is why excessive government debt is generally regarded as a bad thing. It does not burden future generations directly, but only indirectly, insofar as it reduces investment and growth.

So what then does the government accomplish by borrowing money? It can borrow just to smooth out the tax burden over time (since expenditures may rise and fall from year to year), or it may want to engage in countercyclical spending. There is a temptation, however, to borrow just because it represents the course of least political resistance. As Milton Friedman famously observed, "To spend is to tax." And yet to spend is not necessarily to tax right away. Deficit spending allows politicians to spend now while shifting the obligation to tax sometime into the future. In general, people do not like paying taxes because doing so curtails their private consumption. By borrowing, the state can dip into the pool of funds earmarked for investment, which does not depress anyone's rate of consumption in the near term. In the future, of course, the state will have to tax someone in order to repay these funds. So deficit spending does burden future generations, not by imposing a real economic cost on them, but by saddling them with commitments that are politically unpopular to discharge (i.e. either raising taxes or defaulting on the bond).

Thus what Broome and Maltais are proposing is not really to borrow money from future generations to pay for our carbon abatement efforts. Those costs will necessarily be borne in the present. What they are proposing, instead, is that we reduce our savings in order to compensate ourselves for the reduction in consumption that will be required to reduce our emissions. Broome recommends government borrowing only because it would depress the overall level of investment (and thus, implicitly, reduce economic growth). To put it more simply than he does: adopting a climate change mitigation policy is equivalent to increasing our rate of savings, and so we are justified in lowering our rate of traditional capital savings—construed as a sacrifice we are making for future generations—in order to ensure that we come out even.

The problem with Broome's proposal is that, as I argued in Chapter 3, we are not living in anything that resembles the "just savings" world imagined by welfare consequentialists. The savings that we undertake are not a sacrifice we make to benefit future generations; they are driven by our own self-interest. Furthermore, the state is already depressing the rate of savings below what the private rate of savings would be. Thus there are no eleemosynary investment projects that we can cancel in order to compensate ourselves for the costs of climate change mitigation. The "gifts" that we are giving future generations are not really gifts, because they do not cost us anything. They are merely byproducts of our self-interested activities (such as our retirement savings). Because saving is self-interested, we would derive no benefit from increasing government borrowing to the point where it crowds out private investment. This would harm the present generation as well by reducing the return on our own savings.

Because of this, there is only one way that we can extract compensation from future generations for the benefits we would be conferring on them through a carbon abatement policy. This would have to make use of the "overlapping generations" mechanism at the heart of the system of intergenerational cooperation. For instance, it would be possible to grant present generations an unfunded bonus in the state pension scheme (for example, any carbon taxes paid could simply be counted as pension contributions). The unfunded pension liabilities could, in turn, be "rolled over," so that younger workers, who in practice will wind up making up the shortfall, will also receive a pension bonus, which will be made up for by future workers.

This could be phased out gradually, as the benefits of the carbon abatement policy begin to be felt.[73]

The interesting philosophical question is whether any of this is necessary from the standpoint of justice (as opposed to merely being politically expedient). The situation with greenhouse gas emissions, after all, is in at least one respect similar to that of the factory that is pouring mercury into the river, where the cost to the factory owners of restricting emissions is lower than the damages caused to the community living downstream. Imposition of a regulation prohibiting the pollution externality harms the factory owners, but this does not mean that the rationale for the regulation is not essentially Paretian. Efficiency considerations determine whether there should be regulation and, if so, what emissions level it should permit. But as we have seen, the question of whether the factory owners should be compensated after the imposition of a regulation is a distributive justice question. The externality, as Coase observed, is a joint product of the interaction between the factory and the community, so the question of who should bear the cost of remedying it is going to be determined by our sense of which side has a better claim to be pursuing its interests. If the factory was there first, and the residents of the community moved in later, we might be inclined to think that there should be compensation to the factory owners. If, however, the community was long-standing and the factory subsequently moved in and began to pollute the river, we would probably be inclined to say that no compensation should be paid, because the factory owners are the ones whose discretionary actions created the situation.

In the case of climate change, a similar line of reasoning makes it difficult to see why present generations should be owed any compensation for the cost of greenhouse gas regulation. Again, we can set aside the issue of Pareto efficiency, since this is used only to determine the appropriate balance between mitigation and adaptation. When one party is free-riding, resolving the collective action problem does not require that the outcome be a Pareto improvement over the existing state of affairs. In order to justify compensation, there would have to be a distributive justice argument to show that, somehow, it should be people in the future who bear the cost of adapting to climate change. It is not clear how this argument would go. There is a sense in which we are the ones whose discretionary activities are causing climate change; people in the future are just going about their business in perfectly foreseeable ways. This may be why, in the case of diffuse atmospheric externalities, there is seldom any desire to compensate polluters. With the case of sulfur dioxide regulation, for instance, no effort was made to compensate emitters when the US cap-and-trade system was introduced. The judgment was that they should never have been emitting it in such quantities in the first place. It is difficult to see how the case of climate change and greenhouse gases is any different.

These reflections are admittedly rather ad hoc, in part because there is no generally accepted account of how these decisions about the distributive impact of regulatory intervention should be made. Perhaps more significantly, neither Broome nor Stern tries to provide any principled argument in support of the claim that present generations should be compensated. With Broome's analysis, one gets the sense that his primary motive is political. His concern is not that it would be unjust to force present generations to shoulder the cost of climate change mitigation; he just thinks that climate change mitigation efforts would be more likely to be undertaken if present generations were relieved of the cost. Thus he hopes that resistance to carbon taxes could be reduced by offering those who pay them compensating benefits. There are two points worth making with respect to this suggestion. First, if the issue is merely one of political resistance, not distributive justice, it is possible to provide benefits to present people without "charging" future generations. The standard approach, adopted in jurisdictions such as Canada, is to make the carbon taxes revenue-neutral, by rebating all carbon taxes paid in the form of income tax reductions (a so-called green shift in the tax base). Since the income tax rebate is based on the average carbon tax paid, individuals still have an incentive to reduce their own personal carbon emissions. The *political* reality is that very few people care about how much of a burden their generation is shouldering; all they care about is their own taxes. It is not difficult to design a carbon tax-and-rebate system that gives the majority of the population a rebate that exceeds their taxes paid. This should be sufficient to defuse one worry about the willingness of democratic majorities to accept carbon pricing, without the need to transfer the burden forward in time. Second, the mechanism that Broome is proposing to compensate individuals—an increase in consumption financed by government borrowing—is rather abstruse. Even the pension credit that I have suggested is quite abstract and represents a highly deferred benefit. Given the economic sophistication of the average voter, much less the average opponent of climate change mitigation, it is unlikely that such a rebating system would be politically disarming. Much of the opposition to carbon taxes is simply based upon the negative psychological associations surrounding the word "tax." Thus the political intuition driving Broome's suggestion is not a particularly compelling one. It would be better to insist upon the simple and powerful conception of fairness that informs the basic principle of carbon pricing, then leave it up to politicians to ply their trade, by devising ways of convincing the electorate to do the right thing.

6

Positive Social Time Preference

The argument that I have been developing suggests that we should not be thinking about climate change policy in terms of a general obligation to improve the welfare of future generations. Most importantly, we should not be thinking of the bequest that we leave to them as a product of altruism on our behalf, as the traditional "just savings" framework recommends. There is no utilitarian planner in our society, trying to maximize the sum total of human well-being over time; no institution is committed to that purpose, nor does anyone take seriously, in everyday moral life, the obligations that such a project would entail. But this does not mean that we are entitled to ignore future generations. We are all part of an ongoing system of cooperation that extends out indefinitely into the future, and so includes future generations as virtual participants. As a result, thinking about the fairness of existing institutional arrangements requires thinking about the effects that they will have on people in the foreseeable future. And when we examine the current state of affairs, we can easily see that greenhouse gas emissions are a negative externality, the production of which violates a basic principle of fairness, which finds expression in the traditional welfare economist's view that the price of goods should be equal to their social cost. Estimating that cost, and then pricing the externality accordingly, constitutes the appropriate policy response. Because of this, developing an appropriate climate change policy does not require complex or special normative resources, beyond those provided in the standard toolkit used to address issues of environmental regulation.

Unfortunately, it is not quite so easy to escape the intertemporal aspect of the climate change problem. The fact that there is such a large time lag between our emissions activities and the damages that they cause does complicate policy deliberations in one important respect. When calculating the social cost of carbon (SCC), the question arises how we should treat damages that occur at different points in time, and in particular, whether we should worry less about damages that occur in the more distant future. This issue actually arises with every public policy decision, not just climate change. However, with most environmental externalities, both the costs and the

Philosophical Foundations of Climate Change Policy. Joseph Heath, Oxford University Press. © Oxford University Press 2021. DOI: 10.1093/oso/9780197567982.003.0007

benefits occur within a few years of each other, and so this aspect of the cal-culation is not very important. With climate change, however, the costs of abatement are largely upfront, while the benefits are significantly delayed, and thus the way that the temporal aspect of the problem is handled becomes extremely important. In Chapter 3, I argued that climate change was not, first and foremost, an issue of intergenerational justice, but rather a collective ac-tion problem that raises issues of intergenerational justice. The way that this issue of intergenerational justice shows up is in the form of the *discount rate* that is used, in calculating the social cost of carbon.

It is quite standard for economists and policymakers to accommodate temporal considerations by incorporating a discount rate into their assess-ment of costs and benefits, which reduces the present value of future costs, as a function of how far removed they are from the present. Indeed, many countries have an official discount rate, which the state uses for the evalua-tion of public projects such as infrastructure investments. Historically, these rates were determined in a rather ad hoc fashion, in part because they had never been very controversial. Given the timescale of the projects being considered, the choice of discount rate simply did not make a great deal of difference. However, when one considers an issue such as climate change, where the timescale is on the order of centuries, not decades, then suddenly the choice of discount rate becomes highly significant—indeed, it can easily overwhelm almost any other consideration, when it comes to deciding which policies should be undertaken.[1] As a result, the ad hoc justifications have proven no longer sufficient.

The topic of discounting is one that has been the subject of deep and careful analysis by economists over the past century. Although the basic impulse of many was to look for some "positive" measure, or revealed preference, that could be used to pin down the appropriate discount rate for policymaking, this approach has been subjected to powerful criticism.[2] This has given rise to widespread acceptance of the view, articulated by Kenneth Arrow among others, that the choice of a discount rate is fundamentally a "moral" choice.[3] The most difficult question that arises concerns whether the social discount rate should incorporate some form of time preference—whether the mere fact that a particular cost or benefit is to be experienced in the future provides sufficient ground for treating its present value as less than its eventual value. Individuals normally do this when it comes to making their own choices, but from that fact it does not follow that society should do the same. Most economists, however, have assumed that the social discount rate should incorporate some form of time preference, but at the same time have recognized that this position stands in need of moral justification.

One might think that philosophers could be of some help in at least guiding this discussion. Unfortunately, the specific proposal that most philosophers and environmental ethicists have been inclined to make, with respect to the social discount rate, is so far outside the ballpark of what economists and policymakers have been considering, that it has again resulted in them being largely sidelined in the discussion. This is because most philosophers, being of the view that it is morally impermissible to treat future harms as less serious than present ones, move immediately to the conclusion that the only acceptable social rate of time preference is zero. Indeed, this strikes many philosophers as being so obvious as to require little or no argument. The problem with this view is that it has extremely radical implications—indeed, unless some other factor is introduced to raise the overall discount rate above zero, it generates conclusions that are straightforwardly absurd. (This is why most economists, despite being puzzled by what the appropriate discount rate should be, nevertheless consider it essential that it be greater than zero.)[4] As a result, there has been a serious lack of engagement between the two groups, because so many philosophers have again positioned themselves outside the space of plausible policy options.

My goal in this chapter is to show that most philosophers who have specifically addressed this issue (including Derek Parfit, Henry Shue, Simon Caney, Darrel Moellendorf, and Axel Gosseries)[5] have moved too quickly from the idea that the goodness or badness of a beneficial or harmful event is independent of when it occurs, to the conclusion that temporal discounting of benefits and harms in the present is impermissible. As a result, instead of treating the zero-discounting position as a *problem* for whatever normative theory it is derived from, they have tended to treat it as one of the attractions. John Broome has urged a compromise position, arguing that while temporal discounting of welfare is impermissible, discounting the value of commodities may not be, and so the expectation of ongoing economic growth can be appealed to as a way of avoiding the problematic implications of zero time preference.[6] I would like to explore three different avenues of argument that could be adopted, in order to show that discounting of welfare (or interests) may be permissible. I will end with a modest proposal, suggesting that we might use the average *death rate* of the population as a basis for a pure time preference.

6.1. The Case for Temporal Neutrality

I should mention at the outset that the discussion about the social discount rate has, to date, been primarily an intramural discussion among consequentialists, even though it is a problem for a much wider range of views.[7] Indeed, anyone who thinks that, when engaged in normative assessment of some set of policy options, the impact on human welfare of each option should be at least a factor in determining how that option will fare in comparison with the others, is immediately confronted with the question of how to render commensurable, at any given point in time, welfare effects that will occur at different points in time. A radically non-consequentialist normative theory would not have this problem, but as we have seen, these theories have not played much of a role in the specific debate over climate policy, primarily because no one has been able to explain how they could say very much on the subject.[8] Taking action now to mitigate climate change would clearly involve some cost, ranging from insignificant to crippling, depending primarily upon how much carbon abatement was undertaken. Any such investment would be subject to diminishing returns, and so the primary question becomes what cost would be justified given the expected benefits. Some type of SCC calculation, or cost-benefit analysis, seems to impose itself, by the very nature of the problem. Doing this calculation, however, requires deciding how to count costs and benefits that will not be realized for hundreds of years.

Moderate deontological views are no more successful at avoiding this question than consequentialist ones. Consider, for example, the evolution in Simon Caney's thinking on the subject. As mentioned earlier, if one interprets the concept of a right strictly, as Caney initially does, the deontological framework is paralyzing—every policy option will violate the rights of millions of people. The only way to avoid this is to begin trading off one batch of rights violations against some other, in order to determine which is the least bad.[9] This is what Caney winds up doing, suggesting that his rights be interpreted in Razian terms, as a special class of protected interests, which then allows them to be traded off against one another.[10] This naturally raises the question of how to render commensurable, at any given point in time, interest violations that will occur at different points in time (which is, of course, just a different way of posing the basic problem that the discount rate purports to resolve). In the end, Caney winds up having to confront the discounting problem as well, despite being officially committed to a rights framework.

Thus the discounting question is much more difficult to avoid than many recent philosophical discussions might lead one to suspect. This makes it worthwhile to understand the history of the debates over this question. The natural point of departure is to consider the so-called Ramsey formula, which has served as the primary focus of the debate among proponents of CBA. One begins by defining a discount rate (r), which expresses the fraction of its value that a particular outcome loses for every unit of time that it is deferred. The present value (v_p) of some future outcome can then be defined in terms of the value of that outcome (v), discounted for the number of periods of delay before it is realized (t):

$$v_p = (1/1+r)^t v \tag{1}$$

Thus if one were to receive a check for $100 and then notice that it is postdated, so that it can only be cashed in five years, this makes the check less valuable, in the present, than a check that can be cashed right away. Indeed, at an interest rate of 5 percent, one could purchase a five-year bond for only $78.32 that would be worth exactly $100 at maturity, and so the present value of "$100 in five years" is clearly less than "$100 right now." If one were to ask how much less, an obvious suggestion would be to calculate it by setting the discount rate equal to the interest rate. Thus one can imagine the $100 check as losing 5 percent of its value, for every year that one must wait before cashing it. This would make the present value of the $100 check that can be cashed in five years equal to $78.32, a value that can be derived using formula (1), setting $t = 5$, and $r = .05$:

$$v_p = (1/1.05)^5 \$100 = .9523^5 (\$100) = .7832 (\$100) \tag{2}$$

It is worth emphasizing that $78.32 represents the *present value* of the check, at the point when it still cannot be cashed for another five years. The check, however, will still be worth $100 at the time that it is cashed. In other words, future money is not worth as much as present money, even though, in the future, money will be worth just as much as it is in the present (in all these discussions, we assume zero inflation). This may seem like semantic wordplay, but as we shall see, keeping this distinction straight is important when it comes to avoiding certain fallacies that have shown up in the debates over discounting.

Thanks to examples such as this, economists have often just used long-run interest rates as the discount rate for public policy purposes. (This is where the "off-the-shelf" number of 5 percent typically comes from.)[11] Historically this did not attract much controversy, mainly because it did not make all that much difference in most decisions, such as infrastructure projects, which generate returns over a period of decades, not centuries. In the same way that retirement savings accumulate rather slowly on a timescale of 10–20 years, but then begin to take off as the compounding effect kicks in, the discount rate also tends to make little difference on a time-scale of 10–20 years, until the effects of compounding begin to accumulate. On a timescale of a century the effects are massive.

Because of this, climate change policy recommendations are extremely sensitive to the discount rate in a way that other public policies are not. This was dramatized in 2006, with the release of the *Stern Review on the Economics of Climate Change*, which recommended the immediate imposition of carbon taxes an order of magnitude larger than those being contemplated in mainstream policy circles. Critics went through the 700-page report with a fine-toothed comb, trying to find the basis for this rather surprising conclusion. They found that Stern's assessment of the science, as well as the integrated assessment model he used for projecting damages, were all unremarkable. The only real difference, it turned out, was that Stern rejected the mainstream view on social discounting and instead chose to use a rate of only 1.4 percent. As a result, his model simply assigned much greater weight to damages occurring in the distant future and therefore assigned much greater benefit to carbon abatement (keeping in mind that the social cost of carbon specifies the benefit of carbon abatement).

Stern's analysis set off an intense debate over the discount rate. Although Stern was eventually persuaded to raise his own estimate, there was fairly widespread agreement that he was right to regard the discount rate that could be inferred from the interest rate as too high for public policy purposes. The reason is that the annual interest rate is essentially an aggregate measure of how much individuals must be compensated in order to be willing to defer consumption by a year. The problem is that individuals demand this compensation in part because they face the risk of death, which gives them an understandable desire not to delay various satisfactions too far into the future. "Society," however, has no such concern. Similarly, one cannot set the discount rate equal to the marginal rate of return on capital, because this number also incorporates concerns that are inessential from the point of view of society. The marginal productivity of capital is a function of the quantity of capital available for investment, which is determined by the propensity to save of individuals, which is in turn influenced by their own fear of death. Thus the interest rate and the rate of return on capital incorporate a concern over risks that the state is not exposed to.

So while many economists would feel more comfortable if they could find a positive or revealed source for the social discount rate, most have come to the realization that it must, in the end, be specified normatively. The question then becomes, what factors are morally relevant to the determination of the social discount rate? This is where the Ramsey formula comes in. Ramsey suggested that the present value of costs and benefits should be reduced by a factor that reflects four distinct considerations. Although this is not the way that he expressed it, the Ramsey formula is usually written as follows:

$$r = \delta + \eta g \qquad (3)$$

On the right-hand side, in the second term, g represents the rate of growth, and it is multiplied by the so-called elasticity of marginal utility of consumption (η, or *eta*) to reflect the fact that, with respect to welfare, we get less "bang for the buck" out of economic growth as society becomes wealthier. Jointly, the two represent the rate at which the marginal utility of consumption declines over time.[12] The general thought is that we should care less about a particular cost or benefit, measured in monetary terms, if we expect to be richer in the future, because this cost or benefit will have a less significant impact on *welfare* in that future state than it would in the present. It seems natural then to care about it less in the present as well.[13] For instance, the standard critique of consumerism, which claims that economic growth beyond a certain threshold, is absorbed into luxury consumption that does little or nothing to increase human happiness, can be expressed by assigning a high value to *eta*. As we saw in chapter 2, it is a (somewhat perverse) implication of this view that the balance of mitigation and adaptation should be shifted in the direction of adaptation, because the economic losses imposed on future generations will have so little impact on their well-being. This is reflected in the fact that the high value assigned to *eta* will result in a high discount rate, which in turn lowers the SCC value (i.e. it translates into a much lower value being assigned, in the present, to the future costs of climate change). This lowers the cost of adaptation (in the future) relative to mitigation (in the present).

This right-hand term is relatively uncontroversial, in that it simply accounts for economic growth, modified to reflect the declining marginal utility of consumption.[14] It is the first term on the right-hand side that has generated the most commentary and disagreement. The variable δ (or *delta*) normally represents an aggregate of two things, first, a measure of risk, which reflects the probability that the benefit or cost will not actually come about; and second, a *pure time preference*, which reflects a "preference for consumption now, rather than later"[15]—factoring out all effects that can be ascribed to uncertainty and income level. It is the time preference that is most controversial. Henry Sidgwick claimed long ago that such a preference is irrational even at the individual level.[16] On his view, we should be indifferent between happiness achieved now and happiness achieved later, so long as they are in other respects the same. Impatience, as such, is fundamentally irrational.

Economists for the most part have been reluctant to embrace this sort of Olympian rationalism. From a certain perspective, it may be irrational to become impatient, just as it may be irrational to become frustrated, but if people are experiencing impatience or frustration, that looks like a preference, which has just as much claim to being satisfied as any other. Yet even if one were to grant that a time preference is a legitimate preference at the level of the individual, it does not follow that society is allowed to exhibit a preference for welfare that is achieved now over welfare that will be achieved later—particularly not if different people are involved. Many philosophers, I suspect, think that economists who endorse a time preference are committing the utilitarian sin of moving from an individual preference to an aggregate social preference in a way that fails to respect the distinctness of persons. (After all, from the fact that I can make choices that compromise my own future satisfaction, it does not follow that I should be able to compromise someone *else's* future satisfaction.) It has therefore struck many philosophers as obvious that a social welfare function should exhibit *temporal neutrality* (minimally, that if two sets of ordered possible worlds are identical save for the *time* at which benefits or costs are realized, then the social welfare function should be indifferent between the two). Many regard this as sufficiently obvious as to be beyond argument[17]—and so, by way of support, merely cite Ramsey's claim that temporal discounting is "ethically indefensible" or quote Sidgwick's principle, "that the Hereafter *as such* is to be regarded neither less nor more than the Now."[18] But this is all just bald assertion. When one turns to the actual arguments that have been made in support of temporal neutrality, one begins to see that the question is more complicated than it may initially appear.

There are two important arguments against temporal discounting (i.e. a positive pure time preference) that show up in the literature. The first involves what I am inclined to regard as a simple fallacy, and so it can easily be set aside. The second, however, turns on a much more profound difference in metaphysical view, one that is closely tied to the way that consequentialists conceive of our moral obligations. It will therefore require more careful discussion.

6.1.1. Temporal Discrimination

The simplest argument against time preference in social discounting appeals to the equality of persons, claiming that temporal discounting is a case of straightforward discrimination against future persons.[19] For instance, Simon Dietz and Nicholas Stern describe time preference as imposing a system of "discrimination by birth date."[20] A positive temporal discount rate is often described as the view that future welfare is less important than present welfare, or that the welfare of one group of people ("people who live in the future") counts for less than some other group of people ("people living now").[21] Consider the following, particularly compact statement of the view by Geoffrey Heal:

> A PRTP [pure rate of time preference] greater than zero lets us value the utility of future people less than that of present people, *just because they live in the future rather than the present.* They are valued differently even if they have the same incomes. Doing this is making the same kind of judgment as one would make if one valued the utility of people in Asia differently from that of people in Africa, except that we are using different dimensions of the space-time continuum as the basis for differentiation.[22]

Although Heal presents this series of claims as though they were self-evident, it is not difficult to see that the paragraph bristles with philosophical difficulties. The question of what it means to "live in the future," and whether this can productively be compared to living in Africa, will be dealt with in the following section. For now I would like to focus on Heal's rather subtle, but nevertheless important, misrepresentation of how temporal discounting works. What the time preference creates is not a system in which future welfare counts for less than present welfare, but one in which the *present* value of future welfare is less than its eventual value, at the time when it is brought about. It is only so long as the welfare remains in the future that it counts for less. Under a positive rate of time preference, the benefits to all persons are still given exactly the same weight at the time at which they are realized, just as they are all given the same weight when they are temporally removed from the present by the same duration.[23] (In the same way, the check for $100 is still worth $100 at the time that it is cashed.) It may turn out that this constitutes some form of discrimination, but it is certainly not straightforward discrimination. Because *everyone's* (potential) welfare counts for less, the further removed it is from the present, there is no violation of equality, so long as the same policy is applied consistently over time.[24]

For a helpful analogy, it is useful to consider the issue of "age discrimination," which many philosophers believe is not really a form of discrimination (or at least not straightforward discrimination, the way that racial or gender discrimination is).[25] If one were to implement a mandatory retirement policy by requiring that everyone born before the year 1956 stop working, then that would be discriminatory, because it would pick out one cohort for differential treatment (and so it would be "discrimination by birth date"). If, however, there is a standing policy requiring that everyone over the age of 65 stop working, it is not necessarily discriminatory, because age is not a temporally rigid designator, and so the entire population can be expected to flow through that age bracket in the fullness of time. Thus the policy applies equally to all citizens *at the time they reach 65*, regardless of what their birth date is. The policy treats different people differently at different points in their lives, but it treats everyone exactly the same over the *course* of their lives. So no one can complain about discriminatory treatment, or of being treated unequally, because no one can point to any respect in which he or she has been treated differently than anyone else. Thus mandatory retirement is not a system of "discrimination by birth date."

It is only if one takes a snapshot at a particular point in time that there appears to be a violation of equality in the mandatory retirement system. But this is the wrong way to evaluate many policies, not just those that involve discounting. Pensions, for instance, are sometimes mistakenly described as "redistributing" from the young to the old. They only look this way from the snapshot perspective. The correct way to assess their distributive impact is to look at the effects they have over the entire lifetime of participants. It is only then that one can see the true redistributive structure (and thus, the insurance function), which in a defined benefit system involves making transfers from those who die young to those who die old.

Now clearly an age-differentiated policy should not be capricious—preventing everyone from working at age 35, but then allowing them to resume the following year, would be objectionable. It is important to be clear, however, that the objection could not be that this sort of discrimination among persons violates equality; the objection would have to be that the discrimination lacks a rational basis. Similarly, it is important that there be a standing policy. If one were to impose a one-time tax on 35-year olds, to pay for some state venture, this would violate equality, because it would catch only those who happened to be 35 that year. But if the special tax on 35-year-olds were a policy that was consistently applied over time, one could no longer object to it on the grounds that it violated equality, because in the fullness of time everyone would wind up paying it.[26]

So with respect to discounting, all that the equality of persons requires is that there be a *non-capricious* discounting policy that is applied *consistently over time*; it does not prohibit temporal discounting of welfare. When critics of time preference argue, for instance, that a 3 percent rate of discount suggests that "someone born in 1960 should 'count' for roughly twice someone born in 1985,"[27] this is like characterizing mandatory retirement as a policy that prevents people born before 1956 from working. There is clearly some fallacy involved.[28] Part of the problem seems to be that critics of discounting have the wrong implicit picture of how time works. They tend to speak about it as if it were some kind of a plenum, with different individuals "existing" at different temporal locations. A much more appropriate image is that of a conveyor belt, feeding people into the present, with different stations along the belt that perform different operations on people as they pass by.[29] This makes it easier to see how everyone could be treated in exactly the same way over the course of their lives (including their "pre-lives"), even if people are treated differently depending on how far away they are from the present.

These considerations are not sufficient to show that any particular discounting policy is justifiable. They do, however, show that the standard one-line argument, claiming that time preference is inherently discriminatory, or that it clearly violates the equality of persons, is insufficient to establish the need for temporal neutrality. If there is an egalitarian argument to be made against the practice, it will have to be much more subtle than the one that has traditionally been offered.

6.1.2. Space-Time Analogy

The second major argument for temporal neutrality appeals to the intuition that *when* an event occurs is a morally irrelevant property of that event, and so moral evaluation cannot take such features into consideration. We already saw a simple version of this with Heal, and the claim that space and time are just different dimensions of the same "continuum," both of which are irrelevant from the moral point of view. Similarly, Tyler Cowen claims that "we can think of the universe as a four-dimensional block of space-time. We would not discount human well-being for temporal distance per se any more than we would discount well-being for spatial location per se."[30] A more discursive version of this argument can be found in Derek Parfit's discussion, in *Reasons and Persons*, which has been extremely influential. The discussion is very short, and so can be cited almost in its entirety:

> Suppose that I leave some broken glass in the undergrowth of a wood. A hundred years later this glass wounds a child. My act harms this child. If I had safely buried the glass, this child would have walked through the wood unharmed.... Remoteness in time has, in itself, no more significance than remoteness in space. Suppose that I shoot some arrow into a distant wood, where it wounds some person. If I should have known that there might be someone in this wood, I am guilty of gross negligence. Because this person is far away, I cannot identify the person whom I harm. But this is no excuse. Nor is it any excuse that this person is far away. We should make the same claims about effects on people who are temporally remote.[31]

The thought is that, in both cases, I initiate a chain of events that results in some easily foreseeable harm to another person, and so I have done something equally wrong in both cases. The amount of time it takes for this chain of events to unfold, or the amount of space that must be traversed, are both morally irrelevant. Morality is concerned with actions and their consequences, not when or where those consequences occur (or more precisely, it is concerned with qualities of those consequences *other* than when and where they occur). The analogy to space is obviously an invocation of

Peter Singer's well-known argument that "proximity and distance" cannot be considered relevant factors when deciding whether to help someone in distress.[32] Proximity in time, Parfit claims, is just like proximity in space; both are irrelevant from the moral point of view.

One need not reflect upon this argument for long, however, to note some important differences between temporal and spatial distance. First, there is the fact that all persons, regardless of their spatial distance from us, are nevertheless actual persons. If one considers the set of persons who are temporally distant from us, it is obvious that most are not actual persons, in the sense that they do not presently exist. (Furthermore, there is an internal connection between the passage of time and the passage in and out of existence of persons.)[33] For instance, and at the risk of stating the obvious, it would be quite peculiar to describe Napoleon as "living in the past," since he is, at the moment, not living at all, but rather dead. Nevertheless, if one were to take the "space-time continuum" talk seriously, or even just literally, it would imply that we have exactly the same obligations toward dead people that we have toward the living. Of course, one might say that this is irrelevant, because time's arrow makes it impossible to do anything to improve their condition (setting aside the Aristotelian mechanism of posthumous revisionism). It nevertheless seems strange and unparsimonious to assert that the dead have the same moral standing as the living, it is just that, sadly, we are not able to do anything to help them. (It would also sound very strange to criticize someone for discriminating against the dead merely because they "live in the past.")

Turning our attention to the future, it is worth noting that most of those in the future have not yet been born, and are therefore merely possible persons in the present. This seems like a difference that could have some moral significance. Caney has argued that "location in time" is not a "morally relevant feature of persons," and so it is discriminatory to treat people differently because of it.[34] And yet nonexistence clearly is a morally relevant feature of persons. For instance, individuals are usually thought of as acquiring a right to life only sometime *after* conception (and the right to own property, or to vote, only sometime after that). Thus it is clearly false to say that existence is a morally irrelevant feature of persons, since we all acknowledge certain rights or entitlements that are acquired only after a person becomes actual. As a result, "living in the future" is not the same, from the moral point of view, as living in Africa. The simple version of the space-time argument is therefore invalid ab initio. Even if we limit ourselves to the set of possible persons who will someday become actual—call this the set of eventual persons[35]—we might still wonder whether the fact that, at a given point in time, the person who will be affected by our actions does not yet exist is a morally significant fact about that person. Obviously we can treat merely possible people differently from actual people. But is it discriminatory to treat eventual people differently from actual people?

Parfit's argument is more sophisticated, because he does not talk about our treatment of future persons, but rather about causal chains initiated in the present that reach out into the future. It would clearly be sophistry to suggest that just because a person does not exist right now, we are partially or wholly relieved of our moral obligations toward that person. When it comes to the moral assessment of certain actions, it would seem to be not the timing of the *act* that matters, but rather the timing of its *effects*. The fact that, at the time the broken glass is left in the woods, the child who will someday be hurt is not an actual person, is morally irrelevant. What matters is that, at the time she *is* hurt, she is an actual person.[36] So even if we grant that only actual persons can be harmed, eventual persons cannot, we are still obliged to be concerned for the welfare of eventual persons, precisely because they will become actual persons in a foreseeable way, and we are obliged to be concerned with what happens to them at that time. A more economical way of expressing this is just to say that even though we do not need to worry about eventual persons as such, we do need to show concern for the effects that our actions will have, *in the future*, on actual persons. (This also explains why we care more about future people than we do about past people. Neither are actual persons, but the former, unlike the latter, have a chance to become actual, and we care about what happens to actual people.)

There is, however, a more subtle difference between actual and eventual persons, which should have implications for our moral reasoning. The problem with eventual persons, simply put, is that they vastly outnumber us (where "us" is the set of actual persons, alive right now). There are in fact an infinite number of eventual persons, because time—unlike space on this terrestrial sphere—is unbounded (keeping in mind that, because the Ramsey formula distinguishes uncertainty from time preference, the possibility of human extinction is being factored out and treated separately).[37] So if we literally refuse to discriminate, when contemplating the consequences of our actions, between actual and eventual persons, then the interests of eventual persons have the capacity to completely swamp any concern we may have with those currently alive. This can have very perverse consequences.

Ironically, one of the major objections to Singer's argument, over the years, has been that its conclusion is too demanding, in that it suggests that we may be obliged to give to famine relief, or humanitarian assistance more generally, until we ourselves are reduced to a subsistence level of consumption. And yet if this conclusion seems too demanding, expanding the circle of moral concern to include an open-ended stream of eventual people is far more extreme—it suggests that we may not even be entitled to meet our own subsistence requirements. When combined with standard utilitarianism, temporal neutrality implies that we may be obliged to starve ourselves, not just to save future generations from similar torments, but even to provide them with trivial benefits.

This is an observation that was made by Tjalling Koopmans in 1960, in a paper that provided much of the structure for subsequent reflection upon the topic by economists.[38] Fundamentally, the perverse consequence arises because the relationship between people who are alive today and those who will be born sometime in the future does not have the structure of a standard distributive justice problem. As we have seen, we do not simply transfer goods to future people, or send aid into the future, the way that we airlift aid to impoverished parts of the world. Nothing stops us from doing so, of course. Just as Parfit's malfeasor buries glass in the woods, creating a trap that will be sprung in a hundred years, we could pack non-perishable goods into time capsules that will pop open sometime in the distant future. But as we have seen, the standard way that we provide benefits to future generations is by investing some fraction of our economic output, rather than consuming it all. We spend time and energy producing capital goods that will allow us (or them) to produce consumption goods with less effort in the future. There is very little forward transfer of consumption goods—each generation basically produces its own. Earlier generations make productive investments that, in turn, make it easier for later generations to meet their consumption needs (and, of course, allow them to expand their consumption within the scope of their resource constraints). The process is cumulative, so that by investing now we make it easier for future generations, not only to consume more, but to save more as well. This is why economic growth is exponential over time.

All of this is to the benefit of humanity, but it creates a serious problem for anyone endorsing a zero rate of discount. Making a productive investment, rather than consuming, creates a finite loss of welfare in the present, combined with a *stream* of benefits, extending out indefinitely into the future. With a discount rate of zero, the value of these benefits, *no matter how small in any one period,* will outweigh the finite loss of utility in the present *no matter how large*.[39] For a utilitarian, it follows rather immediately that we should be saving as close as possible to 100 percent of our economic output—even if this leaves us unable to meet our present subsistence needs. As Arrow has observed, "Strictly speaking, we cannot say that the first generation should sacrifice everything, if marginal utility approaches infinity as consumption approaches zero. But we can say that given any investment, short of the entire income, a still greater investment would be preferred."[40] Furthermore, there is no obvious point at which we should stop saving, as long as there are productive investments that can be made. This implies that, in effect, almost all consumption should be deferred indefinitely, regardless of the suffering this may create in the present.[41] Koopmans referred to this as the "paradox of the indefinitely postponed splurge."[42] As he put it, "Too much weight given to generations far into the future turns out to be self-defeating,"[43] because it may lead us to *always* privilege the future over the present.

The level of mathematics required to understand Koopmans's paper is nontrivial, which may explain why his argument has largely been passed over by philosophers.[44] Thus it may be helpful to rephrase the basic paradox as a puzzle of the sort that philosophers enjoy. Consider the case of a group of villagers who discover a shmoo. The shmoo, as one may or may not recall, is a fictitious creature introduced by Al Capp in the 1940s comic strip *Li'l Abner*. Round, plump, and vaguely humanoid, the shmoo subsists only on air, reproduces asexually, is delicious to eat and happy to be eaten. (Capp introduced the shmoo as a way of making a point about the threat to capitalism posed by post-scarcity economic conditions, but the creatures can be used to illustrate other themes as well.) The villagers get ready for a feast, preparing to roast the shmoo. But then a small child observes that, if they were to refrain from eating the shmoo for a while, it will reproduce, and so they could eat two shmoos instead of just one. Everyone is impressed by the child's wisdom, and so the feast is postponed. Eventually the time comes when the shmoo has, indeed, reproduced, and so two shmoos are now loitering about the village. The people begin to prepare for a feast. But again, a small child (perhaps a different one this time), observes that, if they were to refrain from eating the shmoos for a while, the creatures will reproduce again, and so the villagers could eat four shmoos instead of just two. Everyone is again impressed, and so the feast is postponed. One can see the problem that is developing.[45] The villagers are now confronting "the paradox of the indefinitely postponed splurge"—no matter how much they have now, they could have more if they waited a bit longer before consuming, but as a result, no one can ever consume anything. Through their far-sighted maximization of future satisfaction, they are making it so that no one ever gets any *actual* satisfaction.

It would be odd, to say the least, to discover that morality requires us to disregard the interests of all actually existing persons, much less all existing persons in perpetuity. And yet these are some of the counterintuitive implications of prohibiting any differential treatment of eventual and actual people. Forced to choose, we may wind up consistently assigning priority to the interests of eventual people over actual people (particularly because there are so many of them, and because they are, in a sense, much easier to help). (As John Maynard Keynes observed, "The 'purposive' man is always trying to secure a spurious and delusive immortality for his acts by pushing his interest in them forward into time. He does not love his cat, but his cat's kittens; nor, in truth, the kittens, but only the kittens' kittens, and so on forward forever to the end of cat-dom. For him jam is not jam unless it is a case of jam to-morrow and never jam to-day.")[46]

This highlights another respect in which the analogy between space and time is fundamentally flawed. The earth is finite, and so refusing to assign any moral significance to spatial "proximity and distance" nevertheless leaves us with moral obligations to a finite set of people. Time, however, extends out indefinitely into the future, and so refusing to assign any moral significance to temporal "proximity and distance" leaves us with obligations to an infinite number of people.[47] Since our moral principles have historically been developed and applied only to finite-set cases, it is hardly surprising to discover that they generate surprising or perverse implications when extended to infinite-set cases.[48] There have, of course, been many attempts by consequentialists to grapple with the "infinite horizon" problem, which I cannot discuss in detail here.[49] My objective has merely been to show that Parfit's "buried glass" argument, which is often appealed to as a decisive refutation of temporal discounting, rests upon a false analogy. Being far away in (terrestrial) space is not the same thing as being far away in time, and no moral theory should treat them as equivalent. So again, while there may be arguments to be made against temporal discounting, they will need to be much more subtle than the ones that are usually appealed to.

6.1.3. Parfit's Nonidentity Problem

There is an additional issue, which has not figured prominently in the discussion so far, but which has had a significant impact on the thinking of consequentialists about the question of obligations to future generations. This is Parfit's "nonidentity problem." While I do not want to go into too much detail on the issue, I would like to point out that this problem actually follows quite closely from the plenum view of space-time and moral obligation, and that part of the reason contractualists have been less troubled by nonidentity concerns is that they are not as attracted to the plenum view.

I began the entire discussion of environmental ethics by noting the importance of Singer's "expanding the circle" arguments. In his "Famine, Affluence, and Morality" paper, he argues that the circle must include all persons regardless of how far removed they are from us in space. In "All Animals Are Equal," he argues that it must be expanded further, to include all sentient beings.[50] When it comes to the question of what we owe to future generations, the dominant impulse among consequentialists has been simply to expand the circle further, to include "future people" (or perhaps "future sentient beings"), in addition to those alive right now. This is, it should be noted, a much more dramatic expansion of the circle than anything proposed by Singer. The important thing about the conceptual shift is that it does not change the nature of our moral obligations, but merely expands the circle of persons (or beings) toward whom these obligations are owed. So if one were to ask why we have an obligation to refrain from performing some action that will hurt some person x in the distant future, the answer is that right now we owe it to that future person x to refrain from imposing that harm. In other words, we owe the duty to some specific person who just happens to occupy a position in the space-time plenum that is some distance removed from us.

The nonidentity problem arises because of a peculiar causal power we have, which is that, through our decisions, we can prevent certain future people from coming into existence, while bringing others in to take their place (think of this as slotting possible people in and out of the eventual person category). This is what it means to change the "identity" of future people. Strangely, even though this looks a lot like killing people (on the plenum view), ordinary morality does not prohibit it, nor is it treated as a wrong against the person who is consigned to oblivion in this way. This generates a puzzle, which is that if one were to do something that harms the person in an eventual person slot, but at the same time, changes the identity of future individuals, in such a way that some other person winds up occupying the slot, then that particular person cannot really claim to have been "harmed" by the action, since the alternative was to have not existed at all. (The standard example is of an act that changes the timing of an act of conception, so that a different person winds up being born than otherwise would have.)

Parfit's argument became quite polarizing, in the sense that it divided moral philosophers into two quite distinct camps: those who worry about the nonidentity problem and those who do not. I would like to suggest that, among those who worry about it, it is their commitment to the plenum view that is generating the concern. The problem is that, if one thinks about future people as merely people who are some distance removed from us in time, then it becomes very difficult to see how it could be permissible for us to act in ways that prevent them from coming into existence. To see this, consider a "presentist" variation on the nonidentity problem. The ratio of abortions to live births in the United States is approximately one to five. Suppose that one could, by waving a magic wand, make it so that one-fifth of the infants conceived in the United States were to disappear, and be replaced (imperceptibly) by those who were aborted. The latter are now to be offered a second chance at life. Assume that it is not immoral to do this. Suppose, however, that as a side effect of the intervention, those who are given this second chance at life have a dramatically reduced life expectancy. Nevertheless, one cannot say that they are "harmed" by this, since the alternative to living with this shortened life-span is to have been aborted. Nevertheless, it does not seem right. Hence the problem.

This thought experiment is rather wild, but I think it shows that there must be something wrong with the plenum view. My inclination is to think that, in this presentist version, the flaw in the argument lies in the early assumption that one does no wrong to the infants who are magically shuffled out of existence and replaced by others. And yet, if future people are just like actual people, save for their location in space-time, this implies that our actions that change the identities of future people are also wrong. But since this is absurd, we can infer by reductio that future people are not "just like actual people save for their location in space-time." More generally, one might conclude that the plenum view is a deeply misleading way of thinking about our obligations to future generations.

Deontologists in general have not been so worried about the nonidentity problem because they see the rightness and wrongness of actions in terms of conformity to rules, and not first and foremost in terms of the effects that they bring about in the world. Within such a framework, it is much easier to arrange things so that which specific persons are affected by one's actions may not matter from the moral point of view. For example, Rawlsian contractualists have been largely untroubled by the nonidentity problem, at least as far as justice is concerned, because they view the basic institutional structure of society as the subject of justice. This means that obligations of justice are associated with institutionally defined *roles*, not concrete, specific individuals. As a result, these obligations are owed to whoever comes to occupy a particular role. In the same way that one has obligations of hospitality toward one's "guests," regardless of which specific individuals show up to the party, one has obligations toward "persons," regardless of the specific identities of those who are born.

6.2. Reflective Equilibrium

Many economists have taken the Koopmans argument as sufficient grounds for adopting a positive rate of time preference. Or, since so many of the problems seem to center around the "infinite horizon" issue, they have introduced various kludges that serve to impose a limit on that horizon in order to render the problem finite. Most obviously, it is possible to rely upon economic growth—the other term in the Ramsey equation—to ensure that in practice none of the counterintuitive implications of a zero time-preference are ever acted upon. One can even change η so that instead of representing the elasticity of marginal utility of consumption, it is taken to represent a measure of inequality aversion. In a positive-growth scenario, this effectively makes the social welfare function a prioritarian one, which in turn puts greater weight upon the present. This approach, however, relies upon the contingent expectation of positive economic growth to avoid absurdity. In the case of climate change, many have felt the need for a framework able to generate plausible recommendations in zero- or negative-growth scenarios as well.

The remaining option on the table—short of outright rejection of the Ramsey formula—is to rely upon uncertainty as a way of avoiding the infinite horizon problem. This is the approach adopted by Stern in his much-debated report.[51] Stern endorses the view that pure time preference is ethically indefensible, yet he assigns δ a value of 0.1 percent by introducing a bit of uncertainty, based on the probability of human extinction.[52] This allows him to avoid certain technical problems, but it does little to avoid the problem of demandingness. If morality requires us to be concerned about an enormous stream of nonexistent people, how can we avoid having this overwhelm our concern for the minority of real, flesh-and-blood persons who happen to exist right now? At Stern's value for δ, it takes more than 4,600 years before the weight that we assign to the welfare of an eventual person declines to less than 1 percent of the present value of the welfare of an actual person. So using that as a cutoff on our horizon of moral concern, and assuming that the current human birthrate remains constant at roughly 2 percent per annum, means that in addition to the roughly seven billion people currently alive, Stern is committed to assigning some weight to the interests of an additional 644.42 billion people (although of course he is discounting, and so not assigning full weight to their interests). At his discount rate, this is equivalent to assigning *equal* weight to the interests of 138.5 billion eventual people. Put in more concrete terms, when we think about the effects that our actions will have upon others, the eventual people that we are obliged to concern ourselves with outnumber actual people by almost 20 to 1. So even though there is no strict absurdity in this way of framing the problem, it is not difficult to imagine circumstances in which the interests of actual people would be drowned out by our concern for those in the future.

At the same time, it must be recognized that if one adopts the economist's preferred solution, of introducing a substantial time preference into the discount rate, it is possible for the interests of the present to completely outweigh those of the (distant) future. The "demandingness" objection seems to be matched by an "undemandingness" objection on the other side.[53] This suggests that what we are dealing with, when we consider the discounting question, is a dilemma: on the one hand, it seems morally impermissible to treat the welfare of eventual persons as less important than the welfare of actual persons, but on the other hand, trying to treat all welfare the same has implications that are deeply problematic.[54] So what to do? At very least, these reflections may help to establish the Socratic wisdom, of knowing that we do not know. It can help us to see that our moral intuitions in this area are unreliable, and so we should be willing to consider all arguments, without prejudging the question of what the appropriate rate of time preference should be. This is, as a matter of fact, exactly the conclusion that Koopmans drew from his model, arguing that the discounting question is too "unfamiliar" to be settled "a priori." Thus he suggested that "one may wish to choose between principles on the basis of the results of their applications."[55]

If we follow up on this suggestion, one way of justifying a positive rate of time preference would be to follow the method of reflective equilibrium, more or less as John Rawls described it, by formulating an abstract principle (the social discount rate), examining the specific consequences of its application to various cases, then making mutual adjustments between the two until an equilibrium is reached.[56] This is a strategy that has been followed most prominently by Arrow. (Despite the supposed popularity of this method among political philosophers, it is difficult to find a single example of a philosopher having followed it. Instead, most proceed as though they were foundationalists, with the commitment to a zero rate of time preference serving as an indubitable proposition.) Again, Arrow begins by pointing out the extreme unreliability of intuitions when it comes to addressing the question.[57] Because it is applied exponentially, tiny differences in the discount rate translate into huge differences with respect to final outcome—the magnitude of which can surprise even seasoned economists, used to working with such models.[58] Thus it makes very little sense, Arrow claims, to apply our moral intuitions to the discount rate itself. No one has moral intuitions fine-grained enough to distinguish a discount rate of 4.5 percent from 5 percent. And yet, in the fullness of time, this difference is one that will make a very large difference.

If the actual rate of discount is too small-scale for us to have useful intuitions about, the implications of these rates over the course of centuries, or millennia, are far too large-scale. It is probably not very useful to consider thought experiments such as the one suggested by Cowen and Parfit: "Imagine finding out that you, having just reached your twenty-first birthday, must soon die of cancer because one evening Cleopatra wanted an extra helping of dessert."[59] Apart from the fact that the example involves consideration of a one-time event, not a standing policy, it is not clear that we have clear and distinct moral intuitions in such cases. Obviously you would be *disappointed* to discover the consequences of Cleopatra's actions. But whether you can attach moral responsibility for your current plight to someone who died in 30 BC, such that you might judge yourself to have been personally *wronged* by the last of the pharaohs, is a different question, one over which, I would suggest, reasonable minds might differ.[60] A more plausible intermediate measure would be helpful as a target for our intuitions— something coarser than the discount rate itself, and yet less coarse than an outcome that will reach fruition only in 2000 years.

Arrow's proposal is to work out the *savings rate* that would be entailed by various social discount rates. (He is working within the utilitarian "just savings" tradition, and so assumes that there is an obligation to maximize welfare over time.) One can do this by taking a simple growth model of the economy, then varying the savings rate until one arrives at a growth trajectory that equalizes the discounted value (according to one's favored rate) of ag-gregate consumption at any time period. In this way, we are able to see just how much consumption we would be obliged to forgo in the present in order to treat all persons—present and future—in the way that a particular rate of time preference prescribed. This has the advantage of permitting us to see what difference it would make in our own lives to take a certain sort of moral obligation toward future generations seriously. It also allows us to put partic-ular proposals into historical perspective, by comparing their implications to the range of saving rates that have been achieved in different societies at different times.

This is the basis on which Arrow develops his criticism of the Stern Review. In his report, Stern wound up recommending very dramatic ac-tion to mitigate climate change, including a rate of carbon taxation approx-imately 10 times higher than the prevailing rate in the states where it has been implemented (or the rate that has been recommended as optimal by economists such as William Nordhaus).[61] Arrow objected to this, arguing that the same discount rate, applied to the more general question of just sav-ings, would require us to adopt a savings rate of just over 97 percent of GDP. This is, needless to say, not a proposal that can be taken seriously.[62] Arrow suggested, therefore, that there was some inconsistency between the way that Stern approached the specific issue of climate change and the way that he would approach other questions of intergenerational justice. His formula for trading off present costs against future costs exhibited a willingness to impose huge costs in the present, in a way that would be self-evidently out-rageous if applied consistently to all areas of either private or public decision-making. For example, as Christian Gollier has observed, Stern's discount rate implies "that one should be ready to give up as much as 60 percent of GDP per capita, now and forever, to raise the growth rate of GDP over the next two centuries from 1% to 2% per year."[63]

Rather than starting then with a value such as the discount rate, over which we have very few clear intuitions, it would make more sense—Arrow suggests—to start with something more tangible, like the savings rate, and then extrapolate the rate of pure time preference implied by that value. For instance, one could start out with some morally attractive principle, such as the one underlying the "golden rule" savings rate—the rate that maximizes steady-state growth of consumption per person—and infer a discount rate from that.

The question is complicated by various factors. First of all, the growth model being implicitly relied upon by Arrow is a relatively simple one, in which the only determinant of growth is the savings rate. If one includes other factors, such as technological change, then it is not obvious that Stern's discount rate implies such a wildly unreasonable rate of savings.[64] Second, there is the simple fact that most savings in our society are not undertaken for eleemosynary purposes, or out of some commitment to justice for future generations. This is obvious in the case of corporations, but equally true of personal savings. A better approach to assessing the plausibility of any particular discount rate would therefore be to examine some of its other policy implications, to see if the same number can be applied consistently in all domains. If we assume a zero rate of pure time preference, for instance, what does this imply for our approach to healthcare funding? For instance, what sort of investments should we be making in malaria eradication?[65] Or what does it imply about our rate of investment in science and technology? What sort of standards should govern infrastructure that is built? It is only by situating the discounting issue within this richer framework of policy issues, and making adjustments between the more abstract and specific levels, that we can apply the method of reflective equilibrium.

If one were to adopt such an approach, and actually work through the implications of a zero rate of time preference, it would become apparent that this commitment is in tension with a very large number of considered judgments. For example, it would certainly turn out that we spend far too much of the healthcare budget on treating sick people and not nearly enough on medical research. Or on a smaller scale, even something as simple as building roads out of asphalt would probably turn out to be impermissible in most cases.[66]

Again, it is important to emphasize that the effects of compounding are *completely unintuitive*, even for seasoned professionals who deal with the phenomenon all the time. Thus it may be wise for philosophers to steer clear of any argument that makes a significant appeal to our moral intuitions with respect to any variable that is to be compounded. If one is going to make an appeal to intuition, it would make a lot more sense to follow Arrow's suggestion and elicit intuitions in areas where they are more likely to be stable and well informed, such as infrastructure projects, or the way that public health investments are made, or even the way that other environmental goods are treated, and then try to infer from this the appropriate social discount rate (making adjustments as necessary in order to achieve consistency among domains).

6.3. Institutionalized Responsibility

Suppose that one is not tempted by the method of reflective equilibrium and would like to treat the principle of temporal neutrality as nonnegotiable. Unless one is also committed to some particularly stark form of act-consequentialism, one might still think that the way one's abstract moral commitments are *institutionally mediated* could give rise to a form of time preference. For example, Singer's "principle of beneficence" imposes an extremely imperfect duty upon the individual. Even if we take into consideration only actual persons, there are literally billions of individuals we could be helping at any given time, e.g. through charitable donation. Narrowing this down to some specific range of persons, in order to produce a concrete duty of assistance to certain named individuals, if carried out perspicuously, would require enormous cognitive resources. And naturally, there is good reason to want to avoid something like the "Baby Jessica" outcome, where thousands of people donate money to the child who happens to fall down a well, resulting in a lavish trust fund for that one child while the greater needs of many others go unmet. Thus there is a lot to be said for having a system of rules that specifies in greater detail exactly who owes what to whom. Spatial and temporal proximity might turn out to be important within such a system.

The relationship between abstract moral commitments and specific rules of conduct could be more or less instrumental. Rule-utilitarians and indirect-strategy consequentialists believe that having everyone follow such rules of conduct is recommended because it will maximize good outcomes over time. Others have argued for a more constitutive status for these rules, on the grounds that our "values" as such are so indeterminate that they cannot guide action without further elaboration. For example, many of our moral commitments are quite general and so require specification in order to be action-guiding. (We may have a duty of "benevolence," but one still needs to know which specific actions count as benevolent.) All of this specification must also be *coordinated* between individuals, so that an appropriate division of moral labor is established, each person is assigned an obligation to help a different person, burdens are distributed fairly, and so on. Thus a set of shared rules is required that will, as John Garthoff puts it, "embody morality by authoritatively proclaiming the content of each individual's obligations and entitlements."[67] We can refer to such a set of rules for convenience as "institutional morality."

From this perspective, it is easy to see how some property of individuals, which has no intrinsic moral significance, might nevertheless *acquire* moral significance, because the system of institutional morality makes use of it as a way of creating determinate moral obligations. With respect to space, for example, one might well agree that from the point of view of the universe, spatial location is a morally irrelevant feature of persons, and yet still think that it has normative significance for real people acting in the world. Consider, for instance, how institutional morality might handle the duty of providing food aid to those who are malnourished. For simplicity, suppose that this obligation is capped, so that once a person has given some reasonable fraction of her food away, she is entitled to keep the rest for her own personal use. The obligation to assist others is then clearly an imperfect duty: the individual is obliged to help some, but not all, persons in need. The question then becomes, which ones? Since there are obvious advantages to having shared rules to determine this, we might expect a set of institutions to develop that determines who has obligations to whom. And given the difficulty of transporting food aid, we might be inclined to adopt a spatial principle, so that people are obliged to help those who are closest to them. In this way, spatial location would become a morally salient feature of persons because of the way that institutional morality parcels out moral obligation.

Suppose that we were to divide the world into a set of "zones" (green zone, red zone, blue zone . . .) and specify that individuals within each zone are responsible for setting up a food distribution system within their territory to ensure that the nutritional needs of everyone within that territory are well met. One might couple this with an undifferentiated obligation to provide emergency relief to people in acute distress in any zone. The result would be a system with a two-tiered structure of obligation, under which each individual is obliged to provide much more in the way of assistance to fellow members of her own zone than to a person living in the neighboring zone. (One might, for instance, have an obligation to strive for equality of condition among all those in one's zone, while having only sufficientarian obligations toward those in other zones.)

It should be noted that, under this scheme, space would still have no intrinsic moral significance. However, space does get *used* as a way of drawing a distinction that becomes, via institutionalization, morally significant. With respect to food aid, the needs of a person generate the same entitlements and obligations, regardless of where that person is located in space. It is just that whom those obligations fall upon is determined by the spatial location of both parties. And as a result, the *specific* obligations that each person has, and whom those obligations are owed to, is determined by spatial location. Thus even though a person in the red zone might reasonably be entitled to a transfer, a person in the green zone might also reasonably refuse to provide it on the grounds that it is not her responsibility. (Parfit's example of shooting an arrow into the woods is unhelpful in this regard. The obligation to refrain from recklessly endangering other people is not something that we have much incentive to territorialize—it wouldn't be the case that it is permissible to shoot an arrow into the woods if those woods happen to be across the border in red zone. But there are many other obligations where it is not so unnatural. If the victim of an archery accident happens to be in the red zone, for instance, it might not be unreasonable that fellow members of the red zone should collectively shoulder the burden of the healthcare system that pays for this person's convalescence.) There can be no doubt that a system of territorialization would generate some outcomes that seem morally arbitrary—simply because the borders are arbitrarily drawn. As Rawls observed, however, to fix on the arbitrariness of borders "is to fix on the wrong thing."[68] It is not difficult to imagine circumstances under which the benefits associated with the assignment of determinate responsibilities would outweigh any cost brought about by arbitrariness in the assignment.

The example of zoned food aid is, of course, nothing but a thinly veiled description of the way that the world we live in is actually organized—instead of zones we have territorially defined states, with elevated levels of obligation among co-nationals. The purpose of the construct is simply to show how spatial proximity might become important in determining what people owe one another. The same is true with respect to time. In order to illustrate this, it is necessary to push the spatial analogy in a somewhat more hypothetical direction. When we think about food aid and our obligations to distant victims of famine, we are accustomed to the idea that assistance can be airlifted from any one region of the globe to another. Suppose instead that both people and goods were confined in their movements. Instead of dividing the world into large territorial zones, suppose that it was divided into small cells, like honeycombs, and individuals for some reason were physically unable to travel outside the cell in which they happened to find themselves. In order to provide food aid to someone in a distant cell, a person would need to travel to the border of his own cell, then pass the goods across the border, with the instruction that they be passed along in a similar fashion until they reach their intended beneficiary. Imagine further that the food never really got passed along for more than one cell, but was consumed by the person who received it, who then found herself with a commensurately larger surplus, some portion of which she would be obliged to pass along to the next person in the direction of the intended beneficiary. In this "honeycomb world," the case to be made for adopting a spatial principle would be even more apparent. One could take an abstract principle of beneficence, for instance, and give it a spatial index, so that when a person found himself in need of food aid, the obligation to provide it would fall on all individuals occupying the same cell, as well as the occupants of all cells that shared an immediate border with it. One might also choose to adopt a spatial discounting principle, so that people sharing the same cell had an obligation to provide a certain transfer, and the amount that others were obliged to transfer declined at some fixed rate the further away they were in terms of number of cells.

This honeycomb model is, of course, precisely analogous to how temporal discounting works, with each cell representing a period of time. It is all well and good to say that every person, throughout all of history, has an obligation to help every other person regardless of what time the beneficiary happens to exist. Nevertheless, there is an enormous amount to be said for making this imperfect duty more determinate. And if we were searching for a principled way of dividing people up into groups, "contemporaneous existence" would seem like an attractive criterion. There are, after all, a number of very practical reasons why we should want to help people who exist right now. In particular, we cannot just airlift goods to the future; the best we can do is pass them along to younger generations (i.e. those with whom we "share a border" temporally). If one were looking for a principle that could be used to divide up time into convenient slices in order to create determinate obligations of assistance, a plausible principle would be to suggest that the existing group of persons should, first and foremost, look after its own, and that its obligation to future generations should decline at some fixed rate, the further removed they are in time. (Imagine how strange it would be to have things organized any other way. What would one say to a person who exhibited no moral concern whatsoever for actual famine victims on the grounds that she was only responsible for some group of future residents of the afflicted country, who were predicted to have plenty of food?) In the same way that it makes sense to territorialize certain moral obligations, it also makes sense to temporalize them.

If this argument is correct, then it is perfectly permissible to introduce a pure time preference when considering the policy questions posed by climate change. Even if one wants to insist that abstract normative principles exhibit temporal neutrality, there is ample room for the introduction of time preference when one turns to the task of elaborating an institutional morality on the basis of such principles.

6.4. Thinking Politically

Finally, it is worth drawing attention to a suppressed premise in the standard Parfitian argument against time preference. Even if it were *morally* impermissible to treat harms and benefits to others as more or less important depending upon when they occur, it does not follow that the state, when formulating public policy, must adhere to a principle of temporal neutrality. In order to get to temporal neutrality of state action, one would need to think that the overriding objective of the state should be to do what is right, from the moral point of view, or that the central objective of law and public policy should be to implement morality. This is a view widely held among utilitarians, but it also has its detractors. In particular, many philosophers have argued that there is an important difference between what you ought morally to do and what the state can legitimately force you to do. The latter set of obligations should be, at very least, a subset of the former. Others have gone further, arguing that because the imperatives of the state are backed by coercion, the principles of justice governing state action are *different* from those governing moral obligation (for example, that they take the form of a system of rights, rather than of moral imperatives).[69]

From this perspective, it is striking that the examples Parfit uses all stem from personal morality. Yet when the state uses a social discount rate to evaluate public policy projects, whether it be a carbon taxation scheme or an infrastructure project, it is contemplating an arrangement that will be coercively imposed. For example, it is a familiar point from the debate over abortion that a person could quite consistently think that having an abortion is immoral and yet still believe that abortion should be legally permitted, or even that prohibition of the procedure would be unjust. Or to use Parfit's example, while it may be immoral to leave broken glass in the woods, in the normal run of circumstances it would not be illegal. Something similar could be true with respect to discounting. Exhibiting time preference in one's actions might be impermissible at the level of personal morality, and yet still be appropriate for the state when it contemplates how to spend tax revenue or regulate environmental externalities.[70] This is not just to say that the state might reasonably concern itself with the opportunity cost of raising tax revenue; it might also legitimately discount the *welfare* of its own future citizens. It is probably worth re-emphasizing that this is not discriminatory, in the ordinary sense of the term, since it applies equally to all citizens. The state may treat its citizens differently at different stages of their lives, or pre-lives, so long as it does so to everyone consistently.

There are two well-known Rawlsian arguments that might be used to generate support for discounting. The first appeals to the fact of reasonable disagreement. In the domain of morality, people can and do agree to disagree on many questions. State action, however, involves decisions that will be binding on everyone. In this situation, the fact that people disagree about what morality requires creates a problem for the idea that the state should simply do what is morally right. People are more likely to disagree about what morality requires the more closely the moral issue is tied to broader cosmological or religious ideas—systems of thought to which individuals may feel a strong sense of attachment, but which are not verifiable in any straightforward way. Beliefs about the distant future obviously fall into this category. With respect to time preference, this has the potential to generate significant disagreement. It is perhaps not an accident that temporal neutrality is largely a preoccupation of atheistic philosophers. For example, concern for future generations is entirely absent from traditional Christian doctrine—the New Testament does not raise the issue and provides no suggestion that anyone has any obligation toward potential people (other than, in some cases, one's own potential progeny). This is partly due to the eschatological structure of Christian beliefs—shared with several other religions—that regards the world as ending sometime in the relatively near future. Belief in reincarnation (a doctrine subscribed to by as much as 25 percent of the world's population) would also appear to justify a strong "presentist" bias, because it suggests that future people are not really new people, just recycled versions of present people. Although the thought experiment becomes mind-bending rather quickly, it does seem to imply that presentism is acceptable because one can discharge one's obligations to future people simply by treating present people in a certain way. (It also blurs the distinction between intrapersonal and interpersonal discounting.)

The general point is just that temporal neutrality does not figure prominently in the moral teachings of *any* major world religion, and many of the latter come freighted with cosmological views that appear to support other principles. This suggests that a certain amount of the presentism in people's moral ideas may not be just self-serving, but may reflect reasonable disagreement about questions of value. At very least, this should encourage greater humility among proponents of temporal neutrality, in recognition of the fact that the position also relies upon something that is, if not exactly a cosmological view, then certainly something that resembles one, and that is, in turn, rejected by many. Liberal neutrality is always most difficult to adhere to when it prohibits the imposition of a view that one considers to be correct.

This is all rather speculative, but there is a second Rawlsian argument that speaks much more clearly in favor of such a conception. It arises in his discussion of the strains of commitment. In order to endorse a particular conception of justice, Rawls claims, the parties must be able to believe that this conception could also be implemented, given human nature roughly as we know it to be (including the extent to which our motivations are modifiable through socialization). One problem with utilitarianism, he claimed, is that it has consequences so extreme that it could never be implemented in practice. Thus contracting parties would not adopt it because, given "general knowledge of human psychology," they would be able to foresee the impossibility of implementation.[71] In Rawls's view, there is nothing wrong with a *moral* code that places extreme demands upon the individual. It is nevertheless important that such actions should not be "demanded as a matter of justice by the basic structure of society."[72]

Utilitarians have often been troubled by the suggestion that their moral code is too demanding because it expects individuals to set aside any special concern for family and friends (much less themselves) and treat everyone impartially. Rule-utilitarians have expended some effort trying to accommodate special obligations within their doctrine. With respect to time preference, however, the same considerations apply. How realistic is it to expect people to show no heightened concern for actual, flesh-and-blood people over abstract eventual people who will be born hundreds of years in the future? How much sacrifice can we realistically expect people to make in order to benefit people who do not exist yet and whom we will never meet? (Many of the arguments advanced over the years in support of "national partiality" in the debates over global justice—which trade on a similar concern about demandingness—work equally well as arguments in support of temporal partiality.)

Of course, if one asks this question as a utilitarian, one risks running afoul of Singer's stricture against confounding what people ought to do with what they are likely to do. Distinguishing the moral from the political, however, offers a way of introducing the more "realist" consideration without confounding the descriptive and the normative. While morality can be as idealized as one likes, when one turns to the question of what the basic structure, or the law, may *require* of people, one cannot be indifferent to the question of what sacrifices people can, as a matter of fact, be persuaded to undertake willingly, because this will affect, in turn, the amount of coercion that must be employed in order to achieve respect for the law. In this context, it is worth asking how the level of carbon taxation consistent with a zero rate of time preference could be implemented right now, at a time when such a policy would produce no significant benefits even for the grandchildren of existing people.

One could follow Dieter Birnbacher and provide a purely "pragmatic" defense of discounting as a concession to the phenomenon of "positive time preference and limited sympathy."[73] This would give it a status similar to the principle of "international Paretianism" advocated by Eric Posner and David Weisbach ("All states must believe themselves better off by their lights as a result of the climate treaty"), which they freely grant "is not an ethical principle but a pragmatic constraint."[74] But the strains of commitment argument is intended to break down this sharp dichotomy by building considerations of motivational feasibility into the choice of principles of justice. In doing so, it obviously raises a host of subtle difficulties. I mention it primarily as a reminder to those involved in the discounting controversy that this argumentation strategy is available and is one way of developing the widely shared intuition that justice lies somewhere *between* the ethical and the pragmatic.

Regardless of the specific argumentation strategy adopted, the decision to treat the choice of discount rate as a political rather than as a moral question suggests the following line of reasoning: despite what one's comprehensive moral view might prescribe, the state might reasonably adopt a principle that imposes a higher level of obligation among contemporaneous generations, in much the same way that individuals have an elevated obligation toward co-nationals.

6.5. Discounting for Deontologists

Making the case for temporal discounting in the abstract is one thing, but in the end we still must answer the more difficult question, which is what the specific rate should be. This is a daunting question, in that there do not seem to be many signposts or ways of orienting ourselves when looking for an answer. We know that individuals have a private rate that is too high to serve as the policy rate because they suffer from both impatience and fear of death, neither of which is a concern shared by the state. We know that zero is too low because of the absurdities that strict temporal neutrality leads to. So we want a number that is somewhere between the two. Ideally, we would also like a rate that does not rely upon the empirical postulate of ongoing economic growth as the *only* way of avoiding absurdity or overdemandingness. Reflective equilibrium could in principle provide us with an answer, but in practice it is likely to be indeterminate because there will be significant inconsistencies in our judgments across different policy domains. So how are we to think about the question?

One approach might be to reconsider the overlapping-generations model of intergenerational cooperation presented in Chapter 3. Cooperation can be sustained in all generations save the last—those who are in the last stage of their lives have no incentive to make any further contribution to the system, as a result of which they are able to defect with impunity. Now of course this occurs in the strictly incentive-based model. However, thinking politically about our own system of cooperation and the strains of commitment that it imposes, one might think it also unreasonable to hold the entire population responsible for events that will occur in the near future, because some fraction of the population will not live that long. Individuals in their last year of life might be thought of as having "paid their dues," and thus as exempt from obligations of intergenerational justice. Taking something like the honeycomb model of intergenerational obligation proposed previously, one can think of a population group "now" having a slightly diminished obligation toward that population group "in one year," and that population group "in one year" having a slightly diminished obligation toward that population group "in two years," and so on, based on the annualized attrition rate in that population. This suggests that we might use the *death rate* as the basis for a pure time preference. Since the annual global rate is around 8 per 1,000, this would produce a pure time preference of 0.8 percent for an international problem like climate change.

The rationale for this would be roughly as follows. The overwhelming majority of us remain participants in the ongoing system of cooperation and, as a result, can be expected to share in its full benefits and burdens. Some small fraction of us, by contrast, are exiting the system and therefore have no further stake in it. The state is governed by principles of justice that are oriented to dividing the benefits and burdens of this system of cooperation. As a result, it is free to discount costs and benefits that will occur in a year's time, on the grounds that only 99.2 percent of the current population will be participating in the cooperative system that will have to bear those costs or be able to enjoy those benefits. With respect to costs and benefits that will occur in two years' time, only 99.2 percent of those who will be alive in one year's time will still be alive in the following year, and so the state should again discount the costs and benefits at a rate that reflects the rate of exit from the cooperative system.

Stern, it may be recalled, introduced a *delta* of 0.1 percent, based on the probability of human extinction. From his perspective—that of the utilitarian planner—society is fundamentally indifferent with respect to which specific individuals get any particular quantity of satisfaction and when. This requires temporal neutrality. On the other hand, if society is going to end through an extinction event, then it is better for us to take our satisfactions prior to that occurrence. This is how he justifies introducing a slight discount to reflect the annual probability of human extinction. In effect, he replaces individual fear of death, which informs individual discount rates, with the probability of societal (or species) death, which is what should inform the social discount rate. Underlying this is the thought that the state should not have a view on how the benefits and burdens of cooperation are distributed across individuals or over time. One can see here how Stern's commitment to temporal neutrality follows from what is widely considered to be the least attractive feature of utilitarianism, viz. the lack of respect for the distinctness of persons. The reason that the state should be unconcerned whether a benefit is realized now or in a hundred years, is that the state is fundamentally unconcerned with *who* receives the benefit—all it cares about is the magnitude of these benefits. Thus sacrificing a generation in order to achieve future prosperity is permissible, on this view, *for the exact same reason* that sacrificing one person, in order to the save the lives of two others, is permissible on the standard utilitarian calculus.

The contractualist perspective, by contrast, does not view the state as a stand-in for the utilitarian planner, but rather treats it as an *institution*, which forms part of an ongoing system of intergenerational cooperation. Its primary obligation is to ensure that its internal affairs are well ordered, which is to say that its rules allocate the benefits and burdens of this system of cooperation in a way that conforms to principles of justice. Because the system of cooperation is intergenerational, this requires some solicitude for the interest of future persons, because new individuals are entering the cooperative system at a steady rate, and the present stability of the system depends upon the expectation that each new generation will be willing to do its part to maintain it. At the same time, it is important to recognize that individuals are exiting the system at a steady rate as well, and there is a sense in which, in the "final round," they are no longer participants in a cooperative system.[75] Because they cannot be expected to enjoy its benefits in the future, they also cannot reasonably be expected to bear its burdens. Thus there is a good case to be made for exempting them from the calculus of costs and benefits governing state policy decisions.

Stern claims that the present value of $1 million one year from now is only 99.9 percent its actual value in one year, because of the slight probability that no one will be around to enjoy it. On the view that I am recommending, the present value is only 99.2 percent of its actual value in one year, because we can be confident that 0.8 percent of the population will not be around to enjoy it, and therefore it would be unreasonable to impose upon them whatever burdens are required in order to achieve it.[76] This might lead one to think that, since only 98.4 percent of the present population will be around in two years, that the discount function should be linear, with the present value of futures costs and benefits dropping to zero in the long term. But this ignores the fact that new generations are born and enter into the cooperative system, effectively replacing those who exited, and the state acquires obligations to them as they enter. Thus it makes sense for the discount function to have the standard exponential form, as the "honeycomb model" suggests.

If we stick with the Stern review assumptions, of 1.3 percent real per capita growth, consumption elasticity of 1 (i.e. a logarithmic utility function), and a 0.1 percent risk of extinction, then adding in the 0.8 percent pure time preference to reflect the death rate yields a social discount rate of 2.2 percent. A more realistic value for *eta* would be around 1.5, which yields a social discount rate of 2.8 percent.[77] This is close to the 3 percent rate that is the median value used in the SCC calculation in Canada (and in the United States during the Obama administration).

6.6. Conclusion

One can see in the debate over the social discount rate that it makes a great deal of difference how we conceive of our obligations toward future generations. The consequentialist sees future generations as inhabiting a currently inaccessible region of a "four-dimensional block of space-time" and believes that we owe them certain moral obligations—such as the obligation to maximize their well-being—merely by virtue of their existence. The contractualist, on the other hand, sees justice as essentially a feature of our institutions and believes that we have obligations toward future generations because of the way that individuals flow through the cooperative system over time. The latter view is, I have suggested, far more congenial to the practice of discounting. This way of conceiving of our obligations, however, is often viewed as inadequate precisely because of the way that it makes obligations of justice depend upon participation in a system of cooperation. It makes them *contingent* in a way that many people find counterintuitive. It invites thought experiments of the form, "What if we were *not* involved in a system of cooperation, could we then do whatever we want to others?" With respect to future generations, in particular, it seems peculiar to suggest that the obligations we have are due to our participation in some relatively recent institutional arrangements, such as the savings-and-investment system.

Tim Mulgan has suggested a thought experiment along these lines, as a way of imposing what he refers to as a "minimal test" on the adequacy of any theory of justice. He imagines a planet inhabited by what he refers to as "Mayfly people," who have a somewhat peculiar life cycle.

> Unlike human beings, the Mayfly people do not have overlapping generations. They live on a planet that takes a hundred of our years to orbit its sun. Each of their four seasons lasts 25 years. The planet is inhabitable in spring, summer and autumn, but not in winter. At the beginning of spring, the Mayfly people's cocoons hatch, and a new generation of Mayfly people are born fully grownup. The previous generation has left behind an established civilization, complete with computers to teach the new generation everything they need to know. This new generation lives for 75 years, adding to the store of culture and knowledge, building new buildings, and so on . . . Then they all die. None of the Mayfly people have particular descendants. They all collectively produce the next batch of eggs.[78]

The question then is what obligations the members of one generation of Mayfly people have toward the next. The "minimal test" that Mulgan proposes is that "no adequate political theory may conclude that the current generation of Mayfly people have no obligations whatever to the next generation."[79]

This minimal test is perhaps a bit too minimal, since it is not difficult to construct a theory that would impose upon the Mayfly people *some* obligation toward the next generation while still claiming that these obligations are significantly less onerous than the ones that we are subject to. As we have seen, contractualist theories typically have a dual structure, whereby duties of justice are linked to the presence of a scheme of cooperation, but where there are other, less onerous duties—such as a weak duty of benevolence or a prohibition on harm—to those who are outside the scheme. Such a theory would prohibit one generation of Mayfly people from destroying their planet—and thus would pass the minimal test—and yet stop short of insisting that they must reproduce and enhance the civilization they have inherited. Thus the interesting question is whether the Mayfly people would have exactly the same obligations toward their descendants that those of us who overlap with them have. Mulgan apparently thinks that they would, since he claims that a plausible political theory must not only "provide a plausible account of our obligations to future generations," but also be able to "derive those obligations from morally significant features of our relationship to those who will live in the future, not from contingent accidents of human biology."[80]

It should be noted that, as far as the debate between consequentialism and contractualism being considered here, Mulgan's example muddies the water a bit by assuming that the current generation of Mayfly people has received a massive cultural inheritance from the previous one. This suggests the presence of a pay-it-forward scheme, which would be a system of cooperation that could serve as a source of obligation. Thus it would be easy to say that, if one generation chose not to pass anything on, they would have failed to do their part, or had violated a duty of fair play. The consequentialist, by contrast, is committed to the view that the Mayfly people would have exactly the same obligations toward their descendants regardless of whether they received any cultural inheritance at all.

Suppose then that the Mayfly people have been leaving nothing for their descendants, have no understanding of the life cycle of the planet, and do not even realize that after they die they will be replaced by a new generation. One day, however, a Mayfly archaeologist discovers traces of past inhabitants of the planet, figures out how to interpret the ruins, laying bare for the first time the structure of the reproductive cycle. Once they have this knowledge, do they suddenly acquire full obligations of justice toward their descendants, identical to our own? Or as the consequentialist imagines, are they obliged to expand the circle of moral concern, so that all of these future people count one-for-one with members of the current generation? Is it plausible to say that the Mayfly people are now obliged to maximize the well-being of future generations?

My inclination is to say that none of this follows. The mere fact that future generations will someday exist does not give them moral entitlements equal to our own. It is fully permissible for the Mayfly people to regard the discovery that they will have descendants as a curiosity, one that has essentially no effect upon their lives. To see this, consider what would happen if humans were to discover the Mayfly planet (something that occurs later on in Mulgan's thought experiment). Since these aliens meet all of the criteria for moral standing, does this mean that we now acquire all of the same moral obligations toward *them* that we have toward our own descendants? Suppose that the harsher conditions on their planet have resulted in their having a standard of living only a fraction as high as our own. Are we now obliged to construct interstellar vessels so that we can share our planetary and civilizational resources with them? Or can we regard their existence as a mere curiosity and remain content to observe it from afar?

It seems to me that the latter is clearly a permissible option. At the same time, I am not particularly confident in this claim, which is one of the reasons that I am not overly partial to these sorts of thought experiments. They are often introduced in the context of discussions of contractualism in order to galvanize opposition to "practice-dependent" conceptions of justice. The goal is to imagine a scenario in which certain social institutions are absent but the moral obligations persist.[81] This is thought to show that the moral obligations must be, in some sense, more fundamental than the institutions. Unfortunately, it is not so easy to construct these scenarios, simply because institutions are such a pervasive feature of human social life. In this respect, it is noteworthy that the institutions of intergenerational cooperation do not entirely disappear in Mulgan's initial presentation of the thought experiment, because he retains the civilizational inheritance. And yet, when one purges the scenario more thoroughly of familiar institutions, the context winds up being so far removed from social life as we know it that it is very difficult to say that our moral intuitions remain the same, or that they have much force.[82]

More importantly, as the thought experiments become stranger and the scenarios more foreign to everyday experience, the moral intuitions that are evoked begin to enter the realm of reasonable disagreement. Moreover, it does not seem plausible to think that our policy response to climate change should hinge upon our reaction to science-fiction thought experiments. We are better off sticking to the institutions we have, along with the widely accepted, well-understood, and minimally controversial conception of justice that they imperfectly embody. Climate change, after all, is a problem that is *endogenous* to the system of cooperation that is currently in place in our society, since it arises only as a consequence of industrialization, economic growth, population increase, etc. It is an undesirable byproduct of the way that the system of production, and in particular the division of labor, is organized in our society.[83] The major trade-offs involved in climate change policy are also internal to this system, since we are mainly contemplating a tension between growth and carbon abatement. Thus there is nothing strange about a policy response to the problem that draws upon normative principles that are also internal to this system.

Of course, we can imagine a completely different scenario, in which there is no system of intergenerational cooperation in place (and thus, no economic growth, no cumulative improvement in living standards, no expectation that future generations will be better off than present ones, etc.). We can imagine environmental harms that might arise in the scenario. From a contractualist perspective, these harms would not constitute an *injustice* toward later generations. Yet this may not be a problem so long as they can be prohibited on other grounds, such as a duty of benevolence or even just a straightforward prohibition on harm. But it is not clear what bearing any of this has on the actual problem that we are confronting. What makes the climate change problem difficult is the fact that it cannot be handled so easily, because the net value of the bequest we are making to future generations is likely to remain positive, and so the entire package does not add up to a harm. In the current situation, as we have seen, duties of benevolence or prohibitions on harm recommend doing nothing to combat climate change. This is why we are obliged to introduce the more onerous obligations of justice if we want to provide any support for a policy response to climate change that involves significant carbon abatement in the present. And it is these principles of justice that are best understood as governing the terms of cooperation within the existing institutional structure of our society.

Conclusion

According to the contractualist, justice represents a response to a fundamental feature of the human condition, viz. the incompossibility of individuals' goals and plans. It is impossible for all of us to get what we want
simultaneously, not just because of natural constraints, but also because each
of us, in pursuing our objectives, *interferes* with others trying to do the same.
We have a tendency to step on each other's toes, get in each other's way, and,
in extreme cases, ruin one another's lives. This can be summed up as the
claim that, in going about our daily affairs, we *harm* others. Furthermore,
there is no natural mechanism that tends to minimize these interferences.
As a result, circumstances often arise in which we can all benefit from imposing upon ourselves a system of rules that will constrain our conduct. It is,
however, impossible to eliminate these harms entirely. When it comes to the
specification of these rules, the question then becomes, when are harms to
others acceptable and when are they not?

Complicating the problem is the fact that what constitutes a harm, and
how grave it is considered, will vary from one person to another, depending
upon how seriously it interferes with the plans that the individual may be
pursuing. There are of course some harms that are so basic, or of such gravity,
that they can be prohibited categorically. But in most cases a compromise is
required. It is difficult, however, to come up with a uniform specification of
how much individuals should be able to interfere with one another. An attractive approach to the problem, therefore, is to let individuals determine
this for themselves, by assigning to them control over features of the environment that are essential to their plans, and then leaving them free to negotiate with one another. A harm will then be considered acceptable if the
person who suffers it is offered a benefit sufficient to outweigh it, where the
"sufficiency" in question is not determined objectively, but rather by the
person in question. Thus the acceptability of an arrangement will be determined, first and foremost, by whether all those affected are willing to accept
it. Furthermore, if this system is comprehensive, it means that individuals
will only be able to produce certain harms if they are willing to offer others

Philosophical Foundations of Climate Change Policy. Joseph Heath, Oxford University Press. © Oxford University Press
2021. DOI: 10.1093/oso/9780197567982.003.0008

compensating benefits sufficient to create an acceptable arrangement. If they are not willing to do so, then they must change their plans in order to reduce or eliminate the harm.

This is why the idea of a contract figures so centrally in this way of thinking about justice. The best evidence that an arrangement is acceptable is that those involved actually accept it. (This is not the only evidence that speaks in favor of or against it, but it is sufficient to generate a strong presumption.) The way that this is institutionalized, first and foremost, is through the classical liberal devices of property and contract law. This is particularly effective when the circumstances of justice (i.e. "moderate scarcity and limited altruism") obtain. People's plans typically involve use of resources of one type or another. This scarcity of resources, combined with a limited willingness on the part of all to sacrifice their interests for those of others, generates incompossibility of plans, and thus the potential for conflict. By giving individuals ownership of resources, and requiring that all transfers be voluntary (i.e. contractual), a mechanism is put in place that will tend to limit interferences, and thus maximize the level of achieved goal-satisfaction.

With the introduction of money as a medium of exchange, the system of bilateral contracts between individuals becomes generalized, and a set of markets emerges, the central feature of which is a system of prices. The prevailing price for a particular resource represents the amount that others must be offered in order to be willing to part with it. It therefore tracks, and in some sense expresses, the amount of harm that is done when the resource is taken out of one use and put to another. The price represents the amount of welfare that others must forgo in order for an individual to enjoy a particular act of consumption. This is what underlies the welfare economist's traditional analysis, in terms of which the ideal price should reflect the full social cost of consumption.

Yet as we know, the system of property rights provides individuals with relatively limited capacity to control aspects of the environment that affect their goals. Because of the requirements of control and excludability, it works well for medium-sized dry goods, including land, but works poorly, or not at all, for more ephemeral, transitory, or mobile goods, as well as terrestrial regions that cannot be easily fenced or controlled, such as the ocean or the atmosphere. Thus it leaves unresolved a multitude of collective action problems, where individuals continue to act in an unconstrained fashion, giving rise to interferences that are unnecessary, in the sense that they could be eliminated without actually making anyone worse off. In these cases, the benefits to the individuals performing the actions are not sufficient to outweigh the harms caused to others, but there is no incentive for them to refrain from performing these actions, because the institutional framework fails to compel them to offer indemnity. It therefore allows arrangements to persist that are unacceptable, and that would not be accepted, were the parties involved actually to contract with one another.

Our first line of response to these circumstances is the tort system, along with civil law more generally, which impose constraints on the damages that individuals can inflict upon one another. This shades over into traditional regulation, in which the state directly imposes rules constraining individual behavior. In most cases, there is no choice but to impose uniform solutions, regardless of the fact that individuals might have contracted to more nuanced arrangements. (Municipal noise bylaws, for instance, impose a standard set of rules on all neighborhoods, even though individuals in particular areas might negotiate quite different arrangements, if they could overcome the transaction costs involved in conducting such negotiations.) In cases where this is justified, it is because having some rule, based on an approximate calculation, is better than having no rule. The exercise in this case is well described by Coase, who argued that the objective of the state, when it imposes a regulation, should be to enforce the outcome that the parties would have contracted to had various contingent circumstances not prevented them from doing so.

Whenever possible, however, there is a great deal to be said for the imposition of a more flexible system, which will allow individuals to retain some discretion in deciding how important their particular goals are, and whether they are willing to do what is necessary to indemnify others for the interferences caused. This is the attraction of having the state engage in *pricing* of particular actions, as a way of controlling the harms that they cause. Control of pricing is a far more powerful policy lever than the imposition of uniform rules because it decentralizes the decision-making, so that instead of the state having to consider policy trade-offs at an aggregate level, individuals are forced to consider the trade-offs in each specific consumption decision. Not only does this result in a vastly more efficient regulatory regime, but it avoids the various injustices that arise as an inevitable byproduct of the imposition of uniform rules.

These are, I have argued, the appropriate "philosophical foundations" for thinking about climate change policy. Climate change is, I have argued, a fairly standard negative externality problem. Because the regime of property rights offers individuals no control over the atmosphere, much less the weather that affects them, markets fail to price greenhouse gas emissions. As a result, the market price for a wide range of goods, but most importantly fossil fuels, fails to reflect the full social cost of their consumption, and thus is lower than it should be. This price distortion is then propagated throughout the entire economy. Because we do not pay for carbon emissions, transport by truck is less expensive than it should be, compared to rail or ship. Building with concrete is less expensive than it should be, compared to stone or wood. Because we do not pay for methane emissions, beef is less expensive than it should be, compared to chicken. Using "brown" electricity, generated from fossil fuel, is less expensive than it should be, compared to "green" electricity. These price distortions affect the relative prices of practically every good in the economy. In every single one of these markets, at the margin, people are consuming more of an environmentally damaging good than they should, or than they would actually be willing to if they had to fully compensate those affected. Often there is an adequate substitute available, which they are failing to choose only because the price distortion gives them no incentive to do so.

For example, of the 500 million tons of rice that are produced every year by farmers around the world, probably the vast majority of it should continue to be produced. The methane emitted by the rice paddies can be thought of as "subsistence emissions" (although the more accurate way of expressing the idea would be to say that, with the inframarginal product, the benefits are clearly greater than the social cost of production, and so the rice would continue to be produced even if the externality were fully internalized). And yet, because the damage caused by the methane is not factored into the cost of production, this means that the price of rice is lower than it should be, and so *at the margin* too much rice is being produced and consumed. The last, say, 10 million tons, and perhaps even more, are being produced only because those who consume it are not having to pay the full cost of their consumption. This means that its production is *unjustified*. We should have grown somewhat less rice and instead produced more of some other crop, such as millet or wheat, which is less environmentally damaging. The same story could be told for almost every single product on the market today, where at the margin we are producing either too much of it or too little of it, relatively speaking.

Seen from this perspective, the complexity of the task of finding an appropriate emissions level may seem overwhelming. And yet it admits of a surprisingly simple solution. By imposing a price on greenhouse gas emissions—first and foremost through the imposition of a carbon tax—the state is able to bring the price into closer alignment with its best estimate of the social cost. The primary goal of this is not to penalize consumers for producing pollution. The importance of the carbon tax is that its effects get propagated throughout the economy, resulting in changes in the relative prices of almost every other good. It does not just increase the price of gasoline and home heating, although these are the most noticeable effects. More imperceptibly, it will raise the price of beef relative to chicken. It will raise the price of local hothouse tomatoes relative to imported field-grown. It will raise the price of air travel compared to train. It will raise the price of traditional delivery service compared to bike couriers, and so on. As a result, millions of consumers will adjust their behavior, at the margin, in millions of slight ways, the *aggregate* effects of which will be a significant reduction in our greenhouse gas emissions. (How large the reduction is will, of course, depend upon how high the carbon tax is set.)

Again, it important to emphasize that the problem is not *that* we are producing greenhouse gas emissions, but that we are producing *too much*. Our current emissions levels are not acceptable and are thus unjustifiable. Why? Because the benefits we are deriving are not great enough to justify the harms. How do we know this? Because if we were forced to pay a price sufficient to compensate those who are harmed, we would choose to cut back our emissions rather than pay it. (And the evidence for that, if we require any, is that carbon taxes change consumer behavior.) Thus carbon pricing is not merely a pragmatic or second-best solution to the climate change problem. It is a way of ensuring that the decisions we each make about our emissions activities are ones that everyone—including future generations—would agree to if we were all to sit down together, *per impossibile*, and come to an agreement about what the terms are that should govern our interactions.

There are, of course, many *practical* features of the climate change problem that make it almost uniquely difficult to solve. The temporal delay caused by inertia in the climate system means that the major constituency that stands to benefit from carbon abatement is not present to press its demands, vote in democratic elections, or participate in deliberative politics. This means that we must rely much more heavily on the power of "ideas" than of "interests" to drive the political process. The dispersion of greenhouse gases means that the problem is global in scope, and so coordinated action among all nations is required. This creates serious international free-rider problems. The situation is, as Stephen Gardiner has observed, something of a "perfect moral storm." There is, however, one aspect of the problem that is fortunate, which is that it is amenable to such a simple and effective regulatory solution. As we have seen, there are many environmental problems that cannot be addressed effectively by pricing, or where the implementation of a pricing system would impose significant compliance or transaction costs. Carbon emissions, by contrast, are almost the ideal scenario for a Pigovian tax—which is why that policy option attracts such broad intellectual consensus.[1] (Methane emissions, by contrast, are probably best addressed through a combination of pricing—e.g. on ruminants—and regulation. In particular, "fugitive emissions," such as those produced through leaks in mining operations, cannot feasibly be priced, and so are probably best addressed through mandated technology and fines.)

Climate change is obviously a very big problem. This makes it incumbent upon us all to refrain from making it bigger than it already is or than it has to be. One of the ways that we can do this is by maintaining a resolute focus on the essential structure of the problem while refusing to be distracted by the many side issues and scenarios that inevitably arise. We must also refrain from allowing our frustration at the slowness of the policy response from states to radicalize either our assessment of the situation or our policy prescriptions. Naturally, the longer the world goes without curtailing GHG emissions, the more dramatic our eventual policy response will have to be. But there is no need to overstate the seriousness of the problem to make this point. What we need instead is to be consistent, firm, and uncompromising with respect to the central message, about the need for a carbon-pricing regime. Beyond that, it is important simply to recognize that collective action is extremely difficult to organize. That is why, even when there is virtual unanimity about what "good policy" amounts to in a particular domain, implementation can remain a challenge. Overcoming these obstacles, however, belongs to the realm of politics. It requires not just the skills of the politician, but even the dark arts of the communications strategist. This is a domain in which, unfortunately, the philosopher has even less than usual to contribute.

Notes

Introduction

1. It would not be incorrect to observe, therefore, that I am not really providing "foundations" either, or at least not in the foundationalist sense.
2. Thus I am not concerned about the issues that one can find discussed in Barry G. Rabe, *Can We Price Carbon?* (Cambridge, MA: MIT Press, 2018).
3. On the latter point, see Joseph Heath, "Rawls on Global Distributive Justice: A Defence," *Canadian Journal of Philosophy*, supp. vol., 35 (2005): 193–226.
4. For contrast, see the papers collected in Clare Heyward and Dominic Roser, eds., *Climate Justice in a Non-ideal World* (Oxford: Oxford University Press, 2016). The major focus in these papers is on the "compliance dilemma" caused by the fact that, once an ideal climate policy is specified, many individuals and nations can be expected to fail to do their part in implementing it. I am concerned rather with the formulation of policy, not the issues that arise at the implementation stage. For a general philosophical analysis of compliance problems, see Allen Buchanan, "Perfecting Imperfect Duties: Collective Action to Create Moral Obligations," *Business Ethics Quarterly*, 6 (1996): 27–42.
5. Ronald Dworkin, "What Is Equality? Part 2: Equality of Resources," *Philosophy and Public Affairs*, 10 (1981): 283–345.
6. Ronald Dworkin, *Justice for Hedgehogs* (Cambridge, MA: Harvard University Press, 2013). Note that, despite the title, Dworkin could more accurately have been described as "the fox who dreamt he was a hedgehog."
7. I am exaggerating somewhat, since Dworkin believed that his criterion allowed for ranking and appeared to be unaware of the difficulties with doing so. See Joseph Heath, "Dworkin's Auction," *Politics, Philosophy and Economics*, 3 (2004): 313–335. Dworkin defended himself against this charge by pointing to the number of very practical, policy-oriented arguments that he made in various popular publications (*Justice for Hedgehogs*, pp. 476–477). It is noteworthy, however, that in all of these publications, he never actually *used* the envy-freeness standard to make any specific claims. Instead he merely appealed to an intuitive conception of what it means to show "equal concern" for all persons.
8. One of the peculiarities of Amartya Sen's book *The Idea of Justice* (Cambridge, MA: Harvard University Press, 2009) is that he spent so much time criticizing John Rawls's work for being overly idealized, while largely ignoring the work of egalitarians such as Dworkin or G. A. Cohen who were defending far more idealized views.
9. John Rawls, *A Theory of Justice*, rev. ed. (Cambridge, MA: Harvard University Press, 1999).

10. G. A. Cohen, *Rescuing Justice and Equality* (Cambridge, MA: Harvard University Press, 2008). Cohen felt that he was "rescuing" justice and equality because Rawls's concerns over incentive-compatibility represented an important threat to these ideals.

11. Jonathan Wolff, *Ethics and Public Policy* (London: Routledge, 2011).

12. Wolff, *Ethics and Public Policy*, p. 2.

13. Donald A. Brown, *Climate Change Ethics* (London: Routledge, 2013), p. 12.

14. Stephen M. Gardiner, *A Perfect Moral Storm* (New York: Oxford University Press, 2011), pp. 20–22; Brown, *Climate Change Ethics*, p. 6.

15. See John Broome, *Counting the Cost of Global Warming* (Cambridge: White Horse Press, 1992); John Broome, *Climate Matters* (New York: Norton, 2012).

16. John M. Taurek, "Should the Numbers Count?," *Philosophy and Public Affairs*, 6 (1977): 293–316.

17. T. M. Scanlon. *What We Owe to Each Other* (Cambridge, MA: Harvard University Press, 1998), p. 232.

18. Frances Kamm, "Equal Treatment and Equal Chances," *Philosophy and Public Affairs*, 14 (1985): 177–194; Iwao Hirose, "Aggregation and Numbers," *Utilitas*, 16 (2004): 62–79; Ben Saunders, "A Defence of Weighted Lotteries in Life Saving Cases," *Ethical Theory and Moral Practice*, 12 (2009): 279–290.

19. Rahul Kumar, "Contractualism on the Shoals of Aggregation," in R. Jay Wallace, Rahul Kumar, and Samuel Freeman, eds., *Reasons and Recognition* (Oxford: Oxford University Press, 2011), pp. 129–154.

20. E.g. see Jussi Suikkanen, "Contractualism and Climate Change," in Marcello Di Paola and Gianfranco Pellegrino, eds., *Canned Heat: Ethics and Politics of Global Climate Change* (New Delhi: Routledge, 2014), pp. 115–128; Matthew Rendall, "Discounting, Climate Change, and the Ecological Fallacy," *Ethics*, 129 (2019): 441–463.

21. See Intergovernmental Panel on Climate Change, *Climate Change 2014: Impacts, Adaptation, and Vulnerability* (Cambridge: Cambridge University Press, 2014).

22. Peter H. Pfromm, "Towards a Sustainable Agriculture: Fossil-Fuel Ammonia," *Journal of Renewable and Sustainable Energy*, 9 (2017): 1–11.

23. Vaclav Smil, "Nitrogen Cycle and World Food Production," *World Agriculture*, 2 (2011): 9–13 at 10. See also Vaclav Smil, *Harvesting the Biosphere* (Cambridge, MA: MIT Press, 2013), p. 248.

24. Pfromm, "Towards a Sustainable Agriculture," p. 8.

25. A. A. K., "Why It Is So Hard to Fix India's Sanitation," *The Economist* (Sept. 25, 2017).

26. This data is taken from https://climateactiontracker.org/ (accessed Feb. 12, 2019). Emissions data is difficult to compile and politically controversial, and so estimates vary from source to source. The numbers presented should be taken to be illustrative of the general dilemma, not as authoritatively correct claims. This is true throughout the book. Data are cited, almost invariably, to prove a philosophical point, where the numbers could be off by a rather substantial degree, and yet the point would still stand.

27. For useful summary of the situation, see Government of Canada, *Crude Oil Facts*, https://www.nrcan.gc.ca/energy/facts/crude-oil/20064 (accessed Feb. 12, 2019).

28. It should also be noted that the European approach to regulation is much more aspi-rational than the American one. European environmental regulation, in particular, is often quite strict on paper, but is then subject to lax and discretionary enforcement. US regulation, by contrast, is typically enforced to the letter, with very severe penalties for noncompliance. (These jurisdictional differences became quite noticeable during the Volkswagen diesel emissions scandal. See "To Stop Carmakers Bending the Rules on Emissions, Europe Must Get Much Tougher," *The Economist* [Jan. 21, 2017].) As a result, it is possible in Europe to win the argument on environmental questions— by getting the appropriate laws enacted—without actually gaining very much on the ground. In the United States, by contrast, it is much more difficult to get legislative action on environmental questions, in part because of the generally shared expecta-tion that these laws, once enacted, will actually be enforced. See Eugene Bardach and Robert A. Kagan, *Going by the Book* (Philadelphia: Temple University Press, 1982). In Canada, regulation is much more in the European style, although the threat of extra-dition to the United States provides a somewhat greater enforcement threat.

29. It has become conventional to distinguish between light, dark, and now bright green positions.

30. Technically, francophone voters in Quebec are a national minority group. What makes them a major locus of political power is the tendency to bloc vote along lin-guistic lines. This in turn makes it extremely difficult to form a federal government without a majority of seats in Quebec. The only effective countervailing power is the (much smaller) Alberta bloc vote, which aligns around oil interests.

31. For an interesting profile of the oil patch, see Chris Turner, *The Patch* (Toronto: Simon & Schuster, 2017).

32. For an early example, see Harold Coward and Thomas Hurka, eds., *Ethics and Climate Change* (Waterloo: Wilfred Laurier University Press, 1993).

33. Rawls, *A Theory of Justice*, p. 155.

34. Rawls, *A Theory of Justice*, p. 4.

35. Joseph Heath, *Morality, Competition and the Firm* (New York: Oxford University Press, 2014), pp. 148–155.

36. In recent philosophical discussions, prioritarianism is often thought of as a variant of utilitarianism, and as a rival to egalitarianism. This is due to the influence of Derek Parfit's article "Equality or Priority?," in M. Clayton and A. Williams, eds., *The Ideal of Equality* (Houndmills: Palgrave, 2000), pp. 81–125. Here I am using the term in the more general sense that Rawls did, to refer to any theory of justice that generates a set of indifference curves that are concave. As John Nash showed long ago, a prioritarian theory of justice can be derived from a joint commitment to equality and efficiency, along with certain other postulates. John Nash, "The Bargaining Problem," *Econometrica*, 18:2 (1950): 155–162. Parfit simply provides another way of deriving a prioritarian social welfare function.

37. David Hume, *A Treatise of Human Nature*, ed. P. H. Nidditch, 2nd ed. (Oxford: Oxford University Press, 1978), p. 477. See also David Gauthier, "David Hume: Contractarian," in his *Moral Dealing* (Ithaca: Cornell University Press, 1990), pp. 45–76.

38. Heath, *Morality, Competition, and the Firm*.

39. *The Economics of Climate Change: The Stern Review* (Cambridge: Cambridge University Press, 2007),
40. Tyler Cowen, "Consequentialism Implies a Zero Rate of Intergenerational Discount," in Peter Laslett and James S. Fishkin, eds., *Justice between Age Groups and Generations* (New Haven: Yale University Press, 1992), pp. 162–168.

Chapter 1

1. The most striking illustration of this was the economist and policy adviser to the UK government Nicholas Stern asking for input on the issue of social discounting. See Nicholas Stern, *The Economics of Climate Change: The Stern Review* (Cambridge: Cambridge University Press, 2007), chap. 2. See also Nicholas Stern, "Ethics, Equity and the Economics of Climate Change, Paper 1: Science and Philosophy," *Economics and Philosophy*, 30 (2014): 397–444 at 434.
2. A good example is Stephen M. Gardiner's *A Perfect Moral Storm* (Oxford: Oxford University Press, 2011), which condemns the "environmental economic" approach to thinking about climate change as facilitating "moral corruption" (p. 298). As a result, when it comes to saying something affirmative about what should be done, Gardiner does little more than list more problems and objections. As for determining the "trajectory" of future emissions, he calls for further research and discussion (pp. 428–431).
3. Darrel Moellendorf, *The Moral Challenge of Dangerous Climate Change* (Cambridge: Cambridge University Press, 2014), p. 17.
4. A typical expression of this point of view: "In order for environmentalists to win our important battles—not just lose a little more slowly, but win—we must end the endless growth economy as we know it." Philip Cafaro, "Taming Growth and Articulating a Sustainable Future," *Ethics and the Environment*, 16 (2011): 1–23 at 9.
5. As Cafaro observes, "Taming Growth," p. 3.
6. Underlying this is the influence of Heidegger's "Question concerning Technology." See Martin Heidegger, *The Question concerning Technology, and Other Essays*, trans. William T. Levitt (New York: Harper & Row, 1977). For discussion, see Kalpita Bhar Paul, "The Import of Heidegger's Philosophy into Environmental Ethics: A Review," *Ethics and the Environment*, 22 (2017): 79–98. Michael E. Zimmerman, "Toward a Heideggerean Ethos for Radical Environmentalism," *Environmental Ethics*, 6 (1983): 99–131.
7. For a representative sample of their works, see John Broome, *Counting the Cost of Global Warming* (Strond: White Horse Press, 1992); John Broome, *Climate Matters* (New York: Norton, 2012); Simon Caney, "Just Emissions," *Philosophy and Public Affairs*, 40 (2012): 255–300; Simon Caney, "Climate Change, Human Rights and Moral Thresholds," in Stephen M. Gardiner, Simon Caney, Dale Jamieson, and Henry Shue, eds., *Climate Ethics: Essential Readings* (Oxford: Oxford University Press, 2010), pp. 163–177; Stephen M. Gardiner, "Ethics and Global Climate

Change," *Ethics*, 114 (2004): 555–600; also Gardiner, *A Perfect Moral Storm*; Catriona McKinnon, *Climate Change and Future Justice: Precaution, Compensation, and Triage* (New York: Routledge, 2011); Henry Shue, *Climate Justice: Vulnerability and Protection* (Oxford: Oxford University Press, 2014).

8. For discussion, see Hans Joachim Schellnhuber, et al. eds., *Avoiding Dangerous Climate Change* (Cambridge: Cambridge University Press, 2006). The technocratic approach typically involves determining a "critical threshold," beyond which climate change becomes "dangerous."

9. Moellendorf, *Moral Challenge*, p. 17.

10. Stephen Gardiner, for instance, has argued that "climate policy is at risk of being driven by hidden assumptions, obscured behind the technical language of other disciplines, such as economics." Stephen M. Gardiner, "Climate Ethics in a Dark and Dangerous Time," *Ethics*, 127 (2017): 430–465 at 431 n. 3.

11. For representative works, see Stern, *Economics of Climate Change*; William Nordhaus, *The Climate Casino* (New Haven: Yale University Press, 2013).

12. See Joseph Heath, *Morality, Competition and the Firm* (New York: Oxford University Press, 2014), p. 148.

13. See, e.g. Byron Williston, *Environmental Ethics for Canadians*, 2nd ed. (Toronto: Oxford University Press, 2016); Andrew Kernohan, *Environmental Ethics* (Buffalo: Broadview Press, 2012); Dale Jamieson, *Ethics and the Environment* (Cambridge: Cambridge University Press, 2008).

14. Kenneth Goodpaster, "On Being Morally Considerable," *Journal of Philosophy*, 75:6 (1978): 308–325; also Peter Singer, *The Expanding Circle: Ethics and Sociobiology* (New York: Farrar, Straus and Giroux, 1982).

15. Katie McShane, "Anthropocentrism vs. Nonanthropocentrism: Why Should We Care?," *Environmental Values*, 16 (2007): 169–185.

16. Following this line of thinking, it is also tempting to conclude that the reason we have so many environmental problems is that we have been working with too small a circle of concern, and so if we expand the circle, it will in turn resolve the practical problems. This is actually a great deal less self-evident than it is normally taken to be.

17. Richard Routley, "Is There a Need for a New, and Environmental, Ethic?," in *Proceedings of the XVth World Congress of Philosophy* (Varna: Sofia Press, 1973), pp. 205–210 at 207. He subsequently changed his last name to Sylvan.

18. For another variant on this argument, see Steven Vogel, *Thinking Like a Mall* (Cambridge, MA: MIT Press, 2015), pp. 160–161.

19. This is a point made by David Schmidtz and Elizabeth Willott, *Environmental Ethics*, 2nd ed. (Oxford: Oxford University Press, 2011).

20. Charles Taylor, *Sources of the Self* (Cambridge, MA: Harvard University Press, 1989), pp. 220–221.

21. David Hume, *A Treatise of Human Nature*, ed. P. H. Nidditch, 2nd ed. (Oxford: Oxford University Press, 1978), pp. 468–469.

22. Hume, *Treatise of Human Nature*, p. 469.

23. See Philip Pettit, "Realism and Response-Dependence," *Mind*, 100 (1991): 587–626.

24. See John O'Neill's excellent discussion in *Ecology, Policy and Politics* (London: Routledge, 1993), p. 10. The example of such a confusion that he cites is from Donald Worster, *Nature's Economy* (Cambridge: Cambridge University Press, 1985).

25. J. Baird Callicott, "Non-anthropocentric Value Theory and Environmental Ethics," *American Philosophical Quarterly*, 21 (1984): 299–309 at 299.

26. J. Baird Callicott, *In Defense of the Land Ethic* (Albany: State University of New York Press, 1989), p. 265.

27. Eric Katz, "A Pragmatic Reconsideration of Anthropocentrism," *Environmental Ethics*, 21 (1999): 377–390.

28. For early formulation, see Donald Vandeveer, "Animal Suffering," *Canadian Journal of Philosophy*, 10 (1980): 463–471 at 465.

29. Peter Singer, "All Animals Are Equal," *Philosophical Exchange*, 1 (1974): 103–116.

30. The phrase was coined by Jan Narveson, "Animal Rights," *Canadian Journal of Philosophy*, 7 (1977): 161–178. For discussion see Julia K. Tanner, "The Argument from Marginal Cases and the Slippery Slope Objections," *Environmental Values*, 18 (2009): 51–66.

31. Peter Singer, *Animal Liberation* (London: Pimlico, 1995), p. 237.

32. Tim Hayward, "Anthropocentrism: A Misunderstood Problem," *Environmental Values*, 6 (1997): 49–63 at 56.

33. Christopher D. Stone, *Should Trees Have Standing?* (Los Altos: William Kaufman, 1974).

34. J. Baird Callicott, "Animal Liberation: A Triangular Affair," *Environmental Ethics*, 2 (1980): 311–338.

35. Paul Taylor, *Respect for Nature* (Princeton: Princeton University Press, 1987); Robin Attfield, "The Good of Trees," *Journal of Value Inquiry*, 15:1 (1981): 35–54.

36. Warwick Fox, *Toward a Transpersonal Ecology* (Albany: SUNY Press, 1995), pp. 162–179. Williston describes the difference between the latter two in the following way: "Biocentrism draws the moral circle around living or biotic things. Ecocentrism goes a step further to draw it around systems that contain both biotic and abiotic elements. Soil and water, although they contain many living things, are not themselves alive. So the primary objects of potential moral concern on this view are ecosystems." *Environmental Ethics for Canadians*, p. 85.

37. Aldo Leopold, *A Sand County Almanac and Sketches Here and There* (London: Oxford University Press, 1949), pp. 224–225. Note that the "biotic community" includes ecological processes, not just life forms, which is why the view is ecocentric and not biocentric.

38. Paul W. Taylor, for instance, interprets the "circle" as involving a judgment of superiority and inferiority (with those inside the circle being superior to those outside it). "The Ethics of Respect for Nature," *Environmental Ethics*, 3 (1981): 197–218 at 211. This allows him to draw a strict analogy between supporters of a previous, discredited form of hierarchy and those who resist moral expansionism. The claim about superiority is introduced, however, out of the blue, and it is quite unclear why anyone might feel obliged to characterize moral "considerability" as following from some form of

superiority. As a result, although the argument is invalid, it reveals a great deal about the way that Taylor is thinking about the issues.

39. Williston, for instance, moves from animal welfare to biocentrism to ecocentrism but shows no concern that there is any sort of a regress problem. *Environmental Ethics for Canadians*, chaps. 2–3. For a similar lack of concern, see Philip Kitcher, *The Ethical Project* (Cambridge, MA: Harvard University Press, 2011), pp. 308–309. For discussion, see Vogel, *Thinking Like a Mall*, pp. 144–164.

40. Sue Donaldson and Will Kymlicka, *Zoopolis* (Oxford: Oxford University Press, 2011), pp. 25–28.

41. Donaldson and Kymlicka, *Zoopolis*, p. 31. Their central response to the concern is to offer assurances that certain canonical cases, such as dogs, horses, and monkeys, fall within the circle. But this is precisely the move that is being denied to the speciesist as a response to the argument from marginal cases. After all, it is easy to point to fully rational, autonomous adult humans as the canonical instance of a moral agent.

42. If one follows Daniel Dennett in thinking that the concept of interest-seeking is an ascribed property, then one will be inclined to think that there is no "fact of the matter" as to which natural systems possess them or not. See Daniel C. Dennett, *The Intentional Stance* (Cambridge, MA: MIT Press, 1987).

43. Fox, *Toward a Transpersonal Ecology*, pp. 179–184. Prior to the advent of concern over climate change, these regress-style arguments were often used to build support for the "Gaia hypothesis"—the view that the earth as a whole should be considered a self-regulating, or autopoetic, system. E.g. James Lovelock, *Gaia* (Oxford: Oxford University Press, 1995). This view has suffered a sudden and dramatic loss of popularity in recent years because the emphasis on self-regulation tends to undermine the basis for concern over climate change. Thus Gaia theory went from being the darling of environmentalists to a political embarrassment almost overnight.

44. The allusion here is to the Sorites Paradox. See Sergi Olms and Elia Zardini, *The Sorites Paradox* (Cambridge: Cambridge University Press, 2019).

45. Taylor, "Ethics of Respect," p. 210.

46. David Wong, *Moral Relativity* (Berkeley: University of California Press, 1984).

47. A position defended by David George Ritchie, *Natural Rights* (London: Swan Sonnenschein, 1916); also Jeff McMahan, "The Meat Eaters," *The Stone, New York Times* (Sept. 19, 2010). The issue came to the forefront with the publication of Tom Regan, *The Case for Animal Rights* (Berkeley: University of California Press, 2004), since many critics found his way of avoiding the problem unpersuasive. See Rainer Ebert and Tibor R. Machan, "Innocent Threats and the Moral Problem of Carnivorous Animals," *Journal of Applied Philosophy*, 29 (2012): 146–159. Incidentally, deontological views are less prone to difficulty in this regard, because the distinction between acts and omissions allows them to say that, even if it is wrong for animals to kill one another, we are not necessarily obliged to do anything about it.

48. Fox, *Toward a Transpersonal Ecology*, p. 195. Views such as this may be one of the reasons that Jonathan Wolff was embarrassed to describe the "state of knowledge" in philosophy on the question of animal welfare.

49. Andrew Brennan, "Moral Pluralism and the Environment," *Environmental Values*, 1 (1992): 15–33.

50. Mark Sagoff, "Animal Liberation and Environmental Ethics: Bad Marriage, Quick Divorce," *Osgoode Hall Law Journal*, 22 (1984): 297–307.

51. Scott R. Loss, Tom Will, and Peter P. Marra, "The Impact of Free-Ranging Domestic Cats on Wildlife of the United States," *Nature Communications*, 4 (2013): 1396.

52. Children raised in US urban areas, for instance, exhibit a pronounced anthropocentric bias in their folk-biological ideas, the result of "insufficient cultural input and a lack of exposure to the natural world." Joseph Henrich, Steven J. Heine, and Ara Norenzayan, "The Weirdest People in the World?," *Behavioral and Brain Sciences*, 33:2–3 (2010): 61–83 at 67–69. "Since such urban environments are highly 'unnatural' from the perspective of human evolutionary history, any conclusions drawn from subjects reared in such informationally impoverished environments must remain rather tentative" (p. 67). This caution should be extended to include whatever moral intuitions these subjects may report about animals and the natural world.

53. Elizabeth Kolbert, *The Sixth Extinction* (New York: Henry Holt, 2014).

54. See Y. S. Lo, "The Land Ethics and Callicott's Ethical System (1980–2001): An Overview and Critique," *Inquiry*, 44 (2001): 331–358.

55. Bryan G. Norton, "Why I Am Not a Nonanthropocentrist: Callicott and the Failure of Monistic Inherentism," *Environmental Ethics*, 17 (1995): 341–358 at 349.

56. Brennan, "Moral Pluralism."

57. Nicole Hassoun, "The Anthropocentric Advantage? Environmental Ethics and Climate Change Policy," *Critical Review of International Social and Political Philosophy*, 14 (2011): 235–257 at 237–238.

58. Richard S. J. Tol, "The Economic Effects of Climate Change," *Journal of Economic Perspectives*, 23 (2009): 29–51.

59. Andrew Dessler, *Introduction to Modern Climate Change*, 2nd ed. (Cambridge: Cambridge University Press, 2016), p. 118.

60. Gordon McGranahan, Deborah Balk, and Bridget Anderson, "The Rising Tide: Assessing the Risks of Climate Change and Human Settlements in Low Elevation Coastal Zones," *Environment and Urbanization*, 19 (2007): 17–37.

61. The planet has in fact been ice-free for most of the past half-billion years, with the past 25 million being a rather prominent exception. See Dessler, *Modern Climate Change*, p. 118. On PETM see pp. 119–120.

62. Vogel, *Thinking Like a Mall*, p. 90, makes the case for eliminating this distinction, on the grounds that all of the environments in which we function are now essentially "built."

63. As Hassoun puts it, "The problem for inclusive environmental ethics just stems from the fact that they are radically incomplete; although they can provide some reasons in favor of particular climate change policies, they also provide reasons not to implement those policies. They do not tell us where the weight of reason lies." "Anthropocentric Advantage," pp. 240–241.

64. Thomas Hobbes, *Leviathan*, ed. Richard Tuck (Cambridge: Cambridge University Press, 1991), p. 39.

65. John Locke, *Second Treatise of Government*, ed. C. B. Macpherson (Indianapolis: Hackett, 1980).

66. This is in contrast to Steven Bernstein, among others, for whom the term has a negative connotation. See Steven Bernstein, *The Compromise of Liberal Environmentalism* (New York: Columbia University Press, 2001).

67. I myself am tempted by this. Perhaps the best example of it in the literature is Bryan Norton, "Ecology and Opportunity: Intergenerational Equity and Sustainability Options," in Andrew Dobson, ed., *Fairness and Futurity* (Oxford: Oxford University Press, 1999), pp. 118–150. He sees the difficulty and therefore develops the argument carefully. See also Daniel A. Farber, *Eco-pragmatism* (Chicago: University of Chicago Press, 1999), pp. 108–110.

68. This neglect of traditional environmental ethics has not gone unnoticed. For a critical discussion, see Katie McShane, "Anthropocentrism in Climate Ethics and Policy," *Midwest Studies in Philosophy*, 60 (2016): 189–204.

69. John Rawls, "Justice as Fairness: Political Not Metaphysical," *Philosophy and Public Affairs*, 14 (1985): 223–251.

70. For discussion of this point, see essays in Gideon Calder and Catriona McKinnon, *Climate Change and Liberal Priorities* (London: Routledge, 2012).

71. See Joseph Heath, "Political Egalitarianism," *Social Theory and Practice*, 34 (2008): 485–516.

72. Ronald Dworkin, "Equality of Resources," in *Sovereign Virtue* (Cambridge, MA: Harvard University Press, 2000), pp. 65–119. He did not invent the idea, but he did a great deal to popularize it. For more general discussion, see William Baumol, *Superfairness* (Cambridge, MA: MIT Press, 1986).

73. Will Kymlicka, *Contemporary Political Philosophy*, 2nd ed. (Oxford: Oxford University Press, 2002), pp. 53–101.

74. Rawls, *A Theory of Justice*, pp. 65–69.

75. Russell Hardin, *Collective Action* (Baltimore: Resources for the Future, 1982).

76. There are many different greenhouses gases, including water vapor and methane. Carbon dioxide, however, is the most important one, not because it is the most powerful, but because it has the unfortunate combination of being produced in very large quantities and being extremely persistent in the atmosphere. (Methane, by contrast, is being produced in large quantities and is a far more powerful greenhouse gas—measured in terms of GWP, or global warming potential—but it is significantly less persistent in the atmosphere.) As a result, carbon dioxide is used as the "benchmark" against which the effects of all other emissions are measured and expressed. Similarly, greenhouse gas emissions are usually measured in "metric tons of carbon dioxide equivalent" (tCO_2e), sometimes abbreviated to just "carbon-ton" (tC), with the understanding that when one talks about "carbon emissions" one is not just talking about carbon dioxide, but rather *all* greenhouse gases, expressed in terms of the equivalent quantity of carbon dioxide based on their warming effects. Also, it should be noted that "carbon-ton" refers to the weight of *carbon dioxide*, which is 3.67 times greater than that of the carbon content alone (since it includes the weight of the two oxygen molecules).

77. For those wondering, the relationship between carbon emissions and temperature increase is essentially linear, because the warming effect of a given quantity of greenhouse gas emissions is independent of the background concentration of those gases. See H. Damon Matthews, Nathan P. Gillett, Peter A. Stott, and Kirsten Zickfeld, "The Proportionality of Global Warming to Cumulative Carbon Emissions," *Nature*, 459 (2009): 829–832. By contrast, the relationship between temperature increase and cost (i.e. damage) is generally assumed to be nonlinear, with the amount of damage from further temperature increases becoming larger at higher temperature levels.

78. E.g., Stephen M. Gardiner and David A. Weisbach, *Debating Climate Ethics* (Oxford: Oxford University Press, 2016).

79. Stephen Gardiner has been the most prominent proponent of the view that it is not a collective action problem on the model of a prisoner's dilemma. See Stephen Gardiner, "The Pure Intergenerational Problem," *Monist* 86 (2003): 481–501; also Gardiner, *A Perfect Moral Storm*. On cutting the cake, see Lukas Meyer and Dominic Roser, "Distributive Justice and Climate Change: The Allocation of Emission Rights," *Analyse & Kritik*, 28 (2006): 223–249 at 226.

80. E.g. see Kok-Chor Tan, *What Is This Thing Called Global Justice?* (Oxford: Routledge, 2017), pp. 120–121.

81. Steve Vanderheiden, *Atmospheric Justice* (Oxford: Oxford University Press, 2008), pp. 104–105.

82. Samuel Randalls, "A History of the 2° Target," *Wiley Interdisciplinary Reviews: Climate Change*, 1 (2010): 598–605.

83. Axel Gosseries, "Cosmopolitan Luck Egalitarianism and the Greenhouse Effect," *Canadian Journal of Philosophy*, supp. vol., 31 (2005): 279–309 at 282.

84. Dale Jamieson, "Adaptation, Mitigation and Justice," in Walter Sinnott-Armstrong and Richard Howarth, eds., *Perspectives on Climate Change* (Oxford: Elsevier, 2005), pp. 217–248 at 231.

85. Anil Agarwal and Sunita Narain, *Global Warming in an Unequal World* (New Delhi: Centre for Science and Environment, 1991); Dale Jamieson, "Climate Change and Global Environmental Justice," in Clark A. Edwards and Paul N. Miller, eds., *Changing the Atmosphere* (Cambridge, MA: MIT Press, 2001), pp. 287–308; Niklas Höhne, Michel den Elzen, and Martin Weiss, "Common but Differentiated Convergence (CDC): A New Conceptual Approach to Long-Term Climate Policy," *Climate Policy*, 6 (2006): 181–199; Peter Singer, "One Atmosphere," in Gardiner et al., *Climate Ethics*, pp. 181–199.

86. See Lukas Meyer, ed., *Climate Justice and Historical Emissions* (Cambridge: Cambridge University Press, 2017). Others have argued, provocatively, that the historical pattern of emissions shows that some people value the atmospheric "good" more than others, and so should be accorded a *greater* share of emissions rights. See Eric Posner and David Weisbach, *Climate Change Justice* (Princeton: Princeton University Press, 2007), p. 136.

87. Gosseries, "Cosmopolitan Luck Egalitarianism," p. 301; Gardiner, *A Perfect Moral Storm*, p. 422.

88. Simon Caney, "Justice and the Distribution of Greenhouse Gas Emissions," *Journal of Global Ethics*, 5 (2009): 125–146 at 130.

89. For an excellent example of this style of reasoning, see Megan Blomfield, "Climate Change and the Moral Significance of Historical Injustice in Natural Resource Governance," in Aaron Maltais and Catriona McKinnon, eds., *The Ethics of Climate Governance* (London: Rowman and Littlefield, 2015), pp. 3–22 at 4; also Brian Elliott, *Natural Catastrophe* (Edinburgh: University of Edinburgh Press, 2016).

90. Dominic Roser and Christian Seidel, *Climate Justice* (Oxford: Routledge, 2012), p. 161.

91. One can find a non-academic version of this argument in Naomi Klein, *This Changes Everything* (New York: Simon & Schuster, 2014). She begins by tying climate change to the problem of economic inequality, then argues that, as a matter of political strategy, activists should focus on fighting inequality, as an indirect strategy for solving the problem of climate change. The net effect is to hijack the climate change agenda in the service of combating inequality—which is a far more difficult problem even to conceptualize, much less solve.

92. Posner and Weisbach, *Climate Change Justice*, p. 86.

93. Posner and Weisbach, *Climate Change Justice*, p. 119.

94. Posner and Weisbach, *Climate Change Justice*, p. 120.

95. This is a point that several economists have observed. See Nicholas Stern, "Ethics, Equity and the Economics of Climate Change, Paper 2: Economics and Politics," *Economics and Philosophy*, 30 (2014): 445–501 at 487.

96. Lawrence H. Goulder and Andrew R. Schein, "Carbon Taxes versus Cap and Trade: A Critical Review," *Climate Change Economics*, 4 (2013): 1–28.

97. Caney, "Justice and the Distribution," 138.

98. There is also a closely related school of thinking that treats it as an issue of "corrective justice." Climate change, on this view, involves one group of people (emitters) *harming* those who will bear the brunt of the costs of global warming. On this view, as Posner and Weisbach observe, "climate justice" is conceived of on the model of a tort, which in turn gives rise to an intense debate about whether or how those who are causing it can be held responsible for these damages. I have not gone into a detailed discussion of it here because this view has not been as important in the philosophical literature. To the extent that it shows up, it is typically subsumed under a more general "distributive justice" framework. I think that Posner and Weisbach offer a compelling set of reasons for not framing things in this way, considered as a stand-alone approach to this issue. *Climate Change Justice*, pp. 105–117.

99. It is fallacious because it ignores the fact that, at the margin, the carbon price gives both rich and poor an incentive to reduce emissions.

100. Some still oppose it, although not as stridently. See Klein, *This Changes Everything*, p. 461.

101. Caney, "Climate Change, Human Rights," p. 165.

102. Caney, "Climate Change, Human Rights," p. 172.

103. Ronald Dworkin, *Taking Rights Seriously* (Cambridge, MA: Harvard University Press, 1977), p. xi.
104. Caney, "Climate Change, Human Rights," pp. 166–168.
105. Caney, "Climate Change, Human Rights," p. 171.
106. Bernadette O'Hare, Innocent Makuta, Levison Chiwaula, and Naor Bar-Zeev, "Income and Child Mortality in Developing Countries: A Systematic Review and Meta-analysis," *Journal of the Royal Society of Medicine*, 105 (2013): 408–414.
107. Daniel Bodansky, in "Climate Change and Human Rights: Unpacking the Issues," *Georgia Journal of International and Comparative Law*, 38 (2010): 511–524, considers it self-evident that "economic and social rights clearly do not always trump other priorities" (pp. 514–515).
108. Caney, "Climate Change, Human Rights," p. 170.
109. This is actually a general problem for such views. See Peter Railton, "Lock, Stock, and Peril: Natural Property Rights, Pollution, and Risk," in Mary Gibson, ed., *To Breathe Freely* (Totowa: Rowman and Allanheld, 1985).
110. Shue, *Climate Justice*, p. 170.
111. Shue, *Climate Justice*, pp. 165–166.
112. Idil Boran and Joseph Heath, "Attributing Weather Extremes to Climate Change and the Future of Adaptation Policy," *Ethics, Policy and Environment*, 19 (2016): 239–255. The example involves the 2010 earthquakes in Chile and Haiti. While the former event was actually stronger than the latter, the death toll was significantly lower, almost entirely because of the higher building standards in Chile, which led to fewer collapsed structures.
113. Shue, for instance, describes rich consumers as "engaged in an orgy of self-indulgent consumerism and unbridled pollution." *Climate Justice*, p. 175.
114. Simon Caney, "Climate Change and the Future: Discounting for Time, Wealth, and Risk," *Journal of Social Philosophy*, 40 (2009): 163–186.
115. Joseph Raz, *The Morality of Freedom* (Oxford: Oxford University Press, 1986), p. 166.
116. Caney, "Climate Change, Human Rights"; Simo Kyllonen and Alessandra Basso, "When Utility Maximization Is Not Enough: Intergenerational Sufficientarianism and the Economics of Climate Change," in Adrian Walsh, Sade Hormio, and Duncan Purves, eds., *The Ethical Underpinnings of Climate Economics* (New York: Routledge, 2017), pp. 65–102.
117. As Jonathan Wolff puts it, "A philosopher becomes famous by arguing for a view that is highly surprising, even to the point of being irritating, but is also resistant to easy refutation. The more paradoxical, or further from common sense, the better." *Ethics and Public Policy* (London: Routledge, 2011), p. 3.
118. For example, Gardiner's *A Perfect Moral Storm* positions itself rhetorically as a middle-of-the-road, sensible discussion of the issues. And yet Gardiner goes on to reject the idea that climate change is a collective action problem, to reject cost-benefit analysis, and to reject social discounting. He winds up taking no clear policy positions, one suspects because his philosophical framework has eliminated all of the plausible alternatives.
119. Farber, *Eco-pragmatism*, p. 40.

120. For an interesting parallel to the case of business ethics see Joseph Heath, Jeffrey Moriarty, and Wayne Norman, "Business Ethics and (or as) Political Philosophy," *Business Ethics Quarterly*, 20 (2010): 427–452.

Chapter 2

1. See William Nordhaus, *A Question of Balance* (New Haven: Yale University Press, 2008), p. 105.

2. Joseph Henrich, *The Secret of Our Success* (Princeton: Princeton University Press, 2015).

3. Dominic Roser and Christian Seidel, *Climate Justice* (New York: Routledge, 2012), p. 69.

4. Chrisoula Andreou, "A Shallow Route to Environmentally Friendly Happiness," *Ethics, Place and Environment*, 13 (2010): 1–10.

5. Eric Neumayer, *Weak versus Strong Sustainability*, 2nd ed. (Cheltenham: Edward Elgar, 2003), Bryan G. Norton, *Sustainability* (Chicago: University of Chicago Press, 2005).

6. The "list of stuff" formulation is due to Bryan Norton, "Ecology and Opportunity: Intergenerational Equity and Sustainable Options," in Andrew Dobson, ed., *Fairness and Futurity* (Oxford: Oxford University Press, 1999), pp. 119–150.

7. See Peter Victor, *Managing without Growth* (Cheltenham: Edward Elgar, 2008); Herman Daly, *Steady-State Economics* (London: W.H. Freeman, 1972); Tim Jackson, *Prosperity without Growth* (New York: Routledge, 2011); Naomi Klein, *This Changes Everything* (Toronto: Knopf, 2014).

8. At the time of writing the global growth rate is relatively low, resulting in a global economy that doubles in size every 24 years.

9. By contrast, philosophers have an almost inexplicable tendency to ignore the issue. See Dan Moller, "Global Justice and Economic Growth: Ignoring the Only Thing That Works," in Jahel Queralt and Bas van der Vossen, eds., *Economic Liberties and Human Rights* (London: Routledge, 2019), pp. 95–113. He notes, as an example, that Peter Singer's book *The Most Good You Can Do* (New Haven: Yale University Press, 2015), contains no discussion of growth. This is like writing a book on public health that fails to mention vaccination.

10. E.g. Tyler Cowen, *Stubborn Attachments* (San Francisco: Stripe, 2018), p. 41.

11. Robert Lucas, "The Industrial Revolution: Past and Future," in *2003 Annual Report Essay* (Minneapolis: Federal Reserve Bank of Minneapolis, 2004).

12. Cowen, *Stubborn Attachments*, pp. 29–30.

13. Angus Deaton, *The Great Escape* (Princeton: Princeton University Press, 2013); Benjamin M. Friedman, *The Moral Consequences of Economic Growth* (New York: Vintage, 2005).

14. John Rawls, *A Theory of Justice*, rev. ed. (Cambridge, MA: Harvard University Press, 1999), pp. 253–254.

15. Frédéric Gaspart and Axel Gosseries, "Are Generational Savings Unjust?," *Politics, Philosophy and Economics*, 6 (2007): 193–217. While the authors remain noncommittal, they certainly fail to treat the conclusion that savings are impermissible as a reductio of the theory that generates it.

16. Robert M. Solow, "Intergenerational Equity and Exhaustible Resources," *Review of Economic Studies*, 41 (1974): 29–46.

17. Gaspart and Gosseries, "Are Generational Savings Unjust?"

18. John Rawls, *A Theory of Justice* (Cambridge, MA: Belknap, 1971), p. 291. Note that this passage is deleted in the 1999 revised edition.

19. Rawls, *A Theory of Justice* (1971), p. 287. The standard that must be met between generations is essentially the same as the "duty of assistance" that he acknowledges in international affairs—to provide assistance to other countries until the point at which they are no longer "burdened." See John Rawls, *The Law of Peoples* (Cambridge, MA: Harvard University Press, 1999), p. 106.

20. Brian Barry, "Sustainability and Intergenerational Justice," in Dobson, *Fairness and Futurity*, pp. 93–117 at 104.

21. Barry, "Sustainability and Intergenerational Justice," p. 104.

22. This is just an inversion of Peter Singer's classic principle of beneficence, which is formulated in terms of avoiding bads rather than promoting goods. Peter Singer, "Famine, Affluence and Morality," *Philosophy and Public Affairs*, 1 (1972): 229–243.

23. Keith Fuglie, James M. MacDonald, and Eldon Ball, "Productivity Growth in U.S. Agriculture," Economic Brief (No EB-9) (Washington, DC: US Department of Agriculture, 2007), p. 2.

24. Ciara Raudsepp-Hearne, Garry D. Peterson, Maria Tengo, Elena M. Bennett, Tim Holland, Karina Benessaiah, Graham K. Macdonald, and Laura R. Pfeifer, "Untangling the Environmentalist's Paradox: Why Is Human Well-being Increasing as Ecosystem Services Degrade?," *BioScience*, 60 (2010): 576–589.

25. Samuel Scheffler has argued that the motivation for much of this production involves the desire to benefit our descendants or to enjoy their regard. Samuel Scheffler, *Death and the Afterlife* (Oxford: Oxford University Press, 2016). This strikes me as rather dubious.

26. This is, of course, purely a hypothetical choice, since the growth in knowledge is strongly correlated with economic growth, mainly because research is expensive in various ways.

27. See Steven Bernstein, *The Compromise of Liberal Environmentalism* (New York: Columbia University Press, 2012).

28. Gru Brundtland, Mansour Khalid, Susanna Agnelli, et al., *Our Common Future* (Oxford: Oxford University Press, 1987).

29. As Alan Holland observes, more perspicuously, "The ability of a given generation to meet its needs or . . . the flow of services yielded by a given stock of natural capital, is a function, first, of the technology, second, of the social arrangements, and, third, of the human needs which pertain at the time." "Natural Capital," in Robin Attfield and Andrew Belsey, eds., *Philosophy and the Natural Environment* (Cambridge: Cambridge University Press, 1994), pp. 169–182 at 172.

30. Jackson, *Prosperity without Growth*, p. 14.

31. Victor, *Managing without Growth*, p. 28. Similarly, Herman Daly states that "the necessary change in vision is to picture the macroeconomy as an open subsystem of the finite natural ecosystem (environment)." *Beyond Growth* (Boston: Beacon Press, 1996), p. 48. Kate Raworth says that the earth should be conceived of as "an open system with constant inflows and outflows of matter and energy." *Doughnut Economics* (White River Junction: Chelsea Green, 2017), p. 64.

32. Victor, *Managing without Growth*, p. 72.

33. Victor, *Managing without Growth*, p. 47.

34. E.g. Clive L. Spash and Clemens Gattringer, "The Ethics Failures of Climate Economics," in Adrian Walsh, Sade Hormio, and Duncan Purves, eds., *The Ethical Underpinnings of Climate Economics* (New York: Routledge, 2017), pp. 162–182, claim that economists, who consider climate change to be merely a market failure, "have not grasped the basic functioning of the economy in biophysical terms" (p. 172).

35. It is, I suppose, worth noting as well that these claims about the "limits of growth" have figured centrally in environmental discourse for over 50 years, and yet growth continues, and the date at which growth is expected to stop keeps getting pushed back. The comparison to a doomsday cult that predicts the end of the world and then must continually revise its forecast as the world fails to end is difficult to avoid. See Leon Festinger, Henry Riecken, and Stanley Schachter, *When Prophecy Fails* (New York: Harper & Row, 1956).

36. Victor, *Managing without Growth*, p. 23.

37. It is uncontroversial, I should note, in environmental circles. In philosophy, by contrast, because of the influence of utilitarianism, many theorists are unfortunately tempted by the view that there is a moral imperative to increase population. See, e.g., John Broome, *Climate Matters* (New York: W. W. Norton, 2013).

38. Jackson, *Managing without Growth*, p. 14. Emphasis added.

39. Jackson, *Managing without Growth*, p. 50. Similarly Donella H. Meadows, Dennis L. Meadows, and Jørgen Randers, *Beyond the Limits* (Post Mills: Chelsea Green, 1992) say, "We will refer to GNP in various figures and tables, because the world's economic data are kept in money terms, not physical terms. But our interest is in what GNP stands for: material flows of capital, industrial goods, services, resources and agricultural products" (p. 35).

40. William Nordhaus, *The Climate Casino* (New Haven: Yale University Press, 2013), p. 88.

41. Jackson, *Managing without Growth*, p. 14.

42. Donella H. Meadows, Dennis L. Meadows, Jørgen Randers, and William W. Behrens III, *The Limits to Growth* (New York: Signet, 1972), p. 29.

43. Nicholas Georgescu-Roegen, *The Entropy Law and the Economic Process* (Cambridge, MA: Harvard University Press, 1971), pp. 277–279.

44. Thomas Princen, using the example of gold as a nonrenewable resource, writes that "to use a gold ring is also to wear down the gold, to send atoms of gold dissipating daily into the environment, never to be recovered by anybody ever again. It is to irreversibly consume the gold. . . . We would take the gold and then watch it entropically

disappear over the years and generations." *The Logic of Sufficiency* (Cambridge, MA: MIT Press, 2005), p. 31. One can see the non sequitur clearly here—reducing something to a lower-energy state is not the same thing as making it "disappear."

45. Padraig Belton, "Could Diesel Made from Air Help Tackle Climate Change?," *BBC News* (Sept. 1, 2015), http://www.bbc.com/news/business-34064072.

46. Vaclav Smil, *Energy and Civilization* (Cambridge, MA: MIT Press, 2017), p. 1.

47. Georgescu-Roegen, *Entropy Law*.

48. Georgescu-Roegen, *Entropy Law*, p. 277.

49. Geological energy is another potential source, due to the fact that the earth still has a molten core. As a source of energy, however, it produces no more than a tiny fraction of what is made available in the form of solar radiation.

50. Daly, *Steady-State Economics*, p. 28.

51. Andrew Dessler, *Introduction to Modern Climate Change*, 2nd ed. (Cambridge: Cambridge University Press, 2016), p. 53. As Dessler observes, somewhat laconically, "This is an immense amount of power. Human society today consumes about 16 TW, so this simple calculation shows why solar energy is the Holy Grail of renewable energy" (p. 53).

52. Jeff Tsao, Nate Lewis, and George Crabtree, "Solar FAQs," in *Report for Office of Basic Energy Sciences* (Washington, DC: US Department of Energy, 2006).

53. Neumayer, *Weak versus Strong Sustainability*, p. 51. Talk of a solar "budget" also ignores the fact that reflectors could be installed in space, to redirect solar energy that would not otherwise have struck the planet. In a more science fiction scenario, a Dyson sphere could be constructed that would capture the entire energy output of the sun (which is 3.8×10^{14} TW).

54. Dieter Helm, *Burn Out* (New Haven: Yale University Press, 2017), p. 8.

55. Thomas Homer-Dixon, *The Ingenuity Gap* (Toronto: Penguin, 2001).

56. Allan Buchanan, "Perfecting Imperfect Duties: Collective Action to Create Moral Obligations," *Business Ethics Quarterly*, 6 (1996): 27–42.

57. Barry G. Rabe, *Can We Price Carbon?* (Cambridge, MA: MIT Press, 2018); Stephen M. Gardiner, *A Perfect Moral Storm* (Oxford: Oxford University Press, 2011).

58. Steven Vanderheiden, *Atmospheric Justice* (Oxford: Oxford University Press, 2008), p. 122. This formulation, I should note, is actually ambiguous between a relative "worsening" and an absolute "worsening."

59. Tim Mulgan, *Ethics for a Broken World* (Montreal: McGill-Queen's University Press, 2011), p. 1.

60. Naomi Oreskes and Erik Conway, *The Collapse of Western Civilization* (New York: Columbia University Press, 2014).

61. Catriona McKinnon, *Climate Change and Future Justice* (New York: Routledge, 2013), p. 1.

62. Clive Hamilton, *Requiem for a Species* (London: Earthscan, 2012).

63. Darrel Moellendorf, *The Moral Challenge of Dangerous Climate Change* (Cambridge: Cambridge University Press, 2014), p. 26.

64. William D. Nordhaus, "A Review of the Stern Review on the Economics of Climate Change," *Journal of Economic Literature*, 45 (2007): 686–702; Christian Gollier, "An

NOTES TO PAGES 83–90 291

Evaluation of Stern's Report on the Economics of Climate Change," IDEI Working Paper no. 464 (2006).

65. Nicholas Stern, *The Economics of Climate Change: The Stern Review* (Cambridge: Cambridge University Press, 2007), p. 179.

66. Stern, *Economics of Climate Change*, p. 180 (Figure 6.6).

67. Stern, *Economics of Climate Change*, p. 179.

68. Stern, *Economics of Climate Change*, p. 104.

69. The obscurity is further exacerbated by Stern's use of BGEs (balanced growth equivalents) to express the damages. This unit "essentially measures the utility generated by a consumption path in terms of the consumption now that, if it grew at a constant rate, would generate the same utility" (*Economics of Climate Change*, p. 190). Translating a reduction in the rate of growth of consumption into a reduced present quantity of consumption that would generate the path without a reduction in the rate of growth, can easily generate the impression that climate change is going to reduce consumption below the present baseline. It is the use of BGEs that allowed Stern to produce misleading verbal formulations such as the following: "If we don't act, the overall costs and risks of climate change will be equivalent to losing at least 5% of global GDP each year, now and forever" (p. iv.) The key phrase, "equivalent to," was overlooked by many commentators, generating the impression that climate change is going to result in a decline in *actual* GDP. See, e.g., Vanderheiden, *Atmospheric Justice*, pp. 113, 122; Donald A. Brown, *Climate Change Ethics* (London: Routledge, 2013), p. 65.

70. Marshall Burke, Solomon M. Hsiang, and Edward Miguel, "Global Non-linear Effect of Temperature on Economic Production," *Nature*, 527 (2015): 235–239.

71. Burke, Hsiang, and Miguel, "Global Non-linear Effect," p. 238.

72. World Bank, "GDP Per Capita Growth," http://data.worldbank.org/indicator/NY.GDP.PCAP.KD.ZG, accessed April 21, 2017.

73. Nordhaus, *The Climate Casino*, p. 97.

74. McKinnon, *Climate Change*, p. 73.

75. McKinnon, *Climate Change*, p. 82.

76. McKinnon, *Climate Change*, p. 76.

77. McKinnon, *Climate Change*, p. 75.

78. McKinnon, *Climate Change*, p. 83.

79. For a detailed analysis of what would be required to maintain a steady state, see Herman E. Daly and John B. Cobb Jr., *For the Common Good* (Boston: Beacon Press, 1989), pp. 416–443.

80. Not everyone assumes this, I should note. The authors of *Beyond the Limits* suggest an income of $350 per capita (1992 dollars)—a shockingly low figure. Meadows, Meadows, and Randers, *Beyond the Limits*, pp. 194–196.

81. Jackson, *Prosperity without Growth*, pp. 55–58. Note that this is approximately equal to the standard of living of the urban population in China at the time of writing.

82. Andreou, "Shallow Route."

83. The "orgy" phrase is from Shue, *Climate Justice*, p. 175.

84. E.g. Tim Kasser, *The High Price of Materialism* (Cambridge, MA: MIT Press, 2002).

85. For discussion, see Joseph Heath, "Political Egalitarianism," *Social Theory and Practice*, 34 (2008): 485–516.

86. Eric Neumayer, "Global Warming: Discounting Is Not the Issue, but Substitutability Is," *Energy Policy*, 27 (1999): 33–43.

87. Norton, *Sustainability*, p. 305.

88. Sade Hormio distinguishes the following types: "natural capital, financial capital, real capital (consumer and investment goods, infrastructure), cultural capital (institutions), social capital (social contacts), human capital (abilities and knowledge, health) and knowledge capital (non-person-bound knowledge)." "Climate Change Mitigation, Sustainability and Non-substitutability," in Walsh, Hormio, and Purves, *Ethical Underpinnings*, pp. 103–121 at 107.

89. Norton, *Sustainability*, p. 313.

90. Robert Solow, "Sustainability: An Economist's Perspective," in Robert Dorfman and Nancy Dorfman, eds., *Economics of the Environment*, 4th ed. (New York: Norton, 1993), p. 181.

91. Solow, "Sustainability," p. 183. See important discussion of the "abstraction" concept in Ronald Dworkin, "What Is Equality Part 3: The Place of Liberty," in his *Sovereign Virtue* (Cambridge, MA: Harvard University Press, 2000).

92. Norton, *Sustainability*, p. 316.

93. Wildred Beckerman, "How Would You Like Your 'Sustainability' Sir? Weak or Strong? A Reply to My Critics," *Environmental Values*, 4 (1995): 169–179.

94. Paul Ekins, Sandrine Simon, Lisa Deutsch, Carl Folke, and Rudolf De Groot, "A Framework for the Practical Application of the Concepts of Critical Natural Capital and Strong Sustainability," *Ecological Economics*, 44 (2003): 165–185 at 169.

95. Ekins et al., "Framework," pp. 173–174.

96. For instance, during the 19th century humans completely depleted the natural stores of accumulated guano in Peru. Wars were fought over access to the dwindling supplies of this important natural resource. The development of the Haber-Bosch process rendered this irrelevant. Nitrogen, after all, is hardly a rare element; the problem is just that it is found mostly in a high-entropy state. To the extent that resources are just negentropic structures, there are *always* candidates for technological substitution.

97. Norton, "Ecology and Opportunity," p. 119.

98. Herman Daly, "On Wilfred Beckerman's Critique of Sustainable Development," *Environmental Values*, 4 (1995): 49–55.

99. See Dolf de Groot, *Functions of Nature* (Groningen: Wolters Noordhoff, 1992).

100. Elizabeth Kolbert, *The Sixth Extinction* (New York: Henry Holt, 2016), p. 18.

101. Martin L. Weitzman, "Fat-Tailed Uncertainty in the Economics of Catastrophic Climate Change," *Review of Environmental Economics and Policy*, 5 (2011): 275–292.

102. This was the substance of Weitzman's critique of the Stern Review. See Martin L. Weitzman, "A Review of the Stern Review on the Economics of Climate Change," *Journal of Economic Literature*, 945 (2007): 703–724.

103. In addition to Weitzman, see Frank Ackerman, Stephen J. DeCanio, Richard B. Howarth, and Kristen Sheeran, "Limitations of Integrated Assessment Models of

Climate Change," *Climatic Change*, 95 (2009): 297–315; or Frank Ackerman, *Can We Afford the Future?* (London: Zed, 2009).

104. Steven Levitt and Stephen Dubner, *Superfreakeconomics* (New York: William Morrow, 2009).

105. For the IPCC position, see Intergovernmental Panel on Climate Change, *Climate Change 2014: Impacts, Adaptation, and Vulnerability* (Cambridge: Cambridge University Press, 2014), pp. 1042–1045, 1079–1080.

Chapter 3

1. For a clear-eyed view, see Dieter Helm, *The Carbon Crunch* (New Haven: Yale University Press, 2012).

2. Government of Canada, *State of Remote/Off-Grid Communities in Canada* (August, 2011), http://www.nrcan.gc.ca/sites/www.nrcan.gc.ca/files/canmetenergy/files/pubs/2013-118_en.pdf (accessed Jan. 28, 2019).

3. Ronald Coase, "The Problem of Social Cost," *Journal of Law and Economics*, 3 (1960): 1–166.

4. Stephen M. Gardiner, "The Pure Intergenerational Problem," *Monist* 86 (2003): 481–501; Stephen M. Gardiner, "A Contract on Future Generations?," in Stephen M. Gardiner and David A. Weisbach, *Debating Climate Ethics* (Oxford: Oxford University Press, 2016), p. 22; Tim Mulgan, "Answering to Future People: Responsibility for Climate Change in a Breaking World," *Journal of Applied Philosophy*, 35 (2018): 532–548 at 535. Note that Gardiner continues to refer to it as a "collective action problem" because he defines this term in a nonstandard way.

5. Walter Sinnott-Armstrong, "It's Not *My* Fault: Global Warming and Individual Moral Obligations," in Walter Sinnott-Armstrong and Richard Howarth, eds., *Perspectives on Climate Change* (Dordrecht: Elsevier, 2005), pp. 221–253. See also Shelly Kagan, "Do I Make a Difference?," *Philosophy and Public Affairs*, 39 (2011): 105–141; Chrisoula Andreou, "Environmental Damage and the Puzzle of the Self-Torturer," *Philosophy and Public Affairs*, 34 (2006): 95–108; Holly Lawford-Smith, "Difference-Making and Individuals' Climate-Related Obligations," in Clare Heyward and Dominic Roser, eds., *Climate Justice in a Non-ideal World* (Oxford: Oxford University Press, 2016), pp. 64–82.

6. John Broome, "Climate Change, Efficiency, Future Generations and the Non-identity Effect" (forthcoming); Edward Page, "Intergenerational Justice and Climate Change," *Political Studies*, 47 (1999): 53–66.

7. One can see the mischief that it causes in John Broome, *Climate Matters* (New York: Norton, 2012), pp. 170–178.

8. Robert William Fogel, *The Escape from Misery and Premature Death, 1700–2100* (Cambridge: Cambridge University Press, 2004).

9. J. Bradford Delong, "Cornucopia: The Pace of Economic Growth in the 20th Century," NBER Working Paper no. 7602 (2000).

10. Delong, "Cornucopia," pp. 3–4.

11. IPCC, *Climate Change 2014: Mitigation of Climate Change* (Cambridge: Cambridge University Press, 2014), states, "Mitigation is a public good; climate change is a case of 'the tragedy of the commons' (*high confidence*)" (p. 211). On the denial that it is a collective action problem, see n. 3.

12. David Gauthier, *Morals by Agreement* (Oxford: Clarendon, 1986), pp. 223–225.

13. Brian Barry, "Circumstances of Justice and Future Generations," in Richard I. Sikora and Brian M. Barry, eds., *Obligations to Future Generations* (Philadelphia: Temple University Press, 1978), pp. 204–248.

14. David Hume, *A Treatise of Human Nature*, ed. P. H. Nidditch, 2nd ed. (Oxford: Oxford University Press, 1978), p. 495; Simon Hope, "The Circumstances of Justice," *Hume Studies*, 36 (2010): 125–148.

15. John Rawls, *A Theory of Justice* (Cambridge, MA: Harvard University Press, 1971), p. 126.

16. Roger Paden, "Reciprocity and Intergenerational Justice," *Public Affairs Quarterly*, 10 (1996): 249–266; Eric Brandstedt, "The Savings Problem in the Original Position: Assessing and Revising the Model," *Canadian Journal of Philosophy*, 47 (2017): 269–289.

17. Gauthier, *Morals by Agreement*; T. M. Scanlon, "Contractualism and Utilitarianism," in Amartya Sen and Bernard Williams, eds., *Utilitarianism and Beyond* (Cambridge: Cambridge University Press, 1982), pp. 103–128. On Gauthier, see Peter Vallentyne, ed., *Contractarianism and Rational Choice* (Cambridge: Cambridge University Press, 1991). On Scanlon, see Matt Matravers, ed., *Scanlon and Contractualism* (London: Frank Cass, 2003).

18. Ann Cudd and Seena Eftekhari, "Contractarianism," in *The Stanford Encyclopedia of Philosophy*, ed. Edward N. Zalta (Summer 2018 ed.).

19. Elizabeth Ashford and Tim Mulgan, "Contractualism," in *The Stanford Encyclopedia of Philosophy*, ed. Edward N. Zalta (Summer 2018 ed.).

20. Brian Barry, *Theories of Justice* (Berkeley: University of California Press, 1989), p. 189.

21. Gardiner, "Contract on Future Generations," pp. 78–86.

22. Daniel Attas, "A Trans-generational Difference Principle," in Axel Gosseries and Lukas H. Meyer, eds., *Intergenerational Justice* (Oxford: Oxford University Press, 2011), pp. 189–218 at 197.

23. Gustaf Arrhenius, "Mutual Advantage Contractarianism and Future Generations," *Theoria*, 65 (1999): 25–35 at 34.

24. John Rawls, *Political Liberalism* (New York: Columbia University Press, 1993), p. 274.

25. See Hugh McCormick, "Intergenerational Justice and the Non-reciprocity Problem," *Political Studies* 57, no. 2 (2009): 451–458, for a very long list of philosophers who have accepted it as a problem. For a more recent example, see Brandstedt, "The Savings Problem in the Original Position."

26. See Drew Fudenberg and Jean Tirole, *Game Theory* (Cambridge, MA: MIT Press, 1991); Roger Myerson, *Game Theory* (Cambridge, MA: Harvard University Press, 1997); Eric Rasmusen, *Games and Information*, 4th ed. (London: Wiley, 2006); also

collected papers in Drew Fudenberg and David K. Levine, *A Long-Run Collaboration on Long-Run Games* (Singapore: World Scientific, 2009).

27. Fudenberg and Tirole, *Game Theory*, p. 171; David J. Salant, "A Repeated Game with Finitely Lived Overlapping Generations of Players," *Games and Economic Behavior*, 3 (1991): 244-259.

28. Raimo Tuomela, *Cooperation: A Philosophical Study* (Dordrecht: Springer, 2000).

29. Rawls, *A Theory of Justice*, p. 4.

30. Fudenberg and Tirole, *Game Theory*, pp. 152-154. The full folk theorem is significantly more complicated. See Drew Fudenberg and Eric Maskin, "The Folk Theorem in Repeated Games with Discounting or with Incomplete Information," *Econometrica*, 54 (1986): 533-554.

31. The "or would be" indicates that the relevant concept of equilibrium is subgame perfection. Unlike Nash equilibrium, which only requires that each strategy be a best response to the other strategies in the equilibrium set, subgame perfection requires that each strategy be a Nash equilibrium, not just of the overall game, but of each proper subgame of the game as well. The goal here is to proscribe from the equilibrium actions that would be non-maximizing if actually played, but which are never played.

32. This is also sometimes known as the "grim" strategy, because it does not allow a return to the cooperative path once a defection has occurred. This may seem extreme, but as Herbert Gintis observes, these strategies are surprisingly common in everyday social interaction. See Herbert Gintis, *Game Theory Evolving* (Princeton: Princeton University Press, 2000), p. 135.

33. Note that all cooperation collapses in this model, but this is not essential, and more targeted punishments can be developed (as in the full folk theorem).

34. It should be noted that defection by any other individual generates a cascade that causes all eight players to defect for the rest of the game. But this is not required in order to sustain the equilibrium. Again, for textbook presentation, see Fudenberg and Tirole, *Game Theory*, pp. 171-172.

35. See Rasmusen, *Games and Information*, p. 132. See also Larry Samuelson, "A Note on Uncertainty and Cooperation in a Finitely Repeated Prisoner's Dilemma," *International Journal of Game Theory*, 16 (1987): 187-195.

36. This is a well-known result in game theory. It is the transformation of an infinitely repeated PD into a finitely repeated one. See Rasmusen, *Games and Information*, pp. 129-130.

37. Axel Gosseries, "Three Models of Intergenerational Reciprocity," in Gosseries and Meyer, *Intergenerational Justice*, pp. 119-146.

38. In Chinese culture (or Confucian ethics), the virtue of "filial piety" involves an upstream flow of benefits, which is why many Westerners find parts of certain texts such as the *Xiaojing* deeply counterintuitive. See *The Chinese Classic of Family Reverence*, trans. Henry Rosemont Jr. and Roger T. Ames (Honolulu: University of Hawaii Press, 2009).

39. This is sometimes known at the Hart-Rawls principle. Hart writes: "When a number of persons conduct any joint enterprise according to rules and thus restrict their

liberty, those who have submitted to these restrictions when required have a right to similar submission from those who have benefited by their submission." H. L. A. Hart, "Are There Any Natural Rights?," *Philosophical Review*, 64 (1955): 175–191 at 185.

40. The assumption that cooperation must have a downstream structure is responsible for the perception, among some philosophers, that all systems of indirect reciprocity require some moral incentive, e.g. Axel Gosseries, *Penser la justice entre les générations* (Paris: Flammarion, 2004), p. 149; Joerg Chet Tremmel, *A Theory of Intergenerational Justice* (Sterling, VA: Earthscan, 2009), p. 194.

41. Marcel Mauss, *The Gift*, trans. W. D. Hall (London: Routledge, 1990).

42. This presentation is somewhat exaggerated, since private pension plans in most jurisdictions are now required by regulation to fund the majority of their liabilities. This is so that the plan can be wound down in an orderly fashion if the need arises. Thus the only pure pay-as-you-go plans are in the public sector. Nevertheless, even funded plans retain significant pay-as-you-go features. These become apparent when the plans suffer significant investment losses, creating a shortfall that *current* employees are expected to make up.

43. Gardiner, "The Pure Intergenerational Problem," p. 490.

44. As Gintis has observed, one can find examples of individuals playing trigger strategies in a wide variety of economic contexts. Obvious examples include employers firing employees who shirk, lenders refusing any further loans to borrowers who have defaulted, and customers refusing to make any further purchases from a firm that has sold them shoddy goods. See *Game Theory Evolving*, p. 135. Gintis refers to these as "contingent renewal contracts"—where the individual either renews the cooperative arrangement, or else discontinues it permanently—and observes that they are "among the most prevalent exchanges in market economies" (p. 136).

45. A more sophisticated criticism would be that the punishment sequence that supports the cooperative equilibrium is not credible because it is not "renegotiation proof" (see Fudenberg and Tirole, *Game Theory*, p. 180). The intuition here is that the punishment would not occur because the offenders could talk their way out of it. But since renegotiation-proofness builds the Pareto principle into the solution concept, it cannot be appealed to in this context.

46. Arrhenius, "Mutual Advantage Contractarianism and Future Generations," pp. 39–40. The model he criticizes is Joseph Heath, "Intergenerational Cooperation and Distributive Justice," *Canadian Journal of Philosophy*, 27 (1997): 361–376.

47. This is basically because uncertainty transforms the finitely repeated PD back into an infinitely repeated PD. See Fudenberg and Tirole, *Game Theory*, pp. 384–385; Rasmusen, *Games and Information*, p. 132.

48. This is why it can be rational to invest in stock market bubbles, and even pyramid schemes. Even though the end is foreseeable, uncertainty about when the end will occur can make it rational to invest—with the intention of getting out before the end occurs.

49. P. D. James, *The Children of Men* (New York: Vintage, 1992).

50. Tim Mulgan, *Future People* (Oxford: Clarendon, 2006), p. 30; Gosseries, *Penser la justice*, pp. 97–98.

51. This is the structure of the argument in Avner de Shalit, *Why Posterity Matters* (London: Routledge, 1995), p. 96; and Arrhenius, "Mutual Advantage Contractarianism."

52. See Fudenberg and Tirole, *Game Theory*, p. 153.

53. Arrhenius, "Mutual Advantage Contractarianism," p. 30.

54. Gardiner, "Contract on Future Generations," p. 104.

55. See, e.g., Gintis, *Game Theory Evolving*, pp. 126–127.

56. For useful discussion, see Gauthier on narrow and broad compliance (*Morals by Agreement*, p. 178).

57. Gardiner, "The Pure Intergenerational Problem," pp. 490–491.

58. Frank P. Ramsey, "A Mathematical Theory of Saving," *Economic Journal*, 38 (1928): 543–549. The paper, it should be noted, was written under the strong influence of John Maynard Keynes and Arthur Cecil Pigou.

59. Tjalling C. Koopmans, "Stationary Ordinal Utility and Impatience," *Econometrica*, 28 (1960): 287–309.

60. One can see the background role that the "socialist state" played in Ramsey's thinking in "Mathematical Theory of Saving," p. 557.

61. David Collard, *Generations of Economists* (London: Routledge, 2011).

62. Robert M. Solow, "Intergenerational Equity and Exhaustible Resources," *Review of Economic Studies*, 41 (1974): 29–46.

63. Rawls, *A Theory of Justice*, p. 291. Again, note that this discussion is deleted in the 1999 revised edition (Cambridge, MA: Harvard University Press).

64. Perhaps the biggest change between the 1971 and 1999 revised edition of *A Theory of Justice* is the insertion of the two-page discussion at pp. 254–255 of the 1999 edition, along with the deletion of the discussion that occurs on p. 291 of the 1971 edition.

65. Rawls, *A Theory of Justice*, rev. ed., p. 255. In John Rawls, *Justice as Fairness* (Cambridge, MA: Harvard University Press, 2001) he revises the position again, arguing that each generation should adopt the rate of savings that it can will all generations, including previous ones, to have saved. See p. 160.

66. A noteworthy exception is Benjamin Franklin, who in his will left $2,000 to the cities of Boston and Philadelphia, with the instruction that half the sum could be withdrawn in 100 years, the remainder in 200 years. His objective was to offer subsequent generations an instructive lesson in the power of compound interest. Fox Butterfield, "From Ben Franklin, A Gift That's Worth Two Fights," *New York Times* (April 21, 1990). The fact that this will is still considered eccentric serves to emphasize the point that individuals seldom have any interest in saving to the benefit of "future generations" generally, but tend to focus on their own living heirs.

67. Martin Feldstein, "Rethinking Social Insurance," *American Economic Review* 95, 1 (2005): 1–24.

68. For fascinating discussion, see Christopher G. Lewin, *Pensions and Insurance before 1800* (East Lothian: Tuckwell, 2003).

69. This is the problem with Gardiner's argument, in "Contract on Future Generations," p. 102, where he suggests that a particular generation might free itself from dependence upon the young by hoarding rather than investing. The availability of this option

does not undermine the contractualist analysis. In order to show that there is a system of intergenerational cooperation in place, one need only show that investment is *better than* hoarding. The fact that investment allows individuals to earn interest on their savings and therefore to consume, during their retirement years, several times more than they actually saved is sufficient to establish this.

70. Contracts with this annuity-like structure were extremely common in Europe before the emergence of capitalism. Farmers would often transfer land title to their children, in return for a lodging and rations, typically specified in great detail. See Lewin, *Pensions and Insurance*. Corrodies—commonly sold by monasteries in order to finance capital projects—also had this structure.

71. While growing up, my older brother benefited from a rather significant, nontransferable inheritance from my grandfather (a peculiar financial instrument, no longer sold, that paid for his university tuition). I had none, because my grandfather died the year before I was born. As is standard, my grandfather's will included provision for both his son and his son's children, but of course it contained no provision for any *future* children that his son might have.

72. A. N. Wilson, *The Victorians* (London: Arrow, 2003).

73. Despite over 50 years of moral argument, the fraction of the US population that is vegetarian remains relatively small at around 5 percent. See R. J. Reinhart, "Snapshot: Few Americans Vegetarian or Vegan," *Gallup* (Aug. 1, 2018).

Chapter 4

1. This is an old idea, recently revived by Paul Mason, *Postcapitalism* (New York: Allen Lane, 2015).

2. Michael Albert, *Parecon: Life after Capitalism* (London: Verso, 2003). Note that the scheme does abolish scarcity pricing for labor, attempting to preserve it only for goods. See also John Roemer, *A Future for Socialism* (Cambridge, MA: Harvard University Press, 1994).

3. Ronald Dworkin, *Sovereign Virtue* (Cambridge, MA: Harvard University Press, 2000), p. 74.

4. Kenneth Arrow and Frank Hahn, *General Competitive Analysis* (San Francisco: Holden-Day, 1971). See also Walter J. Schultz, *The Moral Conditions of Economic Efficiency* (Cambridge: Cambridge University Press, 2009).

5. Joseph Heath, *Filthy Lucre* (Toronto: HarperCollins, 2008), p. 160.

6. Garett Hardin, "The Tragedy of the Commons," *Science*, 162 (1968): 1243–1248.

7. Elinor Ostrom, *Governing the Commons* (Cambridge: Cambridge University Press, 1990).

8. Elinor Ostrom, Joanna Burger, Christopher B. Field, Richard B. Norgaard, and David Policansky, "Revisiting the Commons: Local Lessons, Global Challenges," *Science*, 284 (1999): 278–282.

9. A. C. Pigou, *The Economics of Welfare* (London: Macmillan, 1920).

10. Martin Weitzman, "Prices *vs.* Quantities," *Review of Economic Studies*, 41 (1974): 477–491.
11. Lawrence H. Goulder and Andrew R. Schein, "Carbon Taxes versus Cap and Trade: A Critical Review," *Climate Change Economics*, 4 (2013): 1–28 at 4.
12. Both quotes from Patrick Bigger, "Hybridity, Possibility: Degrees of Marketization in Tradeable Permit Systems," *Environment and Planning A: Economy and Space*, 50:3 (2018): 512–530 at 514. See also Gareth Bryant, *Carbon Markets in a Climate-Changing Capitalism* (Cambridge: Cambridge University Press, 2019); Larry Lohmann, "Financialization, Commodification and Carbon: The Contradictions of Neoliberal Climate Policy," *Socialist Register*, 48 (2012): 85–107.
13. Tim Hayward, "Human Rights versus Emissions Rights: Climate Justice and the Equitable Distribution of Ecological Space," *Ethics and International Affairs*, 21:4 (2007): 431–450 at 433–435.
14. Naomi Klein, *This Changes Everything* (Toronto: Knopf, 2014).
15. John Braithwaite, "The Limits of Economism in Controlling Harmful Corporate Conduct," *Law & Society Review*, 16 (1981–82): 481–504 at 488.
16. Keith Bradsher, *High and Mighty* (New York: PublicAffairs, 2002).
17. Canada's Ecofiscal Commission, *Supporting Carbon Pricing* (June 2017), http://ecofiscal.ca/wp-content/uploads/2017/06/Ecofiscal-Commission-Report-Supporting-Carbon-Pricing-June-2017.pdf (accessed June 18, 2020).
18. This list is distilled from Braithwaite, "Limits of Economism."
19. Goulder and Schein, "Carbon Taxes."
20. Robert Goodin, "Selling Environmental Indulgences," in John Dryzek and David Schlosberg, eds., *Debating the Earth* (Oxford: Oxford University Press, 1994), pp. 237–254; Michael Sandel, "It's Immoral to Buy the Right to Pollute," in his *Public Philosophy: Essays on Morality in Politics* (Cambridge, MA: Harvard University Press, 2005), pp. 93–96.
21. Edward A. Page, "Cashing In on Climate Change: Political Theory and Global Emissions Trading," in Catriona McKinnon and Gideon Calder, eds., *Climate Change and Liberal Priorities* (London: Routledge, 2012), pp. 169–189.
22. Goulder and Schein, "Carbon Taxes," pp. 3–9.
23. Goulder and Schein, "Carbon Taxes," p. 6.
24. For the typical line of reasoning, see Page, "Cashing In," p. 172.
25. Joseph Stiglitz, *The Economic Role of the State*, ed. Arnold Heertje (Oxford: Blackwell, 1989).
26. Dale Jamieson, *Reason in a Dark Time* (New York: Oxford University Press, 2014), p. 9.
27. Jamieson, *Reason in a Dark Time*, p. 143.
28. Henry Shue, "Subsistence Emission and Luxury Emissions," *Law & Policy*, 15 (1993): 39–59.
29. Shue, "Subsistence Emission," p. 55.
30. Shue, "Subsistence Emission," p. 58.
31. David A. Weisbach and Eric Posner, *Climate Change Justice* (Princeton: Princeton University Press, 2007), make a similar point (pp. 85–87). The fact that no political constituency can be mobilized in support of such taxes (indeed, luxury taxes were

abolished in the United States under President Clinton) should give additional pause to those who want to reintroduce them through the back door of climate change policy.

32. There are very few academics who are not in a position to claim that the esoteric, subspecialized field of inquiry in which they have labored anonymously for years has not acquired newfound social relevance because of its connection to climate change. For examples, see John S. Dryzek, Richard B. Norgaard, and David Scholsberg, eds., *Oxford Handbook of Climate Change and Society* (Oxford: Oxford University Press, 2012), or Tahseen Jafry, ed., *Routledge Handbook of Climate Justice* (London: Routledge, 2019).

33. Alisa Smith and J. B. MacKinnon, *The 100-Mile Diet* (Toronto: Random House, 2007). These food movements, it should be noted, are extremely difficult to assess from a policy perspective, because there is seldom a rational connection between the announced objectives and the activities undertaken. For example, the most effective way by far to promote local farming would be to impose tariffs on imported agricultural products, and yet the locavore movement has taken no position on tariff policy. There is a peculiar insistence that the objectives be achieved through self-imposed dietary restrictions, which suggests that there is a more complex psychological phenomenon underlying the trend, perhaps involving an ambivalent relationship to bodily pleasure.

34. Christopher L. Weber and H. Scott Matthews, "Food-Miles and the Relative Climate Impacts of Food Choices in the United States," *Environmental Science & Technology*, 42 (2008): 3508–3513.

35. Wayne Wakeland, Susan Cholette, and Kumar Venkat, "Food Transportation Issues and Reducing Carbon Footprint," in Joyce Boye and Yvew Arcand, eds., *Green Technologies in Food Production and Processing* (New York: Springer, 2012), pp. 211–236 at 220–221.

36. Wakeland, Cholette and Venkat, "Food Transportation Issues," p. 225.

37. Weber and Matthews, "Food-Miles," p. 3512.

38. Peter Scarborough, Paul N. Appleby, Anja Mizdrak, Adam D. M. Briggs, Ruth C. Travis, Kathryn E. Bradbury, and Timothy J. Key, "Dietary Greenhouse Gas Emissions of Meat-Eaters, Fish-Eaters, Vegetarians and Vegans in the U.K.," *Climatic Change*, 125:2 (2014): 179–192.

39. Based on calculations by Weber and Matthews, "Food-Miles." Their data tables show total dollar value and total CO_2-equivalent emissions for these categories. I multiplied the latter figure by the amount of the carbon tax, then expressed the result as a percentage of the former.

40. One must be cautious when attempting to identify the activities that count as "carbon sequestration." In order to be truly sequestered, carbon must be removed entirely from the atmospheric carbon cycle, which normally means that it must be buried or sunk to a depth of at least one kilometer. Most importantly, terrestrial vegetation does not really sequester carbon, because it removes it from the atmosphere only briefly (in the grand scale of things), before dying, decaying, and returning that carbon to the atmosphere. The only important exceptions to this are the extremely dense tropical

rainforests, which lock up so much carbon that they make a tangible impact on the quantity of carbon dioxide in the atmosphere. Thus preservation of Brazilian or Indonesian forest "counts" as carbon sequestration. By contrast, the Canadian boreal forest simply does not contain enough carbon, or hold it for long enough, to make a significant impact on airborne carbon, and so its preservation ought not count as a carbon sink.

41. Simon Caney, "Justice and the Distribution of Greenhouse Gas Emissions," *Journal of Global Ethics*, 5 (2009): 125–146 at 138.

42. For discussion, see Amy Sinden, "Allocating the Costs of the Climate Crisis: Efficiency versus Justice," *Washington Law Review*, 85 (2010): 293–353 at 304–306.

43. For overview of these funds, see Climate Funds Update website: https://climatefundsupdate.org/ (accessed June 5, 2020).

44. In terms of the total quantity of carbon emissions averted, the most successful carbon abatement policy so far enacted is arguably China's "one child" policy ("Curbing Emissions," *The Economist* [Sept. 20, 2014]). Western liberals tend not to like this policy, for reasons having nothing to do with climate change, but instead due to the violation of individual autonomy. The question is then whether this objection could provide legitimate grounds for refusing to count the policy as a contribution to climate change mitigation. It is difficult to see how it could, although it is also not difficult to see how more complex cases could be invented that would create problems in this vein.

45. William Nordhaus, "Climate Clubs: Overcoming Free-Riding in International Climate Policy," *American Economic Review*, 105 (2015): 1339–1370.

Chapter 5

1. See Mark Sagoff, "The Limits of Cost-Benefit Analysis," *Philosophy and Public Policy Quarterly*, 1 (1981): 9–11. The prevailing sentiment among philosophers is perhaps best summed up by the title of Henry Richardson's paper "The Stupidity of the Cost-Benefit Standard," *Journal of Legal Studies*, 29 (2000): 971–1003.

2. HM Treasury, *The Green Book* (2003), p. 1, https://www.gov.uk/government/publications/the-green-book-appraisal-and-evaluation-in-central-governent (accessed February 10, 2021).

3. Richard L. Revesz and Michael A. Livermore, *Retaking Rationality* (Oxford: Oxford University Press, 2008).

4. Donald C. Hubin, "The Moral Justification of Benefit/Cost Analysis," *Economics and Philosophy*, 10 (1994): 169–194 at 170.

5. For example, see Matthew D. Adler and Eric A. Posner, *New Foundations of Cost-Benefit Analysis* (Cambridge, MA: Harvard University Press, 2006); or Eric Schokkaert, "Cost-Benefit Analysis of Difficult Decisions," *Ethical Perspectives*, 2 (1995): 71–84. Also Cass Sunstein, *The Cost-Benefit Revolution* (Cambridge, MA: MIT Press, 2018), inter alia.

6. A point emphasized by Lewis A. Kornhauser, "On Justifying Cost-Benefit Analysis," *Journal of Legal Studies*, 29:S2 (2000): 1037–1057 at 1053.

7. Hubin, "Moral Justification," p. 189.

8. The textbook I happen to own provides a good example of the standard treatment. See Harvery Rosen, Beverly George Dahlby, Roger Smith, and Paul Boothe, *Public Finance in Canada*, 2nd ed. (Toronto: McGraw-Hill, 2002).

9. Cass Sunstein, having started as an academic advocate of CBA and then accepted an administrative position where he was involved in assessing CBAs for a broad range of agencies in the US federal government, has written a number of interesting articles that attempt to describe the differences between textbook CBA and CBA as practiced.

10. Revesz and Livermore, *Retaking Rationality*.

11. Cass Sunstein provides several examples in "The Cost-Benefit State," Coase-Sandor Institute for Law and Economics Working Paper no. 39 (1996).

12. "Executive Order 12866," *Federal Register*, 58 (Oct. 4, 1993).

13. Cass Sunstein, "The Real World of Cost-Benefit Analysis: Thirty-Six Questions (and Almost as Many Answers)," *Columbia Law Review*, 114 (2014): 167–211 at 191 and 193.

14. For purposes of illustration, it is helpful to set aside the issue of interpersonal comparability in the utility scales. Disagreement about what one person's utility is "worth," in terms of the other's, concerns the *slope* of U, but not the basic form of the social welfare function.

15. For an example of this mode of presentation, see Daniel Farber, *Eco-pragmatism* (Chicago: University of Chicago Press, 1999), pp. 44–45.

16. Amartya Sen, "The Discipline of Cost-Benefit Analysis," *Journal of Legal Studies*, 29 (2000): 931–952 at 947.

17. See Rosemary Lowry and Martin Peterson, "Cost-Benefit Analysis and Non-utilitarian Ethics," *Politics, Philosophy and Economics*, 11 (2011): 258–279.

18. For discussion, see Farber, *Eco-pragmatism*, pp. 106–110.

19. One might also describe the exercise as "political" in the Rawlsian sense. See Joseph Heath, "Political Egalitarianism," *Social Theory and Practice*, 34 (2008): 485–516.

20. Jules L. Coleman, "Efficiency, Utility, and Wealth Maximization," *Hofstra Law Review*, 8 (1980): 509–551 at 516.

21. An observation that has been made by David Schmidtz, "A Place for Cost-Benefit Analysis," *Philosophical Issues,* 11 (2001): 148–171, but not defended in detail.

22. Fred Hirsch, *The Social Limits to Growth*, rev. ed. (London: Routledge, 1977), pp. 87–89.

23. Elizabeth Anderson, *Value in Ethics and Economics* (Cambridge, MA: Harvard University Press, 1993), p. 190.

24. Frank Ackerman and Susan Heinzerling, *Priceless* (New York: New Press, 2004), p. 40.

25. Many critics of CBA are either confused on this point or write in a way that encourages significant confusion. Steven Kelman, for example, makes it sound as though CBA involves *actually* assigning prices to things. "Cost-Benefit Analysis: An Ethical Critique," *Regulation*, 33 (Jan. 1981): 33–40 at 39.

26. Anderson, *Value in Ethics and Economics*, p. 193.

27. Anderson, *Value in Ethics and Economics*, p. 193.

28. For another example, Donald Brown argues that "CBAs can be used to allow some people to kill . . . others if the costs to the killer in eliminating the killing behavior are greater than the value of the life of the person who may be killed." *Climate Change Ethics* (London: Routledge, 2013), p. 73. Similarly, he argues the VSL values are too low because "if you ask most people how much money they would accept to give up their life, they would say no amount of money would satisfy them" (p. 73). These claims are so misleading that it is difficult to believe that Brown is arguing in good faith.

29. Jane Jacobs, *The Death and Life of Great American Cities* (New York: Vintage, 1961), p. 94. "In orthodox city planning, neighborhood open spaces are venerated in an amazingly uncritical fashion, much as savages venerated magical fetishes. Ask a houser how his planned neighborhood improves on the old city and he will cite, as a self-evident virtue, More Open Space. Ask a zoner about improvements in progressive codes and he will cite, again as a self-evident virtue, their incentives toward leaving More Open Space . . . More Open Space for what? For muggings? For bleak vacuums between buildings?" (p. 90).

30. Colin Camerer, Samuel Issacharoff, George Loewenstein, Ted O'Donoghue, and Matthew Rabin, "Regulation for Conservatives: Behavioral Economics and the Case for 'Asymmetric Paternalism,'" *University of Pennsylvania Law Review*, 151 (2003): 1211–1254.

31. Ronald Coase, "The Problem of Social Cost," *Journal of Law and Economics*, 3 (1960): 1–44.

32. A. C. Pigou, *The Economics of Welfare*, 3rd ed. (London: Macmillan, 1929), pp. 194–197.

33. This is one of the points that separates the way many economists think about social questions from more commonsense moral perspectives. For example, many economists are not overly concerned about theft because it is *merely* redistributive. If someone sneaks into your house and takes something, the total wealth of the nation is neither increased nor decreased; a portion of it simply passes out of your hands into someone else's. But if you go out and change your locks or install a burglar alarm in response to the theft, then that *is* a concern to the economist, because now real resources are being expended, not to increase the stock of wealth, but merely to ensure that a particular redistribution of that wealth does not occur. Thus the amount spent on theft-deterrence is a loss from the standpoint of efficiency.

34. Cass Sunstein, "Cognition and Cost-Benefit Analysis," *Journal of Legal Studies*, 29 (2000): 1059–1103 at 1090 n. 76. See also Kip Viscusi, "Risk Equity," *Journal of Legal Studies*, 29 (2000): 843–871 at 857–858.

35. Sunstein, "Cognition and Cost-Benefit Analysis," pp. 1063–1064.

36. Peter A. Diamond and Jerry Hausman, "Contingent Valuation: Is Some Number Better Than No Number?," *Journal of Economic Perspectives*, 8:4 (1994): 45–64.

37. Amartya Sen makes the narrower, but nevertheless very significant, point that with any public good, the amount that any individual is willing to pay will be, in part, a function of her beliefs about what other people are going to pay. Thus the type of

question is one better suited for ascertaining the value of goods that are exclusive in consumption. "Discipline of Cost-Benefit Analysis," pp. 948–949.

38. William Desvousges, F. Reed Johnson, Richard W. Dunford, Sara P. Hudson, K. Nicole Wilson, and Kevin J. Boyle, "Measuring Natural Resource Damages with Contingent Valuation: Tests of Validity and Reliability," in Jerry Hausman, ed., *Contingent Valuation: A Critical Assessment* (Amsterdam: North Holland, 1993), pp. 91–164. For discussion see Diamond and Hausmann, "Contingent Valuation," p. 51.

39. See Richard Dorman and Paul Hagstrom, "Wage Compensation for Dangerous Work Revisited," *Industrial and Labor Relations Review*, 52 (1998): 116–135. For discussion see David A. Moss, *When All Else Fails* (Cambridge, MA: Harvard University Press, 2002), pp. 162–169. Cass Sunstein provides a useful summary of studies in *Risk and Reason* (Cambridge: Cambridge University Press, 2002), p. 174.

40. Diamond and Hausman, "Contingent Valuation," p. 58. There are important cases, of course, when the opposite is true. A telling anecdote, for instance, can often overwhelm any consideration of statistical data.

41. Another example is that, prior to the introduction of "valuation of life" based on willingness to pay, US government agencies used a "cost of death" measure consisting of the discounted sum of lost earnings (see Viscusi, "Risk Equity," pp. 854–855). The latter sum is morally arbitrary. Future earnings are of no value to a dead person, and as far as society is concerned, the individual's death is a wash, since that individual's lost contribution to the social product is accompanied by a closely matched decline in consumption.

42. For discussion see Joseph Heath, *Filthy Lucre* (Toronto: HarperCollins, 2008), pp. 218–219.

43. E.g. Anderson, *Value in Ethics and Economics*, pp. 210–216; or Mark Sagoff, "Aggregation and Deliberation in Valuing Environmental Public Goods: A Look beyond Contingent Valuation," *Ecological Economics*, 24 (1998): 213–230.

44. See Douglas Craig Macmillan, Lorna Jennifer Philip, Nick Hanley, and Begoña Alvarez-Farizo, "Valuing the Nonmarket Benefits of Wild Goose Conservation: A Comparison of Interview and Group-Based Approaches," *Ecological Economics*, 43 (2002): 49–59; Begoña Alvarez-Farizo, Nick Hanley, Ramón Barberan, and Angelina Lázaro, "Choice Modeling at the 'Market Stall': Individual versus Collective Interest in Environmental Valuation," *Ecological Economics*, 60 (2007): 743–751. For discussion see Giles Atkinson and Susana Mourato, "Environmental Cost-Benefit Analysis," *Annual Review of Environmental Resources*, 33 (2008): 317–344 at 322.

45. Atkinson and Mourato, "Environmental Cost-Benefit Analysis."

46. Immanuel Kant, *Foundations of the Metaphysics of Morals*, trans. Lewis White Beck (New York: Macmillan, 1990), pp. 51–52.

47. Kelman, "Cost-Benefit Analysis."

48. Michael Walzer, "Political Action: The Problem of Dirty Hands," *Philosophy and Public Affairs*, 2 (1973): 160–180.

49. See Judith Jarvis Thomson, "The Trolley Problem," *Yale Law Journal*, 94 (1985): 1395–1415. See also John Mikhail's analysis in *Elements of Moral Cognition* (Cambridge: Cambridge University Press, 2011).

50. Viscusi, "Risk Equity."

51. Jonathan Wolff, "What Is the Value of Preventing a Fatality?," in Tim Lewens, ed., *Risk: Philosophical Perspectives* (London: Routledge, 2007), pp. 54–67.

52. Wolff provides an excellent example of this involving British Rail, which was under pressure to install a computerized signaling system, at a cost of approximately £6 billion, that would have saved perhaps two lives yearly. See *Ethics and Public Policy* (Oxford: Routledge, 2011), p. 89.

53. E.g. Ackerman and Heinzerling, *Priceless*, p. 87. They repeatedly use the term "priceless" in association with human life, but then deny that they want it to be assigned an infinitely high value (recognizing that this makes it impossible to deal with risk). As a result, what they wind up claiming is simply that the value currently assigned is too low.

54. Viscusi, "Risk Equity," p. 855.

55. Treasury Board of Canada, *Canadian Cost-Benefit Analysis Guide: Regulatory Proposals* (Ottawa: Treasury Board, 2007), p. 20.

56. *Entergy Corp. v. Riverkeep Inc.*, 556 U.S. 208 (2009).

57. John Stuart Mill, *On Liberty*, ed. Elizabeth Rapaport (Indianapolis: Hackett, 1978).

58. For my reflections on a similar set of issues, see Joseph Heath, "Envy and Efficiency," *Revue de philosophie economique*, 13 (2006): 3–30.

59. John Vidal, "The Great Green Land Grab," *The Guardian* (Feb. 13, 2008).

60. Ackerman and Heinzerling, *Priceless*, p. 31.

61. Ackerman and Heinzerling, *Priceless*, p. 36.

62. Ackerman and Heinzerling, *Priceless*, p. 36.

63. Nicholas Stern, "Ethics, Equity and the Economics of Climate Change, Paper 1: Science and Philosophy," *Economics and Philosophy*, 30 (2014): 397–444 at 400.

64. Dominic Roser and Christian Seidel, *Climate Justice* (Oxford: Routledge, 2012), p. 3.

65. E.g. see Richardson, "Stupidity," or Anderson, *Value in Ethics and Economics*, pp. 210–213. Non-liberal philosophers for the most part prefer no procedure at all, but direct enforcement of environmental values.

66. Idil Boran, *Political Theory and Global Climate Action* (London: Routledge, 2019).

67. Nicholas Stern, "Imperfections in the Economics of Public Policy, Imperfections in Markets, and Climate Change," *Journal of the European Economic Association*, 8 (2010): 253–288 at 279; John Broome, "Do Not Ask for Morality," in Adrian Walsh, Sade Hormio, and Duncan Purves, eds., *The Ethical Underpinnings of Climate Economics* (New York: Routledge, 2017), pp. 9–21 at 12.

68. Broome, "Do Not Ask for Morality," p. 12.

69. Broome, "Do Not Ask for Morality," p. 20.

70. Broome, "Do Not Ask for Morality," p. 13.

71. Broome, "Do Not Ask for Morality," p. 20.

72. Aaron Maltais, "Making Our Children Pay for Mitigation," in Aaron Maltais and Catriona McKinnon, eds., *The Ethics of Climate Governance* (London: Rowman and Littlefield, 2015), pp. 91–110 at 91.

73. Maltais, "Making Our Children Pay," pp. 97–99.

Chapter 6

1. This is sometimes referred to as the "tyranny of the discount rate." See Roben Farzard, "Shortcomings of Capitalism: 'Our Grandchildren Have No Value,'" *Businessweek* (March 1, 2012), http://www.businessweek.com/articles/2012-03-01/shortcomings-of-capitalism-our-grandchildren-have-no-value.

2. For summary, see Tyler Cowen and Derek Parfit, "Against the Social Discount Rate," in Peter Laslett and James Fishkin, eds., *Philosophy, Politics, and Society* (New Haven: Yale University Press, 1992), p. 151.

3. Kenneth Arrow, "Intergenerational Equity and the Rate of Discount in Long-Term Social Investment," IEA World Congress presentation (Dec. 1995), http://www-siepr.stanford.edu/workp/swp97005.pdf.

4. Paul R. Portney and John P. Weyant, in their introduction to an influential volume of papers on the subject, write, "Another very important point comes out of these chapters . . . *with one exception, every chapter in this volume suggests that it is appropriate—indeed essential—to discount future benefits and costs at some positive rate.*" Paul R. Portney and John P. Weyant, eds., *Discounting and Intergenerational Equity* (Washington, DC: Resources for the Future, 1999), p. 7.

5. Derek Parfit, *Reasons and Persons* (Oxford: Clarendon, 1986); Henry Shue, "Bequeathing Hazards: Security Rights and Property Rights of Future Humans," in Mohammed Dore and Timothy Mount, eds., *Limits to Markets: Equity and the Global Environment* (Malden: Blackwell, 1998); Simon Caney, "Climate Change and the Future: Discounting for Time, Wealth and Risk," *Journal of Social Philosophy*, 40 (2009): 163–186 at 107–108; Darrel Moellendorf, *The Moral Challenge of Dangerous Climate Change* (Cambridge: Cambridge University Press, 2014); Axel Gosseries, "What Do We Owe the Next Generation(s)?," *Loyola of Los Angeles Law Review*, 35 (2001): 293–354 at 350.

6. John Broome, "Discounting the Future," *Philosophy and Public Affairs*, 23 (1994): 128–156.

7. For discussion, see Marc Fleurbaey, Maddalena Ferranna, Mark Budolfson, Francis Dennig, Kian Mintz-Woo, Robert Socolow, Dean Spears, and Stéphane Zuber, "The Social Cost of Carbon: Valuing Inequality, Risk, and Population for Climate Policy," *The Monist*, 102 (2019): 84–109.

8. See, for example, the puzzlement expressed by Nicholas Stern, "Ethics, Equity and the Economics of Climate Change, Paper 1: Science and Philosophy," *Economics and Philosophy*, 30 (2014): 397–444 at 418.

9. See Moellendorf, *Moral Challenge*, pp. 24–26.

10. Simon Caney, "Human Rights, Climate Change, and Discounting," *Environmental Politics*, 17 (2008): 536–555 at 539.

11. Prior to the debate over climate change, Murray Krahn and Amiram Gafni observed that "most analysts, including those who take great care to measure costs and consequences, pull their discount rates either out of the air or off the shelf, and the lucky number is most often 5%." "Discounting in the Economic Evaluation of Health Care Interventions," *Medical Care*, 31 (1993): 408–418 at 451.

12. Kenneth Arrow, "Discounting and the Public Investment Criteria," in *Collected Papers*, vol. 5, *Production and Capital* (Cambridge, MA: Harvard University Press, 1985), p. 223.

13. This last point may seem like a minor one, but it is actually quite significant, since *eta* is often presented—incorrectly—as a measure of inequality aversion in the present. Whatever aversion to inequality is associated with high values of *eta* is entirely epiphenomenal. High values favor present over future consumption only because consumption in a future state, in which everyone is wealthier, will *actually produce less welfare* than it would in the present. So there is no additional moral standard of evaluation being introduced through *eta* beyond the simple commitment to welfare-maximization. Of course, one could *introduce* some measure of inequality aversion by increasing the value of *eta*, but to the extent that this variable represents the elasticity of marginal utility of consumption, this is not what it is doing.

14. Mancur Olson and Martin J. Bailey, "Positive Time Preference," *Journal of Political Economy*, 89 (1981): 1–25.

15. This discussion follows HM Treasury, *Green Book* (London: HM Treasury, 2003), p. 97.

16. Henry Sidgwick, *The Methods of Ethics* (London: Macmillan, 1884), p. 380.

17. For example, Simon Caney treats several arguments for it, the first of which is simply a statement of the position. See Caney, "Climate Change," p. 166.

18. Frank P. Ramsey, "A Mathematical Theory of Saving," *Economic Journal*, 38 (1928): 543–549 at 543; Sidgwick, *The Methods of Ethics*, p. 381.

19. Frank Ackerman, for instance, asserts that "pure time preference of zero" is to be favored because it "expresses the equal worth of people of all generations." *Can We Afford the Future?* (London: Zed, 2009), p. 26. Similarly, David Pearce claims that "the problem that arises with discounting is that it *discriminates against future generations*." *Economic Values and the Natural World* (London: Earthscan, 1993), pp. 54.

20. Simon Dietz and Nicholas Stern, "Why Economic Analysis Supports Strong Action on Climate Change: A Response to the Stern Review's Critics," *Review of Environmental Economics and Policy*, 2 (2008): 94–113 at 105. See also Nicholas Stern, "Ethics, Equity and the Economics of Climate Change, Paper 2: Economics and Politics," *Economics and Philosophy*, 30 (2014): 445–501 at 461.

21. These two infelicitous formulations are used by John Broome, *Weighing Lives* (Oxford: Oxford University Press, 2006), pp. 92–93. Elsewhere he is more careful in his manner of speaking—since people obviously do not "live in the future" the way that, for example, some people live in China or Belgium. However, one can find similar characterizations throughout the literature, e.g. Richard L. Revesz and Matthew R. Shahabian, "Climate Change and Future Generations," *Southern California Law Review*, 84 (2011): 1099–1164 at 1102 and *passim*; Eric A. Posner and David Weisbach, *Climate Change Justice* (Princeton: Princeton University Press, 2010), p. 151.

22. Geoffrey Heal, "The Economics of Climate Change: A Post-Stern Perspective," *Climatic Change*, 96 (2009): 275–297 at 277.

23. Broome, *Weighing Lives*, pp. 92–93.

24. This seemingly obvious point has eluded, or is ignored by, many authors. For an exception, see Dieter Birnbacher, "Can Discounting be Justified?," *International Journal of Sustainable Development*, 6 (2003): 1–12 at 5.

25. E.g. Norman Daniels, *Am I My Parent's Keeper?* (New York: Oxford University Press, 1998), pp. 18–19, 92. As Onora O'Neill points out, there is a fundamental difference between, say, children and most other oppressed groups, in that whatever discrimination children face, "their main remedy is to grow up." See "Children's Rights and Children's Lives," *Ethics*, 98 (1988): 445–463 at 463. For further discussion, see Claire Breen, *Age Discrimination and Children's Rights* (Leiden: Martinus Nijhoff, 2006), pp. 22–26;

26. This of course leaves open the question of how to introduce such a tax without violating equality. For discussion of these issues, see Michael J. Trebilcock, *Dealing with Losers* (New York: Oxford University Press, 2014).

27. Stephen Gardiner, *A Perfect Moral Storm* (New York: Oxford University Press, 2011), p. 276.

28. For a clear instance of this fallacy, see Dietz and Stern, "Economic Analysis," p. 105.

29. This does, of course, involve a commitment to "presentism" as a metaphysical doctrine. For overview, see Thomas Crisp, "Presentism," in Michael J. Loux and Dean Zimmerman, eds., *Oxford Handbook of Metaphysics* (Oxford: Oxford University Press, 2003), pp. 211–245. As Crisp notes, there are no knockdown arguments in support of this doctrine, but it is the one that accords most closely with common sense.

30. Tyler Cowen, *Stubborn Attachments* (San Francisco: Stripe, 2018), p. 69.

31. Parfit, *Reasons and Persons*, pp. 356–357. For a more extreme version, see Shue, "Bequeathing Hazards," p. 38.

32. Peter Singer, "Famine, Affluence and Morality," *Philosophy and Public Affairs*, 1 (1972): 229–243.

33. For the purposes of this discussion, I will be ignoring the distinction between arguments for discounting that discount for time because they regard the passage of time as morally significant, and arguments that discount for time because time is a good *proxy* for some other phenomenon (such as people passing in and out of existence) that is regarded as morally significant. Derek Parfit and Tyler Cowen claim that arguments of the latter sort do not really count as vindications of temporal discounting ("Against the Social Discount Rate," p. 159), but this strikes me as unproductive hairsplitting.

34. Caney, "Climate Change," p. 168.

35. The Constitution of the Iroquois Confederacy described these people, memorably, as "those whose faces are yet beneath the surface of the ground—the unborn of the future Nation."

36. Mary Warren, "Do Potential People Have Moral Rights?," in R. I. Sikora and Brian Barry, eds., *Obligations to Future Generations* (Philadelphia: Temple University Press, 1978), pp. 28–29.

37. Immanuel Kant famously believed that the boundedness of space was a morally significant feature. See *Metaphysics of Morals*, ed. Mary Gregor (Cambridge: Cambridge

University Press, 1996), pp. 50–51. And so in this regard, the issue of boundedness would serve as the source of a morally significant disanalogy between space and time.

38. Tjalling C. Koopmans, "Stationary Ordinal Utility and Impatience," *Econometrica*, 28 (1960): 287–309. Kenneth Arrow, "Discounting, Morality and Gaming," in Portney and Weyant, *Discounting and Intergenerational Equity*, describes the Koopmans paper as having offered a "crushing answer" to the question, why not have a "zero time perspective"? (p. 14).

39. Stephen Gardiner has attempted to dismiss this concern, arguing that there is nothing special about a rate of zero in this regard. If the discount rate is 5 percent, then any investment that yields a return of 6 percent per annum, indefinitely, will also overwhelm any finite cost in the present (*A Perfect Moral Storm*, p. 294). He fails to note a significant disanalogy, however, which is that if the discount rate is positive, then the size of the returns must grow over time. With a rate of zero, an investment with fixed returns, no matter how small, will still overwhelm any finite present cost. The latter scenario is extremely common, while the former is uncommon, and probably nonexistent.

40. Arrow, "Discounting, Morality, and Gaming," p. 14.

41. It is because of implications such as these that some economists have expressed doubts that "proponents of a zero time preference and an infinite time horizon have understood the implications of their argument," Olson and Bailey, "Positive Time Preference," p. 13.

42. Tjalling C. Koopmans, "Objectives, Constraints, and Outcomes in Optimal Growth Models," *Econometrica*, 35 (1967): 1–15 at 8.

43. Koopmans, "Objectives, Constraints, and Outcomes," p. 9.

44. Consider, for example, the lack of reference to it in Lukas Meyer and Axel Gosseries, eds., *Intergenerational Justice* (Oxford: Oxford University Press, 2009).

45. If the doubling of the population each generation strains credulity, one can consider a version in which only one shmoo reproduces, and so the increase in population is linear. It would not affect the argument.

46. John Maynard Keynes, *The Collected Writings of John Maynard Keynes*, vol. 9, *Essays in Persuasion* (London: Macmillan, 1972), p. 330.

47. Many theorists have noted this distinction between space and time, but then failed to appreciate its full significance. Joseph Mazor, for instance, in "Liberal Justice, Future People, and Natural Resource Conservation," *Philosophy and Public Affairs*, 38 (2010): 380–408, defends the egalitarian view that individuals can consume only a per capita share of nonrenewable resources, but then extends the number of "capitas" to infinity, by including all future generations. He treats this as if it were the same as simply a very large but finite set of individuals ("The situation does not seem to be morally different from one where there are simply a very large number of initial [persons] . . . but no future individuals come into being," p. 406). Thus he states that his scheme will leave individuals only "able to use a miniscule amount of natural resources." In fact, his scheme implies that they will be able to use *no* resources (since, for any proposed consumption plan, one can always show that the plan involves

consuming more than that person is entitled to). The mistake lies in thinking that "one divided by infinity" is equivalent to "one divided by a very large number."

48. On this, see Dale Jamieson, *Reason in a Dark Time* (Oxford: Oxford University Press), p. 166.

49. See, e.g., Peter Vallentyne, "Infinite Utility and Temporal Neutrality," *Utilitas* 6 (1994): 193–199; Luc Van Liedekerke and Luc Lauwers, "Sacrificing the Patrol: Utilitarianism, Future Generations and Infinity," *Economics and Philosophy*, 13 (1997): 159–174. Note that neither produces a general solution to the basic problem.

50. Peter Singer, "All Animals Are Equal," in Tom Regan and Peter Singer, eds., *Animal Rights and Human Obligations* (Oxford: Oxford University Press, 1989), pp. 215–226.

51. Nicholas Stern, *The Economics of Climate Change: The Stern Review* (Cambridge: Cambridge University Press, 2007), p. 53.

52. He is following here the suggestion made by Partha Dasgupta and Geoffrey Heal, *Economic Theory and Exhaustible Resources* (Cambridge: Cambridge University Press, 1979), pp. 273–274.

53. See Graciela Chichilnisky, "An Axiomatic Approach to Sustainable Development," *Social Choice and Welfare*, 13 (1996): 231–257, who suggests that any acceptable solution to the problem must avoid both "the dictatorship of the present" and "the dictatorship of the future." Her conception of sustainability merits a discussion of its own; here it will suffice to note that despite rejecting the traditional discounting approach, she also rejects temporal neutrality.

54. As Daniel Farber puts it, "In deciding how much to sacrifice today for the future, as a general matter we simply *cannot* give future consequences the same weight as current ones." "From Here to Eternity: Environmental law and Future Generations," *University of Illinois Law Review* (2003): 289–336 at 292. This is an important point, which is sometimes not well enough understood. Gardiner, for instance, at one point expresses skepticism about discounting on the grounds that too many different arguments have been offered in support of it. In his view, this suggests that it is "an approach in search of a rationale," and is therefore not "independently motivated" (*A Perfect Moral Storm*, p. 268). He is right about the search for a rationale, but what he fails to appreciate is that the approach is motivated by the apparent absurdity of its alternative.

55. Tjalling Koopmans, "On the Concept of Optimal Economic Growth," *Pontificae Academiae Scientiarum Scripta Varia*, 28 (1965): 225–300 at 226.

56. John Rawls, *A Theory of Justice*, rev. ed. (Cambridge, MA: Harvard University Press, 1999), p. 18.

57. Arrow, "Intergenerational Equity," pp. 6–7.

58. Portney and Weyant write, "No matter how familiar one is with the power of compound interest, it is hard not to be stunned by the small difference that the distant future makes for present-day decision making." *Discounting and Intergenerational Equity*, p. 5.

59. Cowen and Parfit, "Against the Social Discount Rate," p. 145.

60. Thus I agree with Daniel Farber, who writes that in such scenarios "the situation is too far outside of normal human experience for us to have much confidence in our intuitions." "From Here to Eternity," p. 307.

61. William Nordhaus, *A Question of Balance* (New Haven: Yale University Press, 2008), pp. 82–83; also William D. Nordhaus, "A Review of the Stern Review on the Economics of Climate Change," *Journal of Economic Literature*, 45 (2007): 686–702.

62. Partha Dasgupta says the same. "The moral is this: we should be very circumspect before accepting numerical values for parameters for which we have little a priori feel." "Comments on the Stern Review's Economics of Climate Change," *National Institute Economic Review*, no. 199 (2007): 4–7 at 6.

63. Christian Gollier, "An Evaluation of Stern's Report on the Economics of Climate Change," IDEI Working Paper no. 464 (2006), p. 2.

64. See J. Bradford Delong, "Applied Utilitarianism and Global Climate Change," http:// delong.typepad.com/sdj/2006/12/brad_delongs_se.html (accessed Jan. 23, 2013).

65. I mention this specific example because, in cost-benefit terms, investments in carbon abatement must compete with other uses of those resources, and malaria eradication has a particularly favorable benefit-to-cost ratio, making it particularly difficult to compete with.

66. The choice between asphalt and concrete as a road-surfacing material tends to be rather close in terms of long-term cost, and so the decision is often made by the choice of discount rate. (Concrete has a much more significant up-front cost, but lower maintenance costs over the lifetime of the roadway.)

67. John Garthoff, "The Embodiment Thesis," *Ethical Theory and Moral Practice*, 7 (2004): 15–29 at 16.

68. John Rawls, *Law of Peoples* (Cambridge, MA: Harvard University Press, 2001), p. 39.

69. This is, of course, Kant's view. For contemporary versions, see Jürgen Habermas, *Between Facts and Norms*, trans. William Rehg (Cambridge, MA: MIT Press, 1992); Arthur Ripstein, *Force and Freedom* (Cambridge, MA: Harvard University Press, 2009).

70. Something of this structure can be discerned in what is known as the "democratic" argument for social discounting, which is sometimes appealed to by economists looking to make the positive approach sound better. The argument is that, because people discount the future, a democratic state that wishes to respect the will of the people will also discount the future. There are a lot of things wrong with this argument, but one thing that is important about it is the idea that the state's obligation may not simply be to do what is right from the moral point of view.

71. Rawls, *A Theory of Justice*, p. 153.

72. Rawls, *A Theory of Justice*, p. 155.

73. Birnbacher, "Can Discounting Be Justified?," p. 3.

74. Posner and Weisbach, *Climate Change Justice*, p. 6.

75. See Philippe van Parijs, "The Disenfranchisement of the Elderly, and Other Attempts to Secure Intergenerational Justice," *Philosophy and Public Affairs*, 27 (1998): 292–333.

76. Note that this differs from the individual fear of death, which influences the interest rate, because individual savers are likely to have a higher risk of dying than the background probability for the population (largely due to their age).

77. David J. Evans, "The Elasticity of Marginal Utility of Consumption: Estimates for 20 OECD Countries," *Fiscal Studies*, 26 (2005): 197–224.

78. Tim Mulgan, "A Minimal Test for Political Theories," *Philosophia*, 28 (2001): 283–296 at 283–284.

79. Mulgan, "Minimal Test," p. 284.

80. Mulgan, "Minimal Test," p. 294.

81. For a clear example of this, see T. M. Scanlon, *What We Owe to Each Other* (Cambridge, MA: Harvard University Press, 1998), 307–309 (on promising).

82. Walter Sinnott-Armstrong, "Framing Moral Intuitions," in Walter Sinnott-Armstrong, ed., *Moral Psychology*, vol. 2 (Cambridge, MA: MIT Press, 2008), pp. 47–78.

83. As Steven Vogel observes, it is an unintended consequence of our practices. *Thinking like a Mall* (Cambridge, MA: MIT Press, 2015), p. 200.

Conclusion

1. Barry G. Rabe, *Can We Price Carbon?* (Cambridge, MA: MIT Press, 2018), pp. 5–6.

Bibliography

Ackerman, Frank. *Can We Afford the Future?* London: Zed, 2009.

Ackerman, Frank, Stephen J. DeCanio, Richard B. Howarth, and Kristen Sheeran. "Limitations of Integrated Assessment Models of Climate Change." *Climatic Change*, 95:3 (2009): 297–315.

Ackerman, Frank and Susan Heinzerling. *Priceless*. New York: New Press, 2004.

Adler, Matthew D. and Eric A. Posner. *New Foundations of Cost-Benefit Analysis*. Cambridge, MA: Harvard University Press, 2006.

Agarwal, Anil and Sunita Narain. *Global Warming in an Unequal World*. New Delhi: Centre for Science and Environment, 1991.

Albert, Michael. *Parecon: Life after Capitalism*. London: Verso, 2003.

Alvarez-Farizo, Begoña, Nick Hanley, Ramón Barberan, and Angelina Lázaro. "Choice Modeling at the 'Market Stall': Individual versus Collective Interest in Environmental Valuation." *Ecological Economics*, 60:4 (2007): 743–751.

Anderson, Elizabeth. *Value in Ethics and Economics*. Cambridge, MA: Harvard University Press, 1993.

Andreou, Chrisoula. "Environmental Damage and the Puzzle of the Self-Torturer." *Philosophy and Public Affairs*, 34:1 (2006): 95–108.

Andreou, Chrisoula. "A Shallow Route to Environmentally Friendly Happiness." *Ethics, Place and Environment*, 13:1 (2010): 1–10.

Arrhenius, Gustaf. "Mutual Advantage Contractarianism and Future Generations." *Theoria*, 65:1 (1999): 25–35.

Arrow, Kenneth. *Collected Papers*. Vol. 5, *Production and Capital*. Cambridge, MA: Harvard University Press, 1985.

Arrow, Kenneth. "Intergenerational Equity and the Rate of Discount in Long-Term Social Investment." IEA World Congress presentation (Dec. 1995).

Arrow, Kenneth and Frank Hahn. *General Competitive Analysis*. San Francisco: Holden-Day, 1971.

Ashford, Elizabeth and Tim Mulgan. "Contractualism." In Edward N. Zalta, ed., *The Stanford Encyclopedia of Philosophy* (Summer 2018 ed.).

Atkinson, Giles and Susana Mourato. "Environmental Cost-Benefit Analysis." *Annual Review of Environmental Resources*, 33:1 (2008): 317–344.

Attas, Daniel. "A Trans-generational Difference Principle." In Axel Gosseries and Lukas H. Meyer, eds., *Intergenerational Justice*, pp. 189–218. Oxford: Oxford University Press, 2011.

Attfield, Robin. "The Good of Trees." *Journal of Value Inquiry*, 15:1 (1981): 35–54.

Bardach, Eugene and Robert A. Kagan. *Going by the Book*. Philadelphia: Temple University Press, 1982.

Barry, Brian. "Circumstances of Justice and Future Generations." In Richard I. Sikora and Brian M. Barry, eds., *Obligations to Future Generations*, pp. 204–248. Philadelphia: Temple University Press, 1978.

Barry, Brian. "Sustainability and Intergenerational Justice." In Andrew Dobson, ed., *Fairness and Futurity*, pp. 93–117. New York: Oxford University Press, 1999.

Barry, Brian. *Theories of Justice*. Berkeley: University of California Press, 1989.

Basso, Alessandra. "When Utility Maximization is not Enough: Intergenerational Sufficientarianism and the Economics of Climate Change." In Adrian Walsh, Sade Hormio, and Duncan Purves, eds., *The Ethical Underpinnings of Climate Economics*, pp. 65–102. New York: Routledge, 2017.

Baumol, William. *Superfairness*. Cambridge, MA: MIT Press, 1986.

Beckerman, Wilfred. "How Would You Like Your 'Sustainability' Sir? Weak or Strong? A Reply to My Critics." *Environmental Values*, 4:2 (1995): 169–179.

Belton, Padraig. "Could Diesel Made from Air Help Tackle Climate Change?" *BBC News* (Sept. 1, 2015).

Bernstein, Steven. *The Compromise of Liberal Environmentalism*. New York: Columbia University Press, 2001.

Bigger, Patrick. "Hybridity, Possibility: Degrees of Marketization in Tradeable Permit Systems." *Environment and Planning A: Economy and Space*, 50:3 (2018): 512–530.

Birnbacher, Dieter. "Can Discounting Be Justified?" *International Journal of Sustainable Development*, 6:1 (2003): 1–12.

Blomfield, Megan. "Climate Change and the Moral Significance of Historical Injustice in Natural Resource Governance." In Aaron Maltais and Catriona McKinnon, eds., *The Ethics of Climate Governance*, pp. 3–22. London: Rowman and Littlefield, 2015.

Bodansky, Daniel. "Climate Change and Human Rights: Unpacking the Issues." *Georgia Journal of International and Comparative Law*, 38:3 (2010): 511–524.

Boran, Idil. *Political Theory and Global Climate Action*. London: Routledge, 2019.

Boran, Idil and Joseph Heath. "Attributing Weather Extremes to Climate Change and the Future of Adaptation Policy." *Ethics, Policy and Environment*, 19:3 (2016): 239–255.

Bradsher, Keith. *High and Mighty*. New York: PublicAffairs, 2002.

Braithwaite, John. "The Limits of Economism in Controlling Harmful Corporate Conduct." *Law & Society Review*, 16:3 (1981–82): 481–504.

Brandstedt, Eric. "The Savings Problem in the Original Position: Assessing and Revising the Model." *Canadian Journal of Philosophy*, 47:2–3 (2017): 269–289.

Breen, Claire. *Age Discrimination and Children's Rights*. Leiden: Martinus Nijhoff, 2006.

Brennan, Andrew. "Moral Pluralism and the Environment." *Environmental Values*, 1:1 (1992): 15–33.

Broome, John. "Climate Change, Efficiency, Future Generations and the Non-identity Effect." Forthcoming.

Broome, John. *Climate Matters*. New York: Norton, 2012.

Broome, John. *Counting the Cost of Global Warming*. Cambridge: White Horse Press, 1992.

Broome, John. "Discounting the Future." *Philosophy and Public Affairs*, 23:2 (1994): 128–156.

Broome, John. "Do Not Ask for Morality." In Adrian Walsh, Sade Hormio, and Duncan Purves, eds., *The Ethical Underpinnings of Climate Economics*, pp. 9–21. London: Routledge, 2017.

Broome, John. *Weighing Lives*. Oxford: Oxford University Press, 2006.

Brown, Donald A. *Climate Change Ethics*. London: Routledge, 2013.

Brundtland, Gru, Mansour Khalid, Susanna Agnelli, et al. *Our Common Future*. Oxford: Oxford University Press, 1987.

Bryant, Gareth. *Carbon Markets in a Climate-Changing Capitalism.* Cambridge: Cambridge University Press, 2019.

Buchanan, Allen. "Perfecting Imperfect Duties: Collective Action to Create Moral Obligations." *Business Ethics Quarterly*, 6:1 (1996): 27–42.

Burke, Marshall, Solomon M. Hsiang, and Edward Miguel. "Global Non-linear Effect of Temperature on Economic Production." *Nature*, 527:7577 (2015): 235–239.

Butterfield, Fox. "From Ben Franklin, a Gift That's Worth Two Fights." *New York Times* (April 21, 1990).

Cafaro, Philip. "Taming Growth and Articulating a Sustainable Future." *Ethics and the Environment*, 16:1 (2011): 1–23.

Callicott, J. Baird. *In Defense of the Land Ethic.* Albany: State University of New York Press, 1989.

Callicott, J. Baird. "Animal Liberation: A Triangular Affair." *Environmental Ethics*, 2:4 (1980): 311–338.

Callicott, J. Baird. "Non-anthropocentric Value Theory and Environmental Ethics." *American Philosophical Quarterly*, 21:4 (1984): 299–309.

Camerer, Colin, Samuel Issacharoff, George Loewenstein, Ted O'Donoghue, and Matthew Rabin. "Regulation for Conservatives: Behavioral Economics and the Case for 'Asymmetric Paternalism.'" *University of Pennsylvania Law Review*, 151:3 (2003): 1211–1254.

Canada's Ecofiscal Commission. *Supporting Carbon Pricing.* June 2017.

Caney, Simon. "Climate Change and the Future: Discounting for Time, Wealth, and Risk." *Journal of Social Philosophy*, 40:2 (2009): 163–186.

Caney, Simon. "Climate Change, Human Rights and Moral Thresholds." In Stephen Gardiner, Simon Caney, Dale Jamieson, and Henry Shue, eds., *Climate Ethics*, pp. 163–177. Oxford: Oxford University Press, 2010.

Caney, Simon. "Just Emissions." *Philosophy and Public Affairs*, 40:4 (2012): 255–300.

Caney, Simon. "Justice and the Distribution of Greenhouse Gas Emissions." *Journal of Global Ethics*, 5:2 (2009): 125–146.

Chichilnisky, Graciela. "An Axiomatic Approach to Sustainable Development." *Social Choice and Welfare*, 13:2 (1996): 231–257.

The Chinese Classic of Family Reverence. Trans. Henry Rosemont Jr. and Roger T. Ames. Honolulu: University of Hawaii Press, 2009.

Coase, Ronald. "The Problem of Social Cost." *Journal of Law and Economics*, 3 (1960): 1–44.

Cohen, G. A. *Rescuing Justice and Equality.* Cambridge, MA: Harvard University Press, 2008.

Coleman, Jules L. "Efficiency, Utility, and Wealth Maximization." *Hofstra Law Review*, 8:3 (1980): 509–551.

Collard, David. *Generations of Economists.* London: Routledge, 2011.

Coward, Harold and Thomas Hurka, eds. *Ethics and Climate Change.* Waterloo: Wilfred Laurier University Press, 1993.

Cowen, Tyler. "Consequentialism Implies a Zero Rate of Intergenerational Discount." In Peter Laslett and James S. Fishkin, eds., *Justice between Age Groups and Generations*, pp. 162–168. New Haven: Yale University Press, 1992.

Cowen, Tyler. *Stubborn Attachments.* San Francisco: Stripe, 2018.

Cowen Tyler and Derek Parfit. "Against the Social Discount Rate." In Peter Laslett and James Fishkin, eds., *Philosophy, Politics, and Society*, pp. 144–161. New Haven: Yale University Press, 1992.

Crisp, Thomas. "Presentism." In Michael J. Loux and Dean Zimmerman, eds., *Oxford Handbook of Metaphysics*, pp. 211–245. Oxford: Oxford University Press, 2003.

Cudd, Ann and Seena Eftekhari. "Contractarianism." In Edward N. Zalta, ed., *The Stanford Encyclopedia of Philosophy* (Summer 2018 ed.).

Daly, Herman E. *Steady State Economics.* London: W.H. Freeman, 1972.

Daly, Herman E. "On Wilfred Beckerman's Critique of Sustainable Development." *Environmental Values*, 4:1 (1995): 49–55.

Daly, Herman E. *Beyond Growth.* Boston: Beacon Press, 1996.

Daly, Herman E., and John B. Cobb Jr. *For the Common Good.* Boston: Beacon Press, 1989.

Daniels, Norman. *Am I My Parent's Keeper?* New York: Oxford University Press, 1998.

Dasgupta, Partha. "Comments on the Stern Review's Economics of Climate Change." *National Institute Economic Review*, 199:1 (2007): 4–7.

Dasgupta, Partha and Geoffrey Heal. *Economic Theory and Exhaustible Resources.* Cambridge: Cambridge University Press, 1979.

de Groot, Dolf. *Functions of Nature.* Groningen: Wolters Noordhoff, 1992.

de Shalit, Avner. *Why Posterity Matters.* London: Routledge, 1995.

Deaton, Angus. *The Great Escape.* Princeton: Princeton University Press, 2013.

Delong, J. Bradford. "Cornucopia: The Pace of Economic Growth in the 20th Century." NBER Working Paper no. 7602 (2000).

Dennett, Daniel C. *The Intentional Stance.* Cambridge, MA: MIT Press, 1987.

Dessler, Andrew. *Introduction to Modern Climate Change.* 2nd ed. Cambridge: Cambridge University Press, 2016.

Desvousges, William, F. Reed Johnson, Richard W. Dunford, Sara P. Hudson, K. Nicole Wilson, and Kevin J. Boyle. "Measuring Natural Resource Damages with Contingent Valuation: Tests of Validity and Reliability." In Jerry Hausman, ed., *Contingent Valuation: A Critical Assessment*, pp. 91–164. Amsterdam: North Holland Press, 1993.

Diamond, Peter A., and Jerry Hausman. "Contingent Valuation: Is Some Number Better Than No Number?" *Journal of Economic Perspectives*, 8:4 (1994): 45–64.

Dietz, Simon and Nicholas Stern. "Why Economic Analysis Supports Strong Action on Climate Change: A Response to the Stern Review's Critics." *Review of Environmental Economics and Policy*, 2:1 (2008): 94–113.

Donaldson, Sue and Will Kymlicka. *Zoopolis.* Oxford: Oxford University Press, 2011.

Dorman, Richard and Paul Hagstrom. "Wage Compensation for Dangerous Work Revisited." *Industrial and Labor Relations Review*, 52:1 (1998): 116–135.

Dryzek, John S., Richard B. Norgaard, and David Scholsberg, eds. *Oxford Handbook of Climate Change and Society.* Oxford: Oxford University Press, 2012.

Dworkin, Ronald. *Justice for Hedgehogs.* Cambridge, MA: Harvard University Press, 2013.

Dworkin, Ronald. *Sovereign Virtue.* Cambridge, MA: Harvard University Press, 2000.

Dworkin, Ronald. *Taking Rights Seriously.* Cambridge, MA: Harvard University Press, 1977.

Dworkin, Ronald. "What Is Equality? Part 2: Equality of Resources." *Philosophy and Public Affairs*, 10:4 (1981): 283–345.

Ebert, Rainer and Tibor R. Machan. "Innocent Threats and the Moral Problem of Carnivorous Animals." *Journal of Applied Philosophy*, 29:2 (2012): 146–159.

The Economist. "Curbing Emissions" (Sept. 20, 2014).

The Economist. "To Stop Carmakers Bending the Rules on Emissions, Europe Must Get Much Tougher" (Jan. 21, 2017).

The Economist. "Why It Is So Hard to Fix India's Sanitation" (Sept. 25, 2017).

Ekins, Paul, Sandrine Simon, Lisa Deutsch, Carl Folke, and Rudolf De Groot. "A Framework for the Practical Application of the Concepts of Critical Natural Capital and Strong Sustainability." *Ecological Economics*, 44:2–3 (2003): 165–185.

Elliott, Brian. *Natural Catastrophe*. Edinburgh: University of Edinburgh Press, 2016.

Evans, David J. "The Elasticity of Marginal Utility of Consumption: Estimates for 20 OECD Countries." *Fiscal Studies*, 26:2 (2005): 197–224.

Farber, Daniel A. *Eco-pragmatism*. Chicago: University of Chicago Press, 1999.

Farber, Daniel A. "From Here to Eternity: Environmental Law and Future Generations." *University of Illinois Law Review*, 2003:2 (2003): 289–336.

Farzad, Roben. "Shortcomings of Capitalism: 'Our Grandchildren Have No Value.'" *Businessweek* (March 1, 2012).

Feldstein, Martin. "Rethinking Social Insurance." *American Economic Review* 95:1 (2005): 1–24.

Festinger, Leon, Henry Riecken, and Stanley Schachter. *When Prophecy Fails*. New York: Harper & Row, 1956.

Fleurbaey, Marc, Maddalena Ferranna, Mark Budolfson, Francis Dennig, Kian Mintz-Woo, Robert Socolow, Dean Spears, and Stéphane Zuber. "The Social Cost of Carbon: Valuing Inequality, Risk, and Population for Climate Policy." *The Monist*, 102:1 (2019): 84–109.

Fogel, Robert William. *The Escape from Misery and Premature Death, 1700–2100*. Cambridge: Cambridge University Press, 2004.

Fox, Warwick. *Toward a Transpersonal Ecology*. Albany: SUNY Press, 1995.

Friedman, Benjamin M. *The Moral Consequences of Economic Growth*. New York: Vintage, 2005.

Fudenberg, Drew and David K. Levine. *A Long-Run Collaboration on Long-Run Games*. Singapore: World Scientific, 2009.

Fudenberg, Drew and Eric Maskin. "The Folk Theorem in Repeated Games with Discounting or with Incomplete Information." *Econometrica*, 54:3 (1986): 533–554.

Fudenberg, Drew and Jean Tirole. *Game Theory*. Cambridge, MA: MIT Press, 1991.

Fuglie, Keith, James M. MacDonald, and Eldon Ball. "Productivity Growth in U.S. Agriculture." Economic Brief (No. EB-9). Washington, DC: United States Department of Agriculture, 2007.

Gardiner, Stephen M. "Climate Ethics in a Dark and Dangerous Time." *Ethics*, 127:2 (2017): 430–465.

Gardiner, Stephen M. "A Contract on Future Generations?" In Stephen M. Gardiner and David A. Weisbach, *Debating Climate Ethics*, pp. 77–118. Oxford: Oxford University Press, 2016.

Gardiner, Stephen M. "Ethics and Global Climate Change." *Ethics*, 114:3 (2004): 555–600.

Gardiner, Stephen M. *A Perfect Moral Storm*. New York: Oxford University Press, 2011.

Gardiner, Stephen M. "The Pure Intergenerational Problem." *The Monist* 86:3 (2003): 481–501.

Gardiner, Stephen M. and David A. Weisbach. *Debating Climate Ethics*. Oxford: Oxford University Press, 2016.

Garthoff, John. "The Embodiment Thesis." *Ethical Theory and Moral Practice*, 7 (2004): 15–29.

Gaspart, Frédéric and Axel Gosseries. "Are Generational Savings Unjust?" *Politics, Philosophy and Economics*, 6:2 (2007): 193–217.

Gauthier, David. "David Hume: Contractarian." In his *Moral Dealing*, pp. 45–76. Ithaca: Cornell University Press, 1990.

Gauthier, David. *Morals by Agreement*. Oxford: Clarendon, 1986.

Georgescu-Roegen, Nicholas. *The Entropy Law and the Economic Process*. Cambridge, MA: Harvard University Press, 1971.

Gintis, Herbert. *Game Theory Evolving*. Princeton: Princeton University Press, 2000.

Gollier, Christian. "An Evaluation of Stern's Report on the Economics of Climate Change." IDEI Working Paper no. 464 (2006).

Goodin, Robert. "Selling Environmental Indulgences." In John S. Dryzek and David Schlosberg, eds., *Debating the Earth*, pp. 237–254. Oxford: Oxford University Press, 1994.

Goodpaster, Kenneth. "On Being Morally Considerable." *Journal of Philosophy*, 75:6 (1978): 308–325.

Gosseries, Axel. "Cosmopolitan Luck Egalitarianism and the Greenhouse Effect." *Canadian Journal of Philosophy*, supp. vol., 31 (2005): 279–309.

Gosseries, Axel. "Three Models of Intergenerational Reciprocity." In Axel Gosseries and Lukas H. Meyer, eds., *Intergenerational Justice*, pp. 119–146. Oxford: Oxford University Press, 2011.

Gosseries, Axel. "What Do We Owe the Next Generation(s)?" *Loyola of Los Angeles Law Review*, 35:1 (2001): 293–354.

Goulder, Lawrence H. and Andrew R. Schein. "Carbon Taxes versus Cap and Trade: A Critical Review." *Climate Change Economics*, 4:3 (2013): 1–28.

Government of Canada. *State of Remote/Off-Grid Communities in Canada*. August 2011.

Habermas, Jürgen. *Between Facts and Norms*. Trans. William Rehg. Cambridge, MA: MIT Press, 1992.

Hamilton, Clive. *Requiem for a Species*. London: Earthscan, 2012.

Hardin, Garett. "The Tragedy of the Commons." *Science*, 162:3859 (1968): 1243–1248.

Hardin, Russell. *Collective Action*. Baltimore: Resources for the Future, 1982.

Hart, H. L. A. "Are There Any Natural Rights?" *Philosophical Review*, 64:2 (1955): 175–191.

Hassoun, Nicole. "The Anthropocentric Advantage? Environmental Ethics and Climate Change Policy." *Critical Review of International Social and Political Philosophy*, 14:2 (2011): 235–257.

Hayward, Tim. "Anthropocentrism: A Misunderstood Problem." *Environmental Values*, 6:1 (1997): 49–63.

Hayward, Tim. "Human Rights versus Emissions Rights: Climate Justice and the Equitable Distribution of Ecological Space." *Ethics and International Affairs*, 21:4 (2007): 431–450.

Heal, Geoffrey. "The Economics of Climate Change: A Post-Stern Perspective." *Climatic Change*, 96:3 (2009): 275–297.

Heath, Joseph. "Dworkin's Auction." *Politics, Philosophy and Economics*, 3:3 (2004): 313–335.

Heath, Joseph. "Envy and Efficiency." *Revue de philosophie economique*, 7:2 (2006): 3–30.

Heath, Joseph. *Filthy Lucre*. Toronto: HarperCollins, 2008.

Heath, Joseph. "Intergenerational Cooperation and Distributive Justice." *Canadian Journal of Philosophy*, 27:3 (1997): 361–376.

Heath, Joseph. *Morality, Competition and the Firm*. New York: Oxford University Press, 2014.

Heath, Joseph. "Political Egalitarianism." *Social Theory and Practice*, 34:4 (2008): 485–516.

Heath, Joseph. "Rawls on Global Distributive Justice: A Defence." *Canadian Journal of Philosophy*, supp. vol., 35 (2005): 193–226.

Heath, Joseph, Jeffrey Moriarty, and Wayne Norman. "Business Ethics and (or as) Political Philosophy." *Business Ethics Quarterly*, 20:3 (2010): 427–452.

Heidegger, Martin. *The Question concerning Technology, and Other Essays*. Trans. William T. Levitt. New York: Harper & Row, 1977.

Helm, Dieter. *Burn Out*. New Haven: Yale University Press, 2017.

Helm, Dieter. *The Carbon Crunch*. New Haven: Yale University Press, 2012.

Henrich, Joseph. *The Secret of Our Success*. Princeton: Princeton University Press, 2015.

Henrich, Joseph, Steven J. Heine, and Ara Norenzayan. "The Weirdest People in the World?" *Behavioral and Brain Sciences*, 33:2–3 (2010): 61–83.

Heyward, Clare and Dominic Roser, eds. *Climate Justice in a Non-ideal World*. Oxford: Oxford University Press, 2016.

Hirose, Iwao. "Aggregation and Numbers." *Utilitas*, 16:1 (2004): 62–79.

Hirsch, Fred. *The Social Limits to Growth*. Rev. ed. London: Routledge, 1977.

HM Treasury. *Green Book*. London: HM Treasury, 2003.

Hobbes, Thomas. *Leviathan*. Ed. Richard Tuck. Cambridge: Cambridge University Press, 1991.

Höhne, Niklas, Michel den Elzen, and Martin Weiss. "Common but Differentiated Convergence (CDC): A New Conceptual Approach to Long-Term Climate Policy." *Climate Policy*, 6:2 (2006): 181–199.

Holland, Alan. "Natural Capital." In Robin Attfield and Andrew Belsey, eds., *Philosophy and the Natural Environment*, pp. 169–182. Cambridge: Cambridge University Press, 1994.

Homer-Dixon, Thomas. *The Ingenuity Gap*. Toronto: Penguin, 2001.

Hope, Simon. "The Circumstances of Justice." *Hume Studies*, 36:2 (2010): 125–148.

Hormio, Sade. "Climate Change Mitigation, Sustainability and Non-substitutability." In Adrian Walsh, Sade Hormio, and Duncan Purves, eds., *The Ethical Underpinnings of Climate Economics*, pp. 103–121. New York: Routledge, 2017.

Hubin, Donald C. "The Moral Justification of Benefit/Cost Analysis." *Economics and Philosophy*, 10:2 (1994): 169–194.

Hume, David. *A Treatise of Human Nature*. Ed. P. H. Nidditch. 2nd ed. Oxford: Oxford University Press, 1978.

Intergovernmental Panel on Climate Change. *Climate Change 2014: Impacts, Adaptation, and Vulnerability*. Cambridge: Cambridge University Press, 2014.

Jackson, Tim. *Prosperity without Growth*. New York: Routledge, 2011.

Jacobs, Jane. *The Death and Life of Great American Cities*. New York: Vintage, 1961.

Jafry, Tahseen, ed. *Routledge Handbook of Climate Justice*. London: Routledge, 2019.

James, P. D. *The Children of Men*. New York: Vintage, 1992.

Jamieson, Dale. "Adaptation, Mitigation and Justice." In Walter Sinnott-Armstrong and Richard Howarth, eds., *Perspectives on Climate Change*, pp. 217–248. Oxford: Elsevier, 2005.

Jamieson, Dale. "Climate Change and Global Environmental Justice." In Clark A. Edwards and Paul N. Miller, eds., *Changing the Atmosphere*, pp. 287–308. Cambridge, MA: MIT Press, 2001.

Jamieson, Dale. *Ethics and the Environment*. Cambridge: Cambridge University Press, 2008.

Jamieson, Dale. *Reason in a Dark Time*. New York: Oxford University Press, 2014.

Kagan, Shelly. "Do I Make a Difference?" *Philosophy and Public Affairs*, 39:2 (2011): 105–141.

Kamm, Frances. "Equal Treatment and Equal Chances." *Philosophy and Public Affairs*, 14:2 (1985): 177–194.

Kant, Immanuel. *Foundations of the Metaphysics of Morals* Trans. Lewis White Beck. New York: Macmillan, 1990.

Kasser, Tim. *The High Price of Materialism*. Cambridge, MA: MIT Press, 2002.

Katz, Eric. "A Pragmatic Reconsideration of Anthropocentrism." *Environmental Ethics*, 21:4 (1999): 377–390.

Kelman, Steven. "Cost-Benefit Analysis: An Ethical Critique." *Regulation*, 5:1 (1981): 33–40.

Kernohan, Andrew. *Environmental Ethics*. Buffalo: Broadview Press, 2012.

Keynes, John Maynard. *The Collected Writings of John Maynard Keynes*. Vol. 9, *Essays in Persuasion*. London: Macmillan, 1972.

Kitcher, Philip. *The Ethical Project*. Cambridge, MA: Harvard University Press, 2011.

Klein, Naomi. *This Changes Everything*. New York: Simon & Schuster, 2014.

Kolbert, Elizabeth. *The Sixth Extinction*. New York: Henry Holt, 2014.

Koopmans, Tjalling C. "Stationary Ordinal Utility and Impatience." *Econometrica*, 28:2 (1960): 287–309.

Koopmans, Tjalling C. "Objectives, Constraints, and Outcomes in Optimal Growth Models." *Econometrica*, 35:1 (1967): 1–15.

Koopmans, Tjalling C. "On the Concept of Optimal Economic Growth." *Pontificae Academiae Scientiarum Scripta Varia* 28:1 (1965): 225–300.

Kornhauser, Lewis A. "On Justifying Cost-Benefit Analysis." *Journal of Legal Studies*, 29:S2 (2000): 1037–1057.

Krahn, Murray and Amiram Gafni. "Discounting in the Economic Evaluation of Health Care Interventions." *Medical Care*, 31:5 (1993): 408–418.

Kumar, Rahul. "Contractualism on the Shoals of Aggregation." In R. Jay Wallace, Rahul Kumar, and Samuel Freeman, eds., *Reasons and Recognition*, pp. 129–154. Oxford: Oxford University Press, 2011.

Kymlicka, Will. *Contemporary Political Philosophy*. 2nd ed. Oxford: Oxford University Press, 2002.

Lawford-Smith, Holly. "Difference-Making and Individuals' Climate-Related Obligations." In Clare Heyward and Dominic Roser, eds., *Climate Justice in a Non-ideal World*, pp. 64–82. Oxford: Oxford University Press, 2016.

Leopold, Aldo. *A Sand County Almanac and Sketches Here and There*. London: Oxford University Press, 1949.

Levitt, Steven and Stephen Dubner. *Superfreakeconomics*. New York: William Morrow, 2009.

Lewin, Christopher G. *Pensions and Insurance before 1800*. East Lothian: Tuckwell, 2003.

Lo, Y. S. "The Land Ethics and Callicott's Ethical System (1980–2001): An Overview and Critique." *Inquiry*, 44:3 (2001): 331–358.

Locke, John. *Second Treatise of Government*. Ed. C. B. Macpherson. Indianapolis: Hackett, 1980.

Lohmann, Larry. "Financialization, Commodification and Carbon: The Contradictions of Neoliberal Climate Policy." *Socialist Register*, 48 (2012): 85–107.

Loss, Scott R., Tom Will, and Peter P. Marra. "The Impact of Free-Ranging Domestic Cats on Wildlife of the United States." *Nature Communications*, 4:1396 (2013): 1–8.

Lovelock, James. *Gaia*. Oxford: Oxford University Press, 1995.

Lowry, Rosemary and Martin Peterson. "Cost-Benefit Analysis and Non-utilitarian Ethics." *Politics, Philosophy and Economics*, 11:3 (2011): 258–279.

Lucas, Robert. "The Industrial Revolution: Past and Future." *2003 Annual Report Essay*. Minneapolis: Federal Reserve Bank of Minneapolis, 2004.

Macmillan, Douglas Craig, Lorna Jennifer Philip, Nick Hanley, and Begoña Alvarez-Farizo. "Valuing the Nonmarket Benefits of Wild Goose Conservation: A Comparison of Interview and Group-Based Approaches." *Ecological Economics*, 43:1 (2002): 49–59.

Maltais, Aaron. "Making Our Children Pay for Mitigation." In Aaron Maltais and Catriona McKinnon, eds., *The Ethics of Climate Governance*, pp. 91–110. London: Rowman and Littlefield, 2015.

Mason, Paul. *Postcapitalism*. New York: Allen Lane, 2015.

Matravers, Matt, ed. *Scanlon and Contractualism*. London: Frank Cass, 2003.

Matthews, H. Damon, Nathan P. Gillett, Peter A. Stott, and Kirsten Zickfeld. "The Proportionality of Global Warming to Cumulative Carbon Emissions." *Nature*, 459:7248 (2009): 829–832.

Mauss, Marcel. *The Gift*. Trans. W. D. Hall. London: Routledge, 1990.

Mazor, Joseph. "Liberal Justice, Future People, and Natural Resource Conservation." *Philosophy and Public Affairs*, 38:4 (2010): 380–408.

McCormick, Hugh. "Intergenerational Justice and the Non-reciprocity Problem." *Political Studies*, 57:2 (2009): 451–458.

McGranahan, Gordon, Deborah Balk, and Bridget Anderson. "The Rising Tide: Assessing the Risks of Climate Change and Human Settlements in Low Elevation Coastal Zones." *Environment and Urbanization*, 19:1 (2007): 17–37.

McKinnon, Catriona. *Climate Change and Future Justice: Precaution, Compensation, and Triage*. New York: Routledge, 2012.

McMahan, Jeff. "The Meat Eaters." *The Stone, New York Times* (Sept. 19, 2010).

McShane, Katie. "Anthropocentrism in Climate Ethics and Policy." *Midwest Studies in Philosophy*, 40:1 (2016): 189–204.

McShane, Katie. "Anthropocentrism vs. Nonanthropocentrism: Why Should We Care?" *Environmental Values*, 16:2 (2007): 169–185.

Meadows, Donella H., Dennis L. Meadows, and Jørgen Randers. *Beyond the Limits*. Post Mills: Chelsea Green, 1992.

Meadows, Donella H., Dennis L. Meadows, Jørgen Randers, and William W. Behrens III. *The Limits to Growth*. New York: Signet, 1972.

Meyer, Lukas H., ed. *Climate Justice and Historical Emissions*. Cambridge: Cambridge University Press, 2017.

Meyer, Lukas H. and Axel Gosseries, eds. *Intergenerational Justice*. Oxford: Oxford University Press, 2009.

Meyer, Lukas H. and Dominic Roser. "Distributive Justice and Climate Change: The Allocation of Emission Rights." *Analyse & Kritik*, 28:2 (2006): 223–249.

Mikhail, John. *Elements of Moral Cognition*. Cambridge: Cambridge University Press, 2011.

Mill, John Stuart. *On Liberty*. Ed. Elizabeth Rapaport. Indianapolis: Hackett, 1978.

Moellendorf, Darrel. *The Moral Challenge of Dangerous Climate Change*. Cambridge: Cambridge University Press, 2014.

Moller, Dan. "Global Justice and Economic Growth: Ignoring the Only Thing That Works." In Jahel Queralt and Bas van der Vossen, eds., *Economic Liberties and Human Rights*, pp. 95–113. London: Routledge, 2019.

Moss, David A. *When All Else Fails*. Cambridge, MA: Harvard University Press, 2002.

Mulgan, Tim. "A Minimal Test for Political Theories." *Philosophia*, 28:1–4 (2001): 283–296.

Mulgan, Tim. "Answering to Future People: Responsibility for Climate Change in a Breaking World." *Journal of Applied Philosophy*, 35:3 (2018): 532–548.

Mulgan, Tim. *Ethics for a Broken World*. Montreal: McGill-Queen's University Press, 2011.

Mulgan, Tim. *Future People*. Oxford: Clarendon, 2006.

Myerson, Roger. *Game Theory*. Cambridge, MA: Harvard University Press, 1997.

Narveson, Jan. "Animal Rights." *Canadian Journal of Philosophy*, 7:1 (1977): 161–178.

Nash, John. "The Bargaining Problem." *Econometrica*, 18:2 (1950): 155–162.

Neumayer, Eric. "Global Warming: Discounting Is Not the Issue, but Substitutability Is." *Energy Policy*, 27:1 (1999): 33–43.

Neumayer, Eric. *Weak versus Strong Sustainability*. 2nd ed. Cheltenham: Edward Elgar, 2003.

Nordhaus, William D. "Climate Clubs: Overcoming Free-Riding in International Climate Policy." *American Economic Review*, 105:4 (2015): 1339–1370.

Nordhaus, William D. *A Question of Balance*. New Haven: Yale University Press, 2008.

Nordhaus, William D. "A Review of the Stern Review on the Economics of Climate Change." *Journal of Economic Literature*, 45 (2007): 686–702.

Nordhaus, William D. *The Climate Casino*. New Haven: Yale University Press, 2013.

Norton, Bryan G. "Ecology and Opportunity: Intergenerational Equity and Sustainable Options." In Andrew Dobson, ed., *Fairness and Futurity*, pp. 118–150. Oxford: Oxford University Press, 1999.

Norton, Bryan G. *Sustainability*. Chicago: University of Chicago Press, 2005.

Norton, Bryan G. "Why I Am Not a Nonanthropocentrist: Callicott and the Failure of Monistic Inherentism." *Environmental Ethics*, 17:4 (1995): 341–358.

O'Hare, Bernadette, Innocent Makuta, Levison Chiwaula, and Naor Bar-Zeev. "Income and Child Morality in Developing Countries: A Systematic Review and Meta-analysis." *Journal of the Royal Society of Medicine*, 106:10 (2013): 408–414.

Olms, Sergi and Elia Zardini. *The Sorites Paradox*. Cambridge: Cambridge University Press, 2019.

Olson, Mancur and Martin J. Bailey. "Positive Time Preference." *Journal of Political Economy*, 89:1 (1981): 1–25.

O'Neill, John. *Ecology, Policy and Politics*. London: Routledge, 1993.

O'Neill, Onora. "Children's Rights and Children's Lives." *Ethics*, 98:3 (1988): 445–463.

Oreskes, Naomi and Erik Conway. *The Collapse of Western Civilization*. New York: Columbia University Press, 2014.

Ostrom, Elinor. *Governing the Commons*. Cambridge: Cambridge University Press, 1990.

Ostrom, Elinor, Joanna Burger, Christopher B. Field, Richard B. Norgaard, and David Policansky. "Revisiting the Commons: Local Lessons, Global Challenges." *Science*, 284:5412 (1999): 278–282.

Paden, Roger. "Reciprocity and Intergenerational Justice." *Public Affairs Quarterly*, 10:3 (1996): 249–266.

Page, Edward A. "Cashing in on Climate Change: Political Theory and Global Emissions Trading." In Catriona McKinnon and Gideon Calder, eds., *Climate Change and Liberal Priorities*, pp. 169–189. London: Routledge, 2012.

Page, Edward A. "Intergenerational Justice and Climate Change." *Political Studies*, 47:1 (1999): 53–66.

Parfit, Derek. "Equality or Priority?" In Matthew Clayton and Andrew Williams, eds., *The Ideal of Equality*, pp. 81–125. Houndmills: Palgrave, 2000.

Parfit, Derek. *Reasons and Persons*. Oxford: Clarendon, 1986.

Paul, Kalpita Bhar. "The Import of Heidegger's Philosophy into Environmental Ethics: A Review." *Ethics and the Environment*, 22:2 (2017): 79–98.

Pearce, David. *Economic Values and the Natural World*. London: Earthscan, 1993.

Pettit, Philip. "Realism and Response-Dependence." *Mind*, 100:4 (1991): 587–626.

Pfromm, Peter H. "Towards a Sustainable Agriculture: Fossil-Fuel Ammonia." *Journal of Renewable and Sustainable Energy*, 9:3 (2017): 1–11.

Pigou, A. C. *The Economics of Welfare*. London: Macmillan, 1920.

Portney, Paul R. and John P. Weyant, eds. *Discounting and Intergenerational Equity*. Washington, DC: Resources for the Future, 1999.

Posner, Eric A. and David Weisbach. *Climate Change Justice*. Princeton: Princeton University Press, 2007.

Princen, Thomas. *The Logic of Sufficiency*. Cambridge, MA: MIT Press, 2005.

Rabe, Barry G. *Can We Price Carbon?* Cambridge, MA: MIT Press, 2018.

Railton, Peter. "Locke, Stock, and Peril: Natural Property Rights, Pollution, and Risk." In Mary Gibson, ed., *To Breathe Freely*, pp. 187–225. Totowa: Rowman and Allanheld, 1985.

Ramsey, Frank P. "A Mathematical Theory of Saving." *Economic Journal*, 38:152 (1928): 543–549.

Randalls, Samuel. "A History of the 2° Target." *Wiley Interdisciplinary Reviews: Climate Change*, 1:4 (2010): 598–605.

Rasmusen, Eric. *Games and Information*. 4th ed. London: Wiley, 2006.

Raudsepp-Hearne, Ciara, Garry D. Peterson, Maria Tengo, Elena M. Bennett, Tim Holland, Karina Benessaiah, Graham K. Macdonald, and Laura R. Pfeifer. "Untangling the Environmentalist's Paradox: Why Is Human Well-Being Increasing as Ecosystem Services Degrade?" *BioScience*, 60:8 (2010): 576–589.

Rawls, John. "Justice as Fairness: Political not Metaphysical." *Philosophy and Public Affairs*, 14:3 (1985): 223–251.

Rawls, John. *The Law of Peoples*. Cambridge, MA: Harvard University Press, 1999.

Rawls, John. *Political Liberalism*. New York: Columbia University Press, 1993.

Rawls, John, *A Theory of Justice*. Cambridge, MA: Belknap, 1971.

Rawls, John. *A Theory of Justice*. Rev. ed. Cambridge, MA: Harvard University Press, 1999.

Raworth, Kate. *Doughnut Economics*. White River Junction: Chelsea Green, 2017.

Raz, Joseph. *The Morality of Freedom*. Oxford: Oxford University Press, 1986.

Regan, Tom. *The Case for Animal Rights*. Berkeley: University of California Press, 2004.

Reinhart, R. J. "Snapshot: Few Americans Vegetarian or Vegan." *Gallup* (Aug. 1, 2018).

Rendall, Matthew. "Discounting, Climate Change, and the Ecological Fallacy." *Ethics*, 129:3 (2019): 441–463.

Revesz, Richard L. and Michael A. Livermore. *Retaking Rationality*. Oxford: Oxford University Press, 2008.

Revesz, Richard L. and Matthew R. Shahabian. "Climate Change and Future Generations." *Southern California Law Review*, 84:5 (2011): 1099–1164.

Richardson, Henry. "The Stupidity of the Cost-Benefit Standard." *Journal of Legal Studies*, 29:S2 (2000): 971–1003.

Ripstein, Arthur. *Force and Freedom*. Cambridge, MA: Harvard University Press, 2009.

Ritchie, David George. *Natural Rights*. London: Swan Sonnenschein, 1916.

Roemer, John. *A Future for Socialism*. Cambridge, MA: Harvard University Press, 1994.

Rosen, Harvey, Beverly George Dahlby, Roger Smith, and Paul Boothe. *Public Finance in Canada*. 2nd ed. Toronto: McGraw-Hill, 2002.

Roser, Dominic and Christian Seidel. *Climate Justice*. New York: Routledge, 2012.

Routley, Richard. "Is There a Need for a New, and Environmental, Ethic?" In *Proceedings of the XVth World Congress of Philosophy*, pp. 205–210. Varna: Sofia Press, 1973.

Sagoff, Mark. "Aggregation and Deliberation in Valuing Environmental Public Goods: A Look beyond Contingent Valuation." *Ecological Economics*, 24:2–3 (1998): 213–230.

Sagoff, Mark. "Animal Liberation and Environmental Ethics: Bad Marriage, Quick Divorce." *Osgoode Hall Law Journal*, 22:2 (1984): 297–307.

Sagoff, Mark. "The Limits of Cost-Benefit Analysis." *Philosophy and Public Policy Quarterly*, 1:3 (1981): 9–11.

Salant, David J. "A Repeated Game with Finitely Lived Overlapping Generations of Players." *Games and Economic Behavior*, 3:2 (1991): 244–259.

Samuelson, Larry. "A Note on Uncertainty and Cooperation in a Finitely Repeated Prisoner's Dilemma." *International Journal of Game Theory*, 16 (1987): 187–195.

Sandel, Michael. *Public Philosophy: Essays on Morality in Politics*. Cambridge, MA: Harvard University Press, 2005.

Saunders, Ben. "A Defence of Weighted Lotteries in Life Saving Cases." *Ethical Theory and Moral Practice*, 12:3 (2009): 279–290.

Scanlon, T. M. "Contractualism and Utilitarianism." In Amartya Sen and Bernard Williams, eds., *Utilitarianism and Beyond*, pp. 103–128. Cambridge: Cambridge University Press, 1982.

Scanlon, T. M. *What We Owe to Each Other*. Cambridge, MA: Harvard University Press, 1998.

Scarborough, Peter, Paul N. Appleby, Anja Mizdrak, Adam D. M. Briggs, Ruth C. Travis, Kathryn E. Bradbury, and Timothy J. Key. "Dietary Greenhouse Gas Emissions of Meat-Eaters, Fish-Eaters, Vegetarians and Vegans in the U.K." *Climatic Change*, 125:2 (2014): 179–192.

Scheffler, Samuel. *Death and the Afterlife*. Oxford: Oxford University Press, 2016.

Schellnhuber, Hans Joachim, Wolfgang Cramer, Nebojsa Nakicenovic, Tom Wigley, and Gary Yohe, eds. *Avoiding Dangerous Climate Change*. Cambridge: Cambridge University Press, 2006.

Schmidtz, David and Elizabeth Willott. *Environmental Ethics*. 2nd ed. Oxford: Oxford University Press, 2011.

Schmidtz, David. "A Place for Cost-Benefit Analysis." *Philosophical Issues*, 11 (2001): 148–171.

Schokkaert, Eric. "Cost-Benefit Analysis of Difficult Decisions." *Ethical Perspectives*, 2:2 (1995): 71–84.

Schultz, Walter J. *The Moral Conditions of Economic Efficiency*. Cambridge: Cambridge University Press, 2009.

Sen, Amartya. "The Discipline of Cost-Benefit Analysis." *Journal of Legal Studies*, 29:2 (2000): 931–952.

Sen, Amartya. *The Idea of Justice*. Cambridge, MA: Harvard University Press, 2009.

Shue, Henry. "Bequeathing Hazards: Security Rights and Property Rights of Future Humans." In Mohammed Dore and Timothy Mount, eds., *Global Environmental Economics: Equity and Limits to Markets*, pp. 38–52. Malden: Blackwell, 1999.

Shue, Henry. *Climate Justice*. Oxford: Oxford University Press, 2014.

Shue, Henry. "Subsistence Emission and Luxury Emissions." *Law & Policy*, 15:1 (1993): 39–59.

Sidgwick, Henry. *The Methods of Ethics*. London: Macmillan, 1884.

Sikora, Richard I. and Brian Barry. *Obligations to Future Generations*. Philadelphia: Temple University Press, 1978.

Sinden, Amy. "Allocating the Costs of the Climate Crisis: Efficiency versus Justice." *Washington Law Review*, 85:2 (2010): 293–353.

Singer, Peter. "All Animals Are Equal." *Philosophical Exchange*, 5:1 (1974): 103–116.

Singer, Peter. *Animal Liberation*. London: Pimlico, 1995.

Singer, Peter. *The Expanding Circle: Ethics and Sociobiology*. New York: Farrar, Straus and Giroux, 1982.

Singer, Peter. "Famine, Affluence and Morality." *Philosophy and Public Affairs*, 1:3 (1972): 229–243.

Singer, Peter. *The Most Good You Can Do*. New Haven: Yale University Press, 2015.

Singer, Peter. "One Atmosphere." In Stephen M. Gardiner, Simon Caney, Dale Jamieson, and Henry Shue, eds., *Climate Ethics: Essential Readings*, pp. 181–199. Oxford: Oxford University Press, 2010.

Sinnott-Armstrong, Walter. "Framing Moral Intuitions." In Walter Sinnott-Armstrong, ed., *Moral Psychology*, vol. 2, pp. 47–78. Cambridge, MA: MIT Press, 2008.

Sinnott-Armstrong, Walter. "It's Not *My* Fault: Global Warming and Individual Moral Obligations." In Walter Sinnott-Armstrong and Richard Howarth, eds., *Perspectives on Climate Change*, pp. 221–253. Dordrecht: Elsevier, 2005.

Smil, Vaclav. *Energy and Civilization*. Cambridge, MA: MIT Press, 2017.

Smil, Vaclav. *Harvesting the Biosphere*. Cambridge, MA: MIT Press, 2013.

Smil, Vaclav. "Nitrogen Cycle and World Food Production." *World Agriculture*, 2 (2011): 9–13.

Smith, Alisa and J. B. MacKinnon. *The 100-Mile Diet*. Toronto: Random House, 2007.

Solow, Robert M. "Intergenerational Equity and Exhaustible Resources." *Review of Economic Studies*, 41 (1974): 29–46.

Solow, Robert M. "Sustainability: An Economist's Perspective." In Robert and Nancy Dorfman, eds., *Economics of the Environment*. 4th ed. New York: Norton, 1993.

Spash, Clive L. and Clemens Gattringer. "The Ethics Failures of Climate Economics." In Adrian Walsh, Sade Hormio, and Duncan Purves, eds., *The Ethical Underpinnings of Climate Economics*, pp. 162–182. New York: Routledge, 2017.

Stern, Nicholas. *The Economics of Climate Change: The Stern Review*. Cambridge: Cambridge University Press, 2007.

Stern, Nicholas. "Ethics, Equity and the Economics of Climate Change, Paper 1: Science and Philosophy." *Economics and Philosophy*, 30:3 (2014): 397 444.

Stern, Nicholas. "Ethics, Equity and the Economics of Climate Change, Paper 2: Economics and Politics." *Economics and Philosophy*, 30:3 (2014): 445–501.

Stern, Nicholas. "Imperfections in the Economics of Public Policy, Imperfections in Markets, and Climate Change." *Journal of the European Economic Association*, 8:2–3 (2010): 253–288.

Stiglitz, Joseph. *The Economic Role of the State*. Ed. Arnold Heertje. London: Blackwell 1989.

Stone, Christopher D. *Should Trees Have Standing?* Los Altos: William Kaufman, 1974.

Suikkanen, Jussi. "Contractualism and Climate Change." In Marcello Di Paola and Gianfranco Pellegrino, eds., *Canned Heat: Ethics and Politics of Global Climate Change*, pp. 115–128. New Delhi: Routledge, 2014.

Sunstein, Cass. "Cognition and Cost-Benefit Analysis." *Journal of Legal Studies*, 29:S2 (2000): 1059–1103.

Sunstein, Cass. *The Cost-Benefit Revolution*. Cambridge, MA: MIT Press, 2018.

Sunstein, Cass. "The Cost-Benefit State." Coase-Sandor Institute for Law and Economics Working Paper no. 39 (1996).

Sunstein, Cass. "The Real World of Cost-Benefit Analysis: Thirty-Six Questions (and Almost as Many Answers)." *Columbia Law Review*, 114:1 (2014): 167–211.

Sunstein, Cass. *Risk and Reason*. Cambridge: Cambridge University Press, 2002.

Tan, Kok-Chor. *What Is This Thing Called Global Justice?* New York: Routledge, 2017.

Tanner, Julia K. "The Argument from Marginal Cases and the Slippery Slope Objections." *Environmental Values*, 18:1 (2009): 51–66.

Taurek, John M. "Should the Numbers Count?" *Philosophy and Public Affairs*, 6:4 (1977): 293–316.

Taylor, Charles. *Sources of the Self*. Cambridge, MA: Harvard University Press, 1989.

Taylor, Paul W. "The Ethics of Respect for Nature." *Environmental Ethics*, 3:3 (1981): 197–218.

Taylor, Paul W. *Respect for Nature*. Princeton: Princeton University Press, 1987.

Thomson, Judith Jarvis. "The Trolley Problem." *Yale Law Journal*, 94:6 (1985): 1395–1415.

Tol, Richard S. J. "The Economic Effects of Climate Change." *Journal of Economic Perspectives*, 23:2 (2009): 29–51.

Treasury Board of Canada. *Canadian Cost-Benefit Analysis Guide: Regulatory Proposals*. Ottawa: Treasury Board, 2007.

Trebilcock, Michael J. *Dealing with Losers*. New York: Oxford University Press, 2014.

Tremmel, Joerg Chet. *A Theory of Intergenerational Justice*. Sterling, VA: Earthscan, 2009.

Tsao, Jeff, Nate Lewis, and George Crabtree. "Solar FAQs." *Report for Office of Basic Energy Sciences*. Washington, DC: US Department of Energy, 2006.

Tuomela, Raimo. *Cooperation: A Philosophical Study*. Dordrecht: Springer, 2000.

Turner, Chris. *The Patch*. Toronto: Simon & Schuster, 2017.

Vallentyne, Peter, ed. *Contractarianism and Rational Choice*. Cambridge: Cambridge University Press, 1991.

Vallentyne, Peter. "Infinite Utility and Temporal Neutrality." *Utilitas* 6:2 (1994): 193–199.

Vanderheiden, Steve. *Atmospheric Justice*. Oxford: Oxford University Press, 2008.

Vandeveer, Donald. "Animal Suffering." *Canadian Journal of Philosophy*, 10 (1980): 463–471.

Van Liedekerke, Luc and Luc Lauwers. "Sacrificing the Patrol: Utilitarianism, Future Generations and Infinity." *Economics and Philosophy*, 13:2 (1997): 159–74.

van Parijs, Philippe. "The Disenfranchisement of the Elderly, and Other Attempts to Secure Intergenerational Justice." *Philosophy and Public Affairs*, 27:4 (1998): 292–333.

Victor, Peter. *Managing without Growth*. Cheltenham: Edward Elgar, 2008.

Vidal, John. "The Great Green Land Grab." *The Guardian* (Feb. 13, 2008).

Viscusi, Kip. "Risk Equity." *Journal of Legal Studies*, 29:S2 (2000): 843–871.

Vogel, Steven. *Thinking Like a Mall*. Cambridge, MA: MIT Press, 2015.

Wakeland, Wayne, Susan Cholette, and Kumar Venkat. "Food Transportation Issues and Reducing Carbon Footprint." In Joyce Boye and Yvew Arcand, eds., *Green Technologies in Food Production and Processing*, pp. 211–236. New York: Springer, 2012.

Walzer, Michael. "Political Action: The Problem of Dirty Hands." *Philosophy and Public Affairs*, 2:2 (1973): 160–180.

Warren, Mary. "Do Potential People have Moral Rights?" In Richard I. Sikora and Brian Barry, eds., *Obligations to Future Generations*, pp. 28–29. Philadelphia: Temple University Press, 1978.

Weber, Christopher L. and H. Scott Matthews. "Food-Miles and the Relative Climate Impacts of Food Choices in the United States." *Environmental Science & Technology*, 42:10 (2008): 3508–3513.

Weitzman, Martin L. "Fat-Tailed Uncertainty in the Economics of Catastrophic Climate Change." *Review of Environmental Economics and Policy*, 5:2 (2011): 275–292.

Weitzman, Martin L. "Prices *vs.* Quantities." *Review of Economic Studies*, 41:4 (1974): 477–491.

Weitzman, Martin L. "A Review of the Stern Review on the Economics of Climate Change." *Journal of Economic Literature*, 45:3 (2007): 703–724.

Williston, Byron. *Environmental Ethics for Canadians*. 2nd ed. Toronto: Oxford University Press, 2016.

Wilson, A. N. *The Victorians*. London: Arrow, 2003.

Wolff, Jonathan. *Ethics and Public Policy*. London: Routledge, 2011.

Wolff, Jonathan. "What Is the Value of Preventing a Fatality?" In Tim Lewens, ed., *Risk: Philosophical Perspectives*, pp. 54–67. London: Routledge, 2007.

Wong, David. *Moral Relativity*. Berkeley: University of California Press, 1984.

Worster, Donald. *Nature's Economy*. Cambridge: Cambridge University Press, 1985.

Zimmerman, Michael E. "Toward a Heideggerean Ethos for Radical Environmentalism." *Environmental Ethics*, 5:2 (1983): 99–131.

Index